Surveyors of Empire

CARLETON LIBRARY SERIES

The Carleton Library Series publishes books about Canadian economics, geography, history, politics, public policy, society and culture, and related topics, in the form of leading new scholarship and reprints of classics in these fields. The series is funded by Carleton University, published by McGill-Queen's University Press, and is under the guidance of the Carleton Library Series Editorial Board, which consists of faculty members of Carleton University. Suggestions and proposals for manuscripts and new editions of classic works are welcome and may be directed to the Carleton Library Series Editorial Board c/o the Library, Carleton University, Ottawa K1S 5B6, at cls@carleton.ca, or on the web at www.carleton.ca/cls.

CLS board members: John Clarke, Sheryl Hamilton, Jennifer Henderson, Laura Macdonald, Brian McKillop, Stan Winer, Barry Wright

192 *The Blacks in Canada: A History* (second edition)
Robin Winks

193 *A Disciplined Intelligence*
Critical Inquiry and Canadian Thought in the Victoria Era
A.B. McKillop

194 *Land, Power, and Economics on the Frontier of Upper Canada*
John Clarke

195 *The Children of Aataentsic*
A History of the Huron People to 1660
Bruce G. Trigger

196 *Silent Surrender*
The Multinational Corporation in Canada
Kari Levitt

197 *Cree Narrative*
Expressing the Personal Meanings of Events
Richard J. Preston

198 *The Dream of Nation*
A Social and Intellectual History of Quebec
Susan Mann

199 *A Great Duty*
Canadian Responses to Modern Life and Mass Culture, 1939–1967
L.B. Kuffert

200 *The Politics of Development*
Forests, Mines, and Hydro-Electric Power in Ontario, 1849–1941
H.V. Nelles

201 *Watching Quebec*
Selected Essays
Ramsay Cook

202 *Land of the Midnight Sun*
A History of the Yukon
Ken S. Coates and William R. Morrison

203 *The Canadian Quandary*
Harry Johnson
(New edition)

204 *Canada and the Cost of World War II*
The International Operation of the Department of Finance, 1939–1947
Robert B. Bryce
Edited by Matthew Bellamy

205 *Lament for a Nation*
George Grant
(Anniversary edition)

206 *Confederation Debates in the Province of Canada, 1865*
P.B. Waite
(New edition)

207 *The History of Canadian Business, 1867–1914*
R.T. Naylor

208 *Lord Durham's Report*
Based on the Abridgement by Gerald M. Craig
(New edition)

209 *The Fighting Newfoundlander*
A History of the Royal Newfoundland Regiment
G.W.L. Nicholson

210 *Staples and Beyond*
Selected Writings of Mel Watkins
Edited by Hugh Grant and David Wolfe

211 *The Making of the Nations and Cultures*
of the New World
An Essay in Comparative History
Gérard Bouchard

212 *The Quest of the Folk*
Antimodernism and Cultural Selection in
Twentieth-Century Nova Scotia
Ian McKay

213 *Health Insurance and Canadian Public Policy*
The Seven Decisions That Created the Canadian
Health Insurance System and Their Outcomes
Malcolm G. Taylor

214 *Inventing Canada*
Early Victorian Science and the Idea of
a Transcontinental Nation
Suzanne Zeller

215 *Documents on the Confederation of*
British North America
G.P. Browne

216 *The Irish in Ontario*
A Study in Rural History
Donald Harman Akenson

217 *The Canadian Economy in the Great Depression*
(Third edition)
A.E. Safarian

218 *The Ordinary People of Essex*
Environment, Culture, and Economy
on the Frontier of Upper Canada
John Clarke

219 *So Vast and Various*
Interpreting Canada's Regions in
the Nineteenth and Twentieth Centuries
Edited by John Warkentin

220 *Industrial Organization in Canada*
Empirical Evidence and Policy Challenges
Edited by Zhiqi Chen and Marc Duhamel

221 *Surveyors of Empire*
Samuel Holland, J.F.W. Des Barres,
and the Making of The Atlantic Neptune
Stephen J. Hornsby

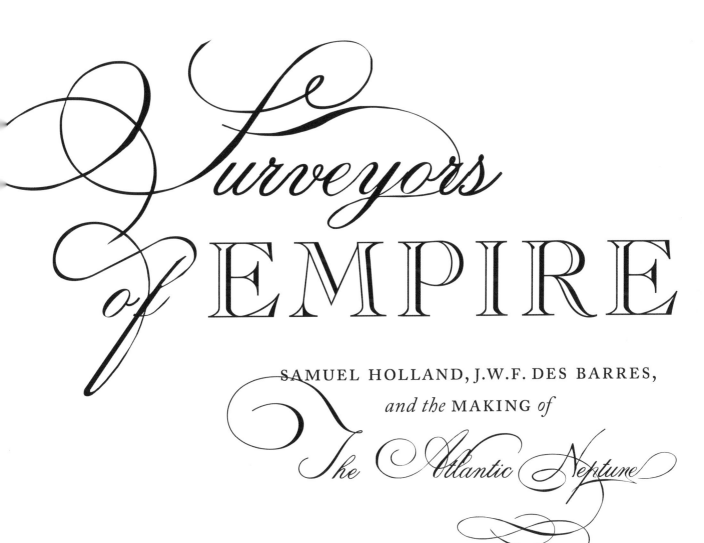

Surveyors of Empire

SAMUEL HOLLAND, J.W.F. DES BARRES, and the MAKING of *The Atlantic Neptune*

STEPHEN J. HORNSBY

With cartography by Hope Stege

CARLETON LIBRARY SERIES 221

McGill-Queen's University Press
Montreal & Kingston · London · Chicago

© McGill-Queen's University Press 2011
ISBN 978-0-7735-3815-3 (cloth)
ISBN 978-0-7735-3848-1 (paper)
ISBN 978-0-7735-8734-2 (ePDF)

Legal deposit second quarter 2011
Bibliothèque nationale du Québec

Reprinted 2016

Printed in Canada on acid-free paper

This book has been published with the help of a grant from the Canadian Federation for the Humanities and Social Sciences, through the Awards to Scholarly Publications Programme, using funds provided by the Social Sciences and Humanities Research Council of Canada. Funding has also been received from the International Council for Canadian Studies through its Publishing Fund and from the University of Maine.

McGill-Queen's University Press acknowledges the support of the Canada Council for the Arts for our publishing program. We also acknowledge the financial support of the Government of Canada through the Canada Book Fund for our publishing activities.

page i: Detail from *A Chart of Nova Scotia* from *The Atlantic Neptune*.

frontispiece: background, Detail from *Sandwich Bay* from *The Atlantic Neptune*; inset image, *West shore of Richmond Isle, near the entrance of the Gut of Canso* from *The Atlantic Neptune*.

Library and Archives Canada Cataloguing in Publication

Hornsby, Stephen J. (Stephen John), 1956–

Surveyors of empire : Samuel Holland, J.F.W. Des Barres, and the making of the Atlantic Neptune / Stephen J. Hornsby ; with cartography by Hope Stege.

(Carleton library series ; 221)
Includes bibliographical references and index.
ISBN 978-0-7735-3815-3 (bnd)
ISBN 978-0-7735-3848-1 (pbk)
ISBN 978-0-7735-8734-2 (ePDF)

1. Holland, Samuel, 1729–1801.   2. Des Barres, Joseph F.W. (Joseph Frederick Wallet), 1729–1824.   3. Nautical charts – Atlantic coast (North America).   4. Atlantic Coast (North America) – Maps.   5. Atlantic Neptune.   6. Cartographers – Great Britain.   I. Stege, Hope   II. Title.   III. Series: Carleton library ; 221

G1108.5.H67 2011   912.1963′44   C2010-906567-0

Set in 11/14 Adobe Caslon Pro with Caslon Open Face and Indenture English Penman
Book design and typesetting by Garet Markvoort, zijn digital

in Memory of

DOUGLAS STIRLING CRAIG

CONTENTS

List of Figures · xi
Acknowledgments · xv

INTRODUCTION · 3

CHAPTER ONE · Surveyors at War · 11

CHAPTER TWO · Surveys · 45

CHAPTER THREE · Surveying · 87

CHAPTER FOUR · Plans and Descriptions · 119

CHAPTER FIVE · Surveyors as Proprietors · 147

CHAPTER SIX · *The Atlantic Neptune* · 163

EPILOGUE · Beyond the Surveys · 199

APPENDIX 1 · Samuel Holland, "List of Plans sent to Government …" · 213
APPENDIX 2 · Cartobibliography of Extant Manuscript Plans and Charts · 215
APPENDIX 3 · Catalogue of the Henry Newton Stevens Collection of *The Atlantic Neptune* · 219

Notes · 227
Bibliography · 251
Index · 265

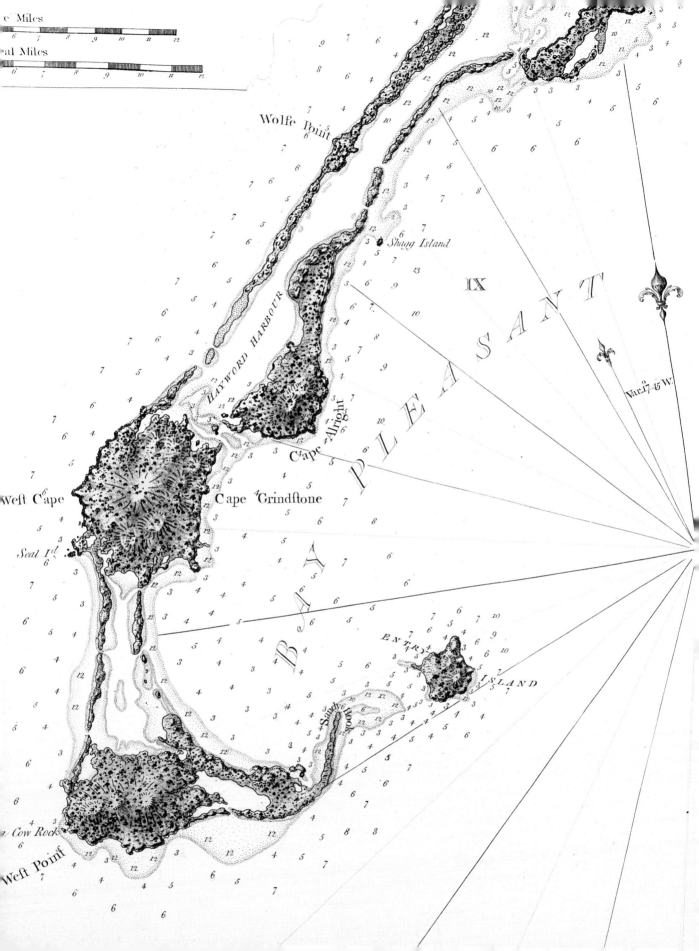

FIGURES

1.1 View near Loch Rannoch · 12

1.2 Movements of Samuel Holland, 1756–1760 · 14

1.3 The provinces of New York, and New Jersey; with part of Pensilvania, and the governments of Trois Rivières and Montreal: Drawn by Captain Holland. Engraved by Thomas Jefferys, geographer to His Majesty · 15

1.4 A plan of Louisbourg … by Samuel Holland Capt Lieut in the 60th Regt. and acting Engineer to Brigr Wolf · 18

1.5 A general view of Quebec from Point Levy · 20

1.6 A view of the Bishop's House with the ruins as they appear in going down the hill from the upper to the lower town · 22

1.7 Movements of J.F.W. Des Barres, 1756–1762 · 24

1.8 Colonel John Montrésor · 26

1.9 Survey of Canada, 1760–1762 · 27

1.10 Town of Quebec, 1761 · 28

1.11 Surveys of North America, 1762–1775 · 35

1.12 The Thames and the City of London from Richmond House · 40

2.1 View of Deptford · 47

2.2 Observation Cove and Fort Amherst, St. John's Island · 51

2.3 Survey of St. John's Island, 1764–1765 · 53

2.4 A north west view of the town and harbour of Louisbourg on the Island of Cape Breton in North America · 54

2.5 Survey of Cape Breton Island, 1765–1766 · 56

2.6   A view of Quebec from the south east · 59

2.7   Survey of the Gulf of St. Lawrence and St. Lawrence River, 1766–1770 · 60

2.8   A view of Cape Torment, part of the Island of Orleans, Isle Madame & Isle au Rot [Ruaux] · 62

2.9   A view of Island Coudre, bearing N.W. distance 2 leagues, taken in crossing the traverse · 63

2.10 Town and harbour of Halifax as they appear from the opposite shore called Dartmouth · 66

2.11 Survey of Nova Scotia, 1764–1772 · 67

2.12 Simeon Perkins House, Liverpool, Nova Scotia · 69

2.13 A view of Portsmouth in New Hampshire taken from the east shore · 73

2.14 Survey of New England, 1770–1775 · 74

2.15 A sketch of the country between New Hampshire and Nova Scotia · 76

2.16 A topographical map of the Province of New Hampshire · 79

2.17 A chart of the harbour of Boston · 82

2.18 Plan of Amboy with its environs from an actual survey · 84

3.1   Lord Halifax and his secretaries · 89

3.2   Board room of the Admiralty · 90

3.3   Lines plan of the *Canceaux*, February 1771 · 102

3.4   Ship model of cargo schooner, circa 1780 · 104

3.5   A view from the camp at the east end of the naked sand hills, on the south east shore of the Isle of Sable · 106

3.6   West shore of Richmond Isle, near the entrance of the Gut of Canso · 107

3.7   Field survey of Grand River, St. John's Island, 1765 · 108

3.8   Heath & Wing trade card · 110

3.9   Holland's astronomical clock made by George Graham, circa 1750 · 111

3.10 Latitudes and longitudes · 113

3.11 A plan of the inhabited part of the province of Quebec · 115

4.1   Mud flats at low tide, St. John's Island · 122

4.2   A plan of the Island of St. John … Survey'd by Capt. Holland, 1775 · 127

4.3   Principal civil divisions on St. John's Island · 128

4.4   Political place names on St. John's Island · 130

4.5   Military and other place names on St. John's Island · 133

4.6   Principal civil divisions and place names on Cape Breton Island · 136

4.7   King's Bay [and] Lunenburg · 138

5.1   Holland's lot 28 on St. John's Island · 152

5.2   Hollandville, New Hampshire · 153

5.3   Holland's land holdings in 1775 · 155

5.4   Castle Frederick, Falmouth Township, Hants Co., Nova Scotia · 156

5.5   Elysian Fields, Minudie Marsh, Chignecto · 157

5.6   Des Barres's land holdings in 1775 · 159

6.1   John Montagu, 4th Earl of Sandwich · 165

6.2   White Haven · 166

6.3   Sandwich Bay · 168

6.4   Admiral of the Fleet Richard Howe, 1st Earl Howe · 170

6.5   Denmark Street (showing number 21) and Soho Square, London · 173

6.6   Title page from *The Atlantic Neptune* · 174

6.7   Imprint dates of *The Atlantic Neptune* · 176

6.8   Small-scale coverage of *The Atlantic Neptune* · 178

6.9   Large-scale coverage of *The Atlantic Neptune* · 179

6.10 Port Jackson, Nova Scotia · 180

6.11 The Magdalen Islands in the Gulph of St. Lawrence · 182

6.12 Buzzards Bay and Vineyard Sound · 183

6.13 Missing harbours on Mount Desert Island, Maine · 184

6.14 Delineation of undersea banks on A chart of Nova Scotia · 186

6.15 Coastal profiles on a Chart of New York Harbour · 189

7.1   French squadron entering Newport, 8 August 1778 · 200

7.2   John Knight Esq. Rear Admiral of the White Squadron · 201

7.3   The Town of Falmouth burnt by Captain Moet, Octr. 18, 1775 · 203

7.4   Penobscot River and Bay, with the operations of the English fleet, under Sir George Collyer, against the division of Massachusetts troops acting against Fort Castine, August 1779 · 204

7.5   A view of the ruins of the fort at Cataraqui · 206

7.6   Encampment of the Loyalists at Johnston, a new settlement on the banks of the River St. Lawrence in Canada · 207

7.7   Areas surveyed by Samuel Holland and his deputies, 1756–1801 · 209

7.8   Joseph Frederick Wallet Des Barres · 210

7.9   Founding of Sydney, Cape Breton Island, 1785 · 211

TABLES

Table 6.1   Charts supplied for His Majesty's Service sent to the Commander in Chief of His Majesty's ships in America, 24 July 1779 · 193

Table 6.2   Distribution of *A sketch of the operations before Charlestown the capital of South Carolina*, 1780 · 194

ACKNOWLEDGMENTS

The surveys of Samuel Holland and J.F.W. Des Barres have interested me since the early 1980s when I was doing research on the historical geography of nineteenth-century Cape Breton. Among the first documents I looked at in the National Archives of Canada (as it was then called) was D.C. Harvey's compilation, *Holland's Description of Cape Breton Island and Other Documents*, published by the Public Archives of Nova Scotia in 1935. The mass of material dealt with Holland's survey of Cape Breton Island in the 1760s. To a callow graduate student who knew little about early Canada, Holland's report hardly seemed relevant to understanding Cape Breton in the nineteenth century and thus could safely be ignored. But the report still seemed important and stuck with me through the years. I was also intrigued by the island's place names. In particular, Richmond County seemed an odd name set amid a mass of Scottish place names, especially when seen in conjunction with Lennox Passage to the north of Isle Madame. Having been at school in Chichester, West Sussex, I made the immediate connection to the local grandee, the Duke of Richmond and Lennox, whose country house was at Goodwood, almost within sight of the cathedral city, but the link to Cape Breton seemed far-fetched and coincidental. The place names of neighbouring Prince Edward Island also interested me, again for their echoes of England, but there were no obvious explanations. Andrew Hill Clark's historical geography of the island, written in the 1950s, said virtually nothing about them. The superb British maps of northeastern North America also attracted my attention. On an Eastern Historical Geography Association field trip to Newport, Rhode Island, in 1989, Martyn Bowden talked excitedly about the Blaskowitz map of Newport, but nobody seemed to know who Blaskowitz was or why the British had produced such a spectacular map of the city on the eve of the Revolution.

In seeking answers to these and many other questions about the surveys of Holland and Des Barres, I have benefitted considerably from the help and wisdom of numerous people. At an early stage of the research, I contacted Fred Thorpe, who, I had been told, knew more about Samuel

Holland than anyone else. With great generosity, Fred provided a file of references to Holland in the Colonial Office papers. I followed up these references in Library and Archives Canada, where Patricia Kennedy was an immense help in explaining the complexities of the British colonial records. It soon became clear, however, that extended research in London was needed, and for making that possible I must thank Nigel Rigby and the National Maritime Museum. A Caird (Short-Term) Fellowship from the museum allowed six weeks research at Greenwich, where I was made most welcome and greatly assisted by Gillian Hutchinson, Brian Thynne, Gloria Clifton, Janet Norton, and the staff of the Caird Library. I know of no other library in London that is so congenial for research.

Staff at the following institutions and collections have also been helpful: National Archives (UK); British Library; British Museum; National Portrait Gallery, London; UK Hydrographic Office; West Sussex County Record Office; Goodwood Collection; Library and Archives Canada; Canadian Museum of Civilization; Public Archives and Records Office of Prince Edward Island; Public Archives of Nova Scotia; Art Gallery of Nova Scotia; New Brunswick Museum; Royal Ontario Museum; Library of Congress, Geography and Map Division; Detroit Institute of Arts; John Carter Brown Library at Brown University; New Hampshire Division of Archives and Records Management; and Special Collections, Fogler Library, University of Maine. Cheney J. Schopieray provided a most useful transcription of several personal papers held in the William L. Clements Library at the University of Michigan. Sophie Forgan, Chairman of the Trustees of the Captain Cook Memorial Museum, Whitby, supplied the image of the Earl of Sandwich and has shown continued interest in the project. Scott Baltjes kindly provided the photograph of the Perkins house in Liverpool, Nova Scotia.

I have also benefitted from comments and suggestions on the project from Andrew Cook, Max Edelson, Michael Heffernan, Elizabeth Mancke, and Graeme Wynn. I am extremely grateful to Cole Harris and Anne Kelly Knowles for reading the entire manuscript and for offering numerous penetrating comments and helpful suggestions. I appreciate the reviews made by two readers of the manuscript for McGill-Queen's University Press. Abigail Davis and Micah Pawling provided sterling assistance in acquiring images and permissions, and Hope Stege was a pleasure to work with in developing the maps. Michael Hermann and Craig Harris also assisted with maps and images. A Canadian Embassy Research Grant greatly eased the costs of research and image acquisition. For the Embassy's support, I am especially grateful to Dan Abele. As always, Philip Cercone, Joan McGilvray, and the staff of McGill-Queen's University Press have been encouraging and supportive. Garet Markvoort designed the book with style, wit, and sensitivity. The endowment fund of the Canadian-American Center, University of Maine, assisted with the publication subvention.

Surveyors of Empire is dedicated to my late uncle, Doug Craig. Born and raised in Duncan, British Columbia, Doug joined the Royal Marines at the outbreak of the Second World War and served as a career officer. I first knew him soon after his retirement in the early 1960s, a great bear of a man, sporting a military moustache, a crew cut, and often a Cowichan sweater. He was full of North American generosity and was immensely good to me. I think he would have enjoyed reading a book about the British army and navy surveying eastern Canadian waters.

overleaf · *The coast of Nova Scotia, New England, New York, Jersey, the Gulph and River of St. Lawrence. The Islands of Newfoundland, Cape Breton, St. John, Antecosty, Sable &c …* (detail) from *The Atlantic Neptune*. © National Maritime Museum, Greenwich, London. HNS 9i.

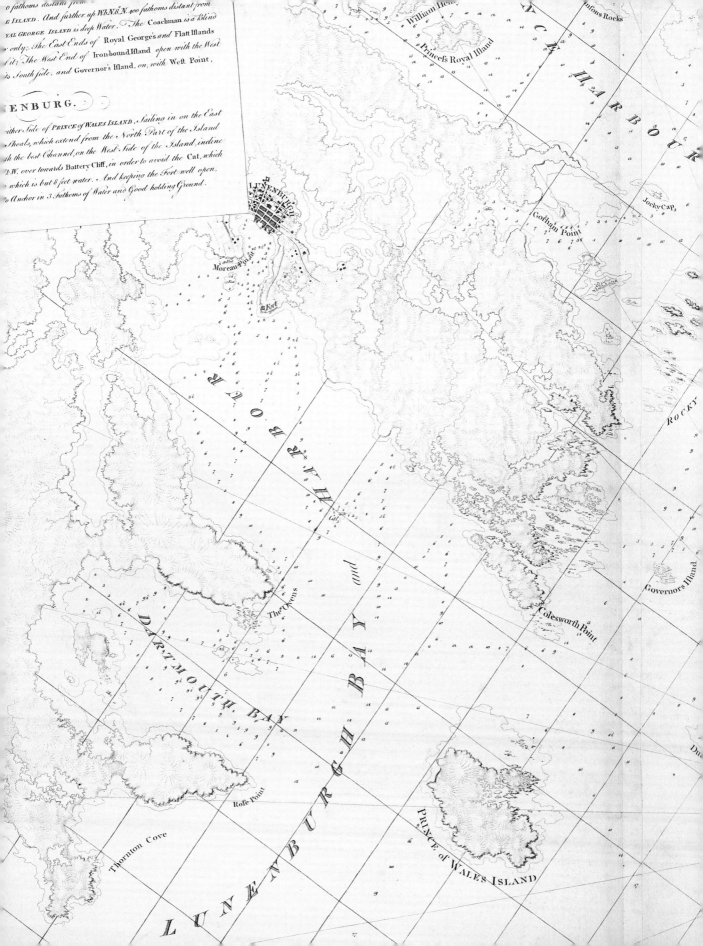

INTRODUCTION

In the wake of the Seven Years War (1756–1763), Britain emerged as a global military and scientific power. At the Treaty of Paris in 1763, Britain massively enlarged its empire at the expense of France and Spain, acquiring new territories in South Asia, North America, and the Caribbean. The empire now reached, discontinuously, from the Bay of Bengal to the Mississippi. At the same time, Britain began to wrench scientific leadership away from France, particularly in the realms of navigation, astronomy, exploration, and cartography. Clockmaker John Harrison solved what has been called the greatest scientific problem of the age, namely how to establish longitude at sea, while the Royal Society contributed mightily to solving another major scientific problem – calculating the distance between the earth and the sun – by sponsoring several expeditions to observe the transit of Venus in 1769. The most ambitious of these expeditions, Cook's voyage to Tahiti in 1768, ushered in a decade of Pacific exploration that enormously expanded scientific knowledge of the globe. Britain was also engaged in surveying and mapping its newly acquired territories. In South Asia, the East India Company directed James Rennell to survey and map Bengal. In North America, the British government instituted coastal surveys that stretched from southern Labrador to the Gulf of Mexico, and river surveys that reached into the heart of the continent along the St. Lawrence, Ohio, and Mississippi. These surveys were the first attempts by any European power to systematically survey extensive areas overseas at a large scale.

The most significant of the North American surveys were the General Survey of the Northern District and the Survey of Nova Scotia conducted by army officers Samuel Holland and Joseph Frederick Wallet Des Barres. Between 1764 and 1775, the two surveyors produced large-scale detailed surveys of more than 15,000 miles of coast from Quebec to Newport, Rhode Island, much of it among the most intricate and complex in the world. In the course of the surveys, Holland undertook cadastral and land-use surveys of St. John's Island (modern Prince Edward Island) and Cape Breton Island, while Des Barres produced numerous offshore soundings and sailing directions. All the surveying and mapping were done to the highest scientific standards. The surveys generated manuscript maps and accompanying reports and correspondence that created an archive of information about northeastern North America that the British government drew upon to administer and settle the region over the following decades. Most of the maps were later engraved, printed, and published as *The Atlantic Neptune*, generally considered the greatest maritime atlas of the age. Arranged in four volumes covering the Gulf of St. Lawrence, Nova Scotia, New England, and from New York southward to the Gulf of Mexico, the atlas comprised more than a hundred maps at different scales, as well as many views and coastal profiles. In a review published in 1784, the French journal *L'Espirit des Journeaux* considered the atlas "one of the most remarkable products of human history that has ever been given to the world through the arts of printing and engraving … the most splendid collection of charts, plans and views ever published."[1] The French recognized the British achievement.

Despite their importance, the surveys have largely been ignored by scholars.[2] In American and British imperial historiographies, the 1760s and early 1770s are dominated by the run-up to the American Revolution and Cook's opening of the Pacific, rather than by what was going on during the same years in what is today eastern Canada.[3] Canadian historiography, too, has overlooked the surveys. Recent scholarship, dominated by nationalist concerns, has focused on the French Acadians displaced by the Seven Years War and their replacement by New England planters, rather than on imperial designs for the newly acquired territories.[4] The surveys of Holland and Des Barres, the extraordinary cadastral division of St. John's Island, and the superb charts and views of *The Atlantic Neptune* scarcely fit narratives of revolution, oceanic exploration, and founding peoples. And yet the General Survey and the Survey of Nova Scotia provide windows into the operations of the British imperial state during those critical years between the end of the Seven Years War and the outbreak of the American Revolution, and also reveal more general characteristics of the developing British Empire.

In the short term, the General Survey and the Survey of Nova Scotia exposed weaknesses in the imperial bureaucracy, particularly the lack of a centralized mapping agency. The Board of Trade and the Admiralty cobbled together survey teams from existing resources, including army surveyors, navy ships, and civilian crews. The shortcomings of this *ad hoc* approach quickly became apparent, and contributed to the Admiralty's decision to create its own hydrographic office in 1795. At the policy level, the surveys revealed the shortsightedness of government officials in Whitehall. Although the surveys contributed significantly to the British government's intelligence gathering in North America, producing information on coasts, terrain, and settlements, this geographic knowledge had little relevance to the conflicts that erupted up and down the eastern seaboard in the 1760s and 1770s. In the wake of the Stamp Act riots in 1765, General Gage, commander-in-chief of the British army in North America, recognized the need for maps of the interior of the continent, but failed to have the surveys redirected towards the seaboard colonies. Government officials in London continued to instruct Holland and Des Barres to survey parts of the continent captured in the Seven Years War rather than other areas of North America that were increasingly sites of civil unrest. The government's immense investment in surveying and mapping the coastlines of North America, probably the most ambitious and sophisticated mapping operation in the world at the time, had little impact on the American conflict that broke out in 1775. The maps were used by the navy during the war but they were of little help to the army fighting on land. The British government's focus on the continent's coastlines and waters failed to anticipate the outbreak of a land-based conflict in the American colonies. In many ways, the story of the surveys serves as a parable of imperial power.

In the longer term, the surveys reveal several important features of the developing British imperial state. First, the metropolitan government was directly involved in the surveys. The Board of Trade, through its agent, the surveyor general, gathered information about the potential of the newly acquired territories for commerce and settlement, and directed the cadastral division of the largely unsettled spaces of St. John's Island and Cape Breton Island. This direct metropolitan involvement reflected Britain's more general tightening of control over its North American colonies in the 1760s, which would be repeated in different parts of the empire after the end of the War of American Independence. Second, metropolitan control was exercised through the military. Although the British military played little role in the thirteen colonies before the mid-1750s, it was critically important to the administration and defence of British possessions elsewhere in the Atlantic world.[5] After 1763, the government relied on the army and the navy to survey the new frontiers of empire, and continued to do so as the empire expanded in the nineteenth century. Third, the surveys demonstrated Britain's growing scientific mastery. The surveys were done using the latest scientific techniques and equipment and to the highest possible standards; in today's parlance, the surveys were "big science."[6] Although recent scholarship has traced the interaction of science and empire to James Cook and Joseph Banks, the surveys of North America revealed the British government's investment in science as a tool to govern empire even before Cook and Banks sailed for the Pacific.[7] Surveyor general Holland set the standard for imperial surveys on land; marine surveyor Des Barres contributed to the navy's growing proficiency at hydrographic surveying. Indeed, John Knight and Thomas Hurd, two of the naval officers involved in the surveys, went on to become leading marine surveyors, with Hurd becoming the first naval officer appointed as Hydrographer of the Navy. Science and empire were inextricably intertwined. Fourth, the surveys fixed and arranged the geographic space of northeastern North America in preparation for colonial settlement. Through triangulation, the surveys delineated the boundary between land and sea, and through astronomical observation, tied it to a universal grid of latitude and longitude. The geography of empire was placed on a rational, scientific basis. Furthermore, surveyor general Holland created a pattern of systematic survey and imperial naming in St. John's Island and Cape Breton Island that reflected the hierarchy of British political and military authority. The surveyor would replicate this pattern in surveys of Upper and Lower Canada after the end of the American war, creating a generic colonial landscape marked by allegiance to Crown, government, and military. This pattern of survey would be applied in other parts of the world as the British Empire expanded in the late eighteenth and nineteenth centuries. Finally, the General Survey, the Survey of Nova Scotia, the other coastal surveys, and the publication of *The Atlantic Neptune* reflected Britain's increasing global ambition. To survey the coastline of eastern North America from southern Labrador to the mouth of the Mississippi and then represent it in maps at a large scale demonstrated the government's enormous confidence in its military and scientific capabilities. The engraved maps and their many copies created a cartography of empire based on scientific principles and tied to the Greenwich meridian that increasingly arranged the world around Britain. Cook's charts of the Pacific were part of this global geographic imagining; so, too, were Holland and Des Barres's charts of northeastern North America. If the General Survey of the Northern District of North America and the Survey of Nova Scotia failed in the short-term to affect British policy towards the thirteen colonies, they made a considerable contribution to the long-term development of the British Empire.

In geographical terms, the surveys are also a useful reminder that Canada and Nova Scotia remained an important focus of the British government in the years immediately after the end of the Seven Years War. In many ways, the British government was working out through these surveys, as well as those of Cook in Newfoundland, the best method to survey extensive coastlines and territories overseas. In a sense, the surveys in the Gulf of St. Lawrence and around the coasts of Nova Scotia were like shakedown cruises, in which the problems of marine and land surveying were sorted out.[8] The resolution of these problems paved the way for the later successes of Cook, Vancouver, Flinders, Fitzroy, and many other naval surveyors in the late eighteenth and early nineteenth centuries.

These historiographical and geographical arguments bear on more general literatures in the history of cartography, science, and the Atlantic world. In the 1980s and early 1990s, geographer Brian Harley shifted the study of maps away from antiquarian concerns with description and cataloguing towards explaining the role that maps play in society. Influenced by such thinkers as Michel Foucault, Jacques Derrida, and Edward Said, Harley examined the relationship between knowledge and power in cartography, deconstructing a wide range of maps to reveal the hidden agendas of map makers.[9] According to Harley, maps were far from neutral or natural, but encoded the power of the cartographer and the institutions that supported map-making. The task of the scholar, he argued, "is to search for the social forces that have structured cartography and to locate the presence of power – and its effects – in all map knowledge."[10] For Harley, a great deal of cartographic power was concentrated in the hands of the state. *Surveyors of Empire* clearly reveals the power of the British state, particularly the military, to conduct systematic surveys and create large-scale topographic maps, which shaped the cartographic representation and governance of northeastern North America in the late eighteenth and early nineteenth centuries. But this study also reveals the shortcomings of state power. For all the resources of the British state in the 1760s and early 1770s, the failure of government policy makers to direct the General Survey to map the American colonies during a period of increasing civil unrest suggests that state power was far from all-knowing. The fact that many of the imperial names that Holland and Des Barres bestowed on their maps failed to take hold among Acadian and British settlers suggests that the state was far from all-powerful. Maps may have contained information that the state could deploy to survey and control territory but until maps were put to use, their power always remained latent.

Surveyors of Empire also demonstrates that science, in this case cartographic science, is historically and geographically contingent. Recent literature in the sociology of scientific knowledge has focused on the construction of scientific knowledge in spaces, places, and regions. Scholars have intensively scrutinized the making of science in such sites as laboratories, academies, courts, and ships.[11] Much of this scholarship has focused on the Enlightenment, a seminal period in Western science, when important advances were made in such subjects as astronomy, navigation, cartography, geography, and botany.[12] Yet for all the recent interest in geographies of scientific knowledge during the Enlightenment, there is a tendency for such work to float above particular places and to stand aside from the flow of time.[13] The British surveys of North America were not simply interesting examples of the entwining of Enlightenment science and military power in the late eighteenth century, but directly influenced British understanding of specific parts of the globe and helped shape government policy at a critically important juncture. The surveys were instrumental in Britain's engagement with North America in those crucial years between the end of the Seven Years War in 1763 and the beginning of the American Revolution in 1775.

Lastly, *Surveyors of Empire* contributes to our understanding of the British Atlantic world.[14] At one level, this study of marine surveyors is a useful reminder of the physical reality of the Atlantic as a major ocean. In much Atlantic scholarship, the sea rarely intrudes.[15] Both the General Survey and the Survey of Nova Scotia charted the Atlantic's northwestern rim, recording in considerable detail its intricate coastline, shelving seabed, and shifting winds and currents. These physical features were represented by the engraved maps and sailing directions in *The Atlantic Neptune*, the first scientific atlas of any part of the ocean.[16] The marine surveying was so exact that geologists today use maps from the atlas as base-lines to measure rises in sea-level.[17]

At another level, *Surveyors of Empire* demonstrates that the Atlantic was not just a physical space that could be surveyed and mapped but also an ocean traversed by flows of men, equipment, and information. Drawing on the methodological writings of social theorist Bruno Latour, *Surveyors of Empire* examines both metropole and periphery by following the "traces" of men, ships, scientific equipment, knowledge, and information back and forth across the Atlantic.[18] This circulation of people, objects, and knowledge created a veritable cat's cradle of linkages and networks across the ocean.[19] As historian Tony Ballantyne has argued, empires consisted of "assemblages of networks, complex threads of correspondence and exchange that linked distant components together … While some of these networks were weak and perished quickly, there is no doubt that from the 1760s the British empire increasingly functioned as a system, albeit an often inefficient and uncertain one."[20] The General Survey and the Survey of Nova Scotia were significant parts of these developing imperial networks. Science, imperialism, cartography, and the Atlantic world were bound up together in the British surveying and mapping of northeastern North America in the third quarter of the eighteenth century.

The first chapter of *Surveyors of Empire* sketches the military background of the surveys. It looks at the role army engineers played in the Seven Years War in North America, and follows Holland and Des Barres through some of the principal campaigns. Even before the war was finished, the British army instituted the Survey of Canada, the first large-scale topographic survey of any part of North America. Samuel Holland played a central part in the survey and helped produce maps for George III, wartime prime minister William Pitt, and leading generals. Arguably the most qualified surveyor in the British army, certainly among those in North America, Holland gained the attention of the Board of Trade in London and was invited to prepare a proposal for surveying the newly captured territories. At the same time, the Admiralty instituted surveys of Newfoundland, Nova Scotia, and the Gulf Coast, appointing J.F.W. Des Barres, another army surveyor, to lead the survey of Nova Scotia. By 1764, an immense British effort to survey much of the eastern seaboard of North America was underway.

The course of the two surveys is the subject of chapter two. Holland's surveys of St. John's Island, Cape Breton Island, the Gulf of St. Lawrence, and the coast of New England, as well as Des Barres's surveys of Nova Scotia and Sable Island, are treated in detail. If Cook's swings across the Pacific represented the exploratory phase of British imperialism, Holland and Des Barres's surveys in North America represented the next phase of detailed coastal surveys and systematic large-scale topographic surveying in preparation for colonial settlement. The immense logistical exercise of mounting the surveys is discussed in chapter three. It considers the personnel, scientific equipment, and Native and settler knowledge that were all essential to the success of the two surveys. Chapter four focuses on the manuscript maps produced by the surveys, and the importance of cadastral surveying and naming to British settlement and cultural appropriation of the northeast. The survey maps were not simply texts for reading in London, but documents that shaped the material and cultural landscapes of several British colonies. The relationship between surveying and land speculation is the subject of chapter five. Both Holland and Des Barres speculated in frontier land. Indeed, Des Barres got into such financial difficulties that he saw the publication of *The Atlantic Neptune* as a way to save his extensive land holdings. Chapter six examines the production and publication of *The Atlantic Neptune* in London and its distribution to the British navy during the American war. An epilogue considers the careers of the leading surveyors after the end of the war, and assesses their impact on the development of British colonial surveying.

Unlike Cook and Vancouver, Holland and Des Barres did not publish their surveying journals. Holland kept a journal of his military campaigns during the Seven Years War, which he intended to publish, but it was stolen, along with his baggage, from the back of his *post chaise* near London in the early 1760s.[21] If Holland and Des Barres kept journals of their respective surveys, they do not appear to have survived. As a result, the primary source for this study is the official correspondence from the surveyors to the Board of Trade and the Admiralty. In quoting from this correspondence, spelling and punctuation have been left as in the original, but capitalization has been modernized. Quebec refers to the town that is now Quebec City; Canada and Canadians refer to the former French colony along the St. Lawrence and its French-speaking inhabitants.

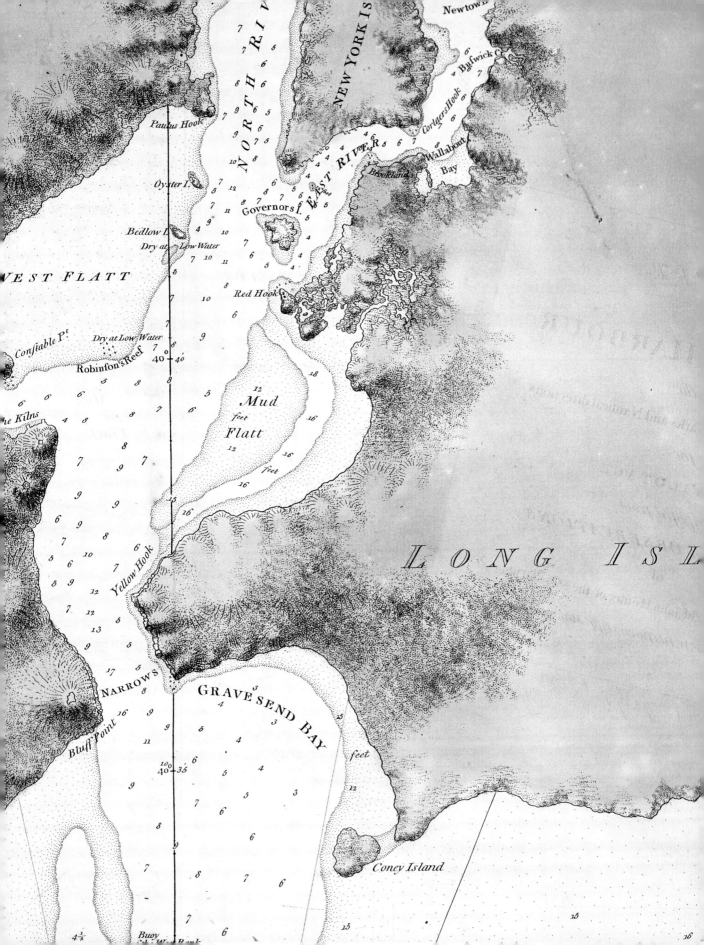

Surveyors at War

1

The Seven Years War (1756–1763) was a pivotal moment in Britain's rise to global pre-eminence. Britain's victory over France and Spain confirmed its military superiority, notably in battles at Louisbourg, Quebec, Quiberon Bay, Havana, Manila, and Plassey. British forces were especially dominant at sea, destroying French fleets and facilitating the transfer of the army across the world's oceans. Such military victories secured a vastly expanded empire at the Treaty of Paris (1763), particularly in eastern North America and northeastern India.[1] As William Pitt, the wartime prime minister, boasted in the House of Commons, the British "had over-run more world" in three campaigns than the Romans had "conquered in a century."[2] The war also marked Britain's rise as a scientific power. Before the war, France was the undoubted scientific leader in Europe. In such sciences as military engineering, surveying, and cartography, the work of Vauban and the Cassini family set the standard. Yet by the end of the Seven Years War, the British had caught up with their French rivals and, in some respects, surpassed them. British military engineers had not only mastered complex siege warfare, but were able to conduct large-scale topographic surveys of captured territory. Moreover, British military capabilities to survey and map territory were massively enhanced during the 1760s and early 1770s, helping to push Britain into the first rank of European science in the last decades of the eighteenth century.

THE SEVEN YEARS WAR, 1756–1763

At the outbreak of the Seven Years War, British military engineering was still in its infancy. The Board of Ordnance, which handled munitions for the army and the navy, had responsibility for military engineering and had created a Corps of Engineers separate from the regular army in 1717.³ The Board had also established the Royal Military Academy at Woolwich in 1741 to train military engineers. The Corps of Engineers was responsible for constructing and maintaining military infrastructure, such as fortifications, barracks, roads, and bridges; conducting sieges; and surveying, sketching, and mapping military buildings and topography. By the mid-1750s, the Corps' greatest achievement had been the pacification of the Highlands of Scotland after the crushing of the Jacobite Rebellion at Culloden in 1746. The Corps worked diligently in the Highlands during the late 1740s and

early 1750s, laying out a network of roads and bridges, building new forts and barracks, and conducting a Military Survey of Scotland, the first large-scale topographic surveying and mapping of any part of the British Isles (figure 1.1).[4]

Even so, the British had only a handful of engineers in North America at the beginning of the Seven Years War, and many more would be needed if the army was to capture the great French fortresses at Ticonderoga, Louisbourg, and Quebec.[5] To bolster the regular army and to provide additional engineers in North America, the British government decided in January 1756 to create a new unit, the Royal American Regiment (60th Regiment of Foot), with officers recruited from Europe, particularly from Protestant Germany and Switzerland, and regular soldiers from the German population of Pennsylvania.[6] Among the continental officers commissioned were Frederic Haldimand and Henry Bouquet, who were to make outstanding contributions to the British war effort, as well as several Swiss engineers. Late in 1756, Lord Loudoun, commander-in-chief in North America, wrote to the Duke of Cumberland, captain general of the British forces, expressing his disappointment in the Swiss engineers, only to be reminded by the duke that he should consider "whether a *Vauban*, or a *Coehorne* would have come a Captain or a Subaltern into an *American* service."[7] The British were making do with whoever they could get, but among the new recruits were military engineers Samuel Johannes Holland and Joseph Frederick Wallet Des Barres.

Born in the small Dutch town of Deventer in 1729, not far from the modern German border, Samuel Holland grew up in the bloodied "cockpit of Europe."[8] In 1745, he entered the Dutch artillery as a cadet and served during the War of Austrian Succession (1744–1748), a global conflict between Britain, the Netherlands, and Austria on one side, and France, Spain, and Prussia on the other. Holland was part of the Dutch force, supported by English troops, that defended Bergen-op-Zoom and other Scheldt River towns against French attack. At the end of the war, he was engaged in surveying fortified towns along the Dutch border. In the course of that work, he met Charles Lennox, 3rd Duke of Richmond. Richmond was an officer in the 20th Regiment of Foot, serving under Lieutenant Colonel James Wolfe, and made a tour of fortified towns in the Low Countries in 1753, in the company of his military tutor, the young Guy Carleton.[9] With few prospects in the small Dutch army, Holland accepted Richmond's offer of a lieutenancy in the fourth battalion of the Royal American Regiment and thereafter always considered the duke to be his patron and protector. After the siege of Louisbourg in 1758, Wolfe wrote to Richmond observing that Holland "looks upon himself, as in some measure under your protection, & upon my word he deserves it."[10]

View near Loch Rannoch, by Paul Sandby, 1749. A draftsman in the Survey of Scotland, Sandby shows an army survey team in the field. A survey team consisted of a surveyor, a non-commissioned officer, and six soldiers who provided two chain men, two flag men, a carrier of instruments, and a batman. Similar teams worked in the Survey of Canada and the General Survey of the Northern District. © The British Library Board, Map Library, K. Top. 50.83.2.

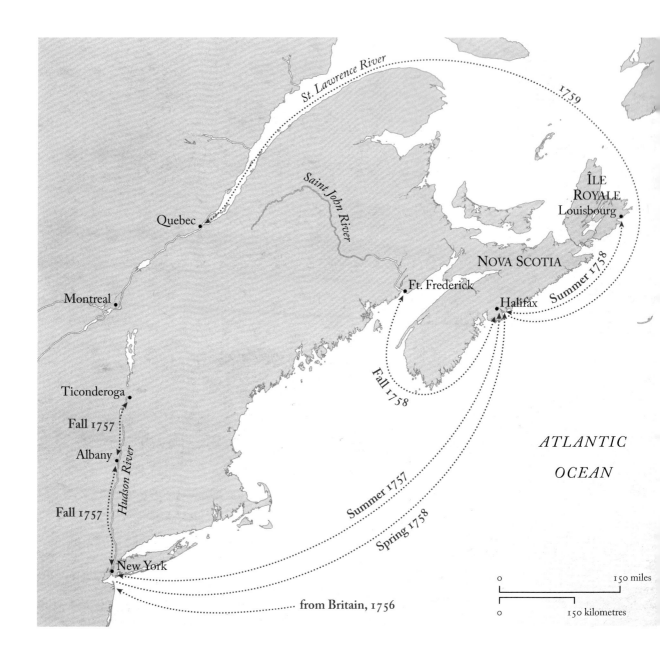

Richmond introduced Holland to Lord Loudoun, colonel-in-chief of the Royal American Regiment, who appointed him as the regiment's acting engineer and draftsman. After his arrival in New York with Loudoun in July 1756 (figure 1.2), Holland's first task was to compile and draft a map of the province of New York (figure 1.3). As the Hudson Valley served as the primary overland route into Canada, the British needed a good map of the province. Holland did not have time or resources to produce a sys-

1.2 · Movements of Samuel Holland, 1756–1760

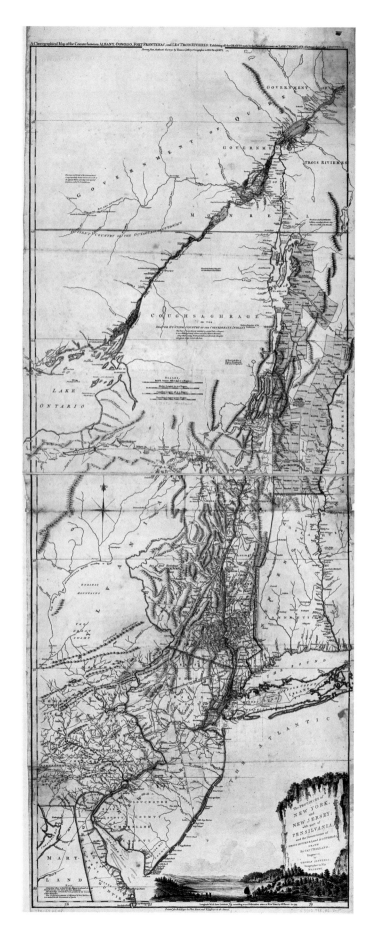

1.3 · *The provinces of New York, and New Jersey; with part of Pensilvania, and the governments of Trois Rivières and Montreal: Drawn by Captain Holland. Engraved by Thomas Jefferys, geographer to His Majesty.* Holland's first map of any part of North America. Library of Congress, Geography and Map Division, G3800 1768.H6 Vault.

tematic survey but instead compiled a map from materials supplied by "the Gentlemen of the Province, as well as proprietors & surveyors."[11] The map was later engraved and printed by the London map publishers Sayer and Jefferys in 1768, much to the annoyance of Holland who, by then, not only considered the map "extreamly erroneous" but also resented his name being used as a "catchpenny."[12] Nevertheless, on the strength of the New York map, he was promoted to a captain lieutenancy in May 1757.[13]

Holland attended Loudoun on his expedition to Halifax, Nova Scotia, in June 1757 in preparation for launching an attack on Louisbourg, the great French fortress town located near the southeastern tip of Île Royale (Cape Breton Island). But bad weather prevented the operation from taking place, and Holland returned to New York in August. From there, he and two battalions of the Royal American Regiment were transferred up the Hudson to Albany, the assembly point for launching assaults up the Hudson-Champlain corridor into Canada. In November 1757, Lord George Howe, the new commander of the Royal Americans, ordered Holland to scout and sketch the environs of Fort Carillon at Ticonderoga, which guarded the southern entrance to the Champlain Valley.[14] Soon after, Holland was transferred back to Halifax to take part in another attempt at capturing Louisbourg, thus avoiding the disastrous British attack on Ticonderoga in July 1758.

For Holland, the switch from the overland campaign up the Champlain Valley to the seaward campaign via the Gulf of St. Lawrence had important repercussions. Early in 1758, Loudoun had been replaced as commander-in-chief by Major General Abercromby, and younger, more dynamic officers had been sent out from Britain to take charge of the faltering campaigns. Among them was Jeffery Amherst, who took command of the assault on Louisbourg planned for that summer, and James Wolfe, his acting brigadier. Louisbourg commanded the southern entrance to the Gulf of St. Lawrence and had to be captured before any seaborne assault on Quebec. The success of the maritime strategy depended considerably on how well British engineers and artillery batteries performed in the siege of the French fortress city.

Wolfe, by far the most effective senior officer on Amherst's staff, quickly recognized Holland's abilities.[15] According to Holland, "General Wolfe did him the honor to admit him one of his family" (i.e., his inner group of officers) by appointing him acting engineer.[16] After the successful landing by British troops on the coast near Louisbourg, the siege of the fortress city began. Wolfe's force was particularly aggressive. Troops established batteries at the North East Harbour in order to bombard the town, and drove siege trenches towards the town's West Gate. As Wolfe's engineer, Holland directed the digging of trenches, driving them to within fifty or sixty yards of the fortress's *glacis*, well within French fire (figure 1.4).[17] On 25 July, scaling ladders were sent to the trenches in anticipation of an assault on the walls but they were never used.[18] The next day, their position hopeless and much of the town in ruins, the French surrendered. On the 28th, Wolfe wrote to the Duke of Richmond declaring that "Hollandt the Dutch engineer has been with me the whole siege, & a brave active fellow he is, as ever I met with; he shou'd have been killed a hundred times, his escape is a miracle." Wolfe went on to express the hope that Howe would take Holland into the Corps of Engineers and "when there is any business to be done, he will find him the most useful man in it." Wolfe finished his encomium by assuring the duke that the surveyor would send "a plan of the attack, & I do believe it will amaze you."[19] Holland most likely sent a plan to the duke but he would never again serve with Howe.[20] The greatly esteemed officer had been killed in a skirmish near Ticonderoga a few days earlier.

If the British siege of Louisbourg followed a carefully scripted and logical progression that culminated in the French surrender, the meeting of Samuel Holland and James Cook on the beach at Kennington Cove the day after the siege ended was entirely fortuitous but had similarly far-reaching consequences. In preparation for making his map of the battlefield (figure 1.4), Holland surveyed Kennington Cove, the principal landing place of British troops, and during the work was approached by Cook, sailing master of the *Pembroke*, one of the warships supporting the siege.[21] Cook expressed interest in Holland's use of a plane table and the following day was given his first lesson in basic surveying techniques. John Simcoe, captain of the *Pembroke*, sent his regrets at not being able to join the surveying party but invited Holland to dine with him onboard ship that evening. Holland found him a "truly scientific gentleman" and a "most intimate and friendly acquaintance." The next day, Cook and two midshipmen accompanied Holland in surveying White Point.

Although the British had originally planned to advance on Quebec after the capture of Louisbourg, the season was too far advanced; instead, forces were sent into the Gulf of St. Lawrence to destroy the French fishery at Gaspé and round up French settlers on Île Saint Jean (modern Prince Edward Island), as well as into the Bay of Fundy to attack French settlements along the Saint John River. Holland accompanied Brigadier Monckton's force to the mouth of the Saint John in early September to assist in the capture of the French fort.[22] They landed without opposition, and Holland immediately set about repairing the fort and building barracks for 300 men. Work on the new Fort Frederick was largely complete by November 1758 when Holland returned to Halifax.

Over the winter months, Holland renewed his acquaintance with Simcoe and Cook onboard the *Pembroke* in Halifax Harbour. In a letter

written in 1792 to Simcoe's son, John Graves Simcoe, the newly appointed lieutenant-governor of Upper Canada (modern Ontario), Holland explained the routine: "Whenever I could get a moment of time from my duty, I was on board the *Pembroke* where the great cabin, dedicated to scientific purposes and mostly taken up with a drawing table, furnished no room for idlers."[23] Under Captain Simcoe's supervision, Holland and Cook began preparing two charts of the Gulf and River St. Lawrence, which the navy needed in order to convey the army to Quebec the following summer. According to Holland, both charts were sent to Jefferys in London for engraving and some copies were available prior to the fleet leaving Halifax in May 1759.[24] Given Holland's demonstrated skills as a cartographer, much of the work on these two charts must have been his hand, although he gives credit to Simcoe and Cook: "By the drawing of these plans under so able an instructor [i.e., Simcoe], Mr. Cook could not fail to improve and thoroughly brought in his hand as well in drawing as in protracting, etc." Holland also credited Simcoe with recognizing the need for accurate surveys of the coast. Finding latitudes and longitudes of Newfoundland and the Gulf of St. Lawrence greatly inaccurate on existing maps, Simcoe talked to "several of his friends in power [about] the necessity of having surveys of these parts

1.4 · *A plan of Louisbourg ... by Samuel Holland Capt Lieut in the 60th Regt. and acting Engineer to Brigr Wolf* (detail). Holland's plan shows the trenches dug under his direction towards the West Gate of the fortress, and White Point, where Holland instructed Cook in surveying. With permission of the Royal Ontario Museum. © ROM, 949.39.23.

and astronomical observations made as soon as peace was restored." Among senior naval officers in Halifax that winter was Alexander Colvill.[25] After the capture of Quebec, Colvill was made commodore on the Halifax station (commander-in-chief of naval forces in North America), and instigated Des Barres's survey of Nova Scotia in 1764. Conceivably, the original idea for the survey came from Simcoe.

The working relationship that developed among Holland, Cook, and Simcoe in the great cabin of the *Pembroke* was of considerable significance. At an institutional level, it marked the close cooperation that existed between the army and navy during the North American campaign and provided the cartographic basis for the expedition to Quebec in 1759. It also revealed the navy's dependence on the army for cartographic expertise. At the time, the navy did not have a hydrographic office responsible for producing nautical charts, and instead left the task to sailing masters, who compiled information from marine surveys, and to commercial map dealers, who engraved and published charts. At an individual level, the interaction was immensely important for Cook, who received his training in land surveying and cartography from Holland, but it was also significant for Holland. Although he hardly could have thought at the time of preparing large-scale surveys of northeastern North America, Holland's interaction with the two navy men prepared him for his close relationship with the senior service after the war.

After the success of the Louisbourg campaign and the failure of Abercromby to capture Ticonderoga, the British government shuffled command of the army. Amherst was promoted to commander-in-chief and given the campaign up the Champlain Valley; Wolfe was promoted to major-general and given the assault on Quebec. Robert Monckton, James Murray, and George Townshend served as Wolfe's brigadiers. The attack on Quebec required an even more ambitious combined operation by the army and the navy than at Louisbourg.[26] Naval fleets from Halifax and Britain were to rendezvous at Louisbourg to pick up the bulk of the troops and then convey them through the Gulf of St. Lawrence and up the St. Lawrence River to Quebec. Vice-Admiral Saunders, the naval commander, brought the fleet from England, while his second in command, Durrell, led a small squadron from Halifax to secure the mouth of the St. Lawrence and prepare the way for the larger force.

There was water deep enough for the fleet to get through the Gulf of St. Lawrence and along the St. Lawrence River as far as Cap Tourmente, just east of Île d'Orléans, but thereafter vessels had to make the "Traverse" from the deepwater channel on the north side of the river to one on the south side. With tides to contend with and the St. Lawrence valley a natural funnel for prevailing westerly winds, the "Traverse" was not the place for

navigational errors. To ensure success, Durrell's squadron captured several French river-pilots, and also sent small boats ahead to sound. Cook and other sailing masters were involved in the sounding, but Holland's role is less clear. He and Cook worked together correcting charts as they made their way upriver, suggesting that he was in Durrell's squadron, but it is unlikely that he was involved in the soundings. Most likely, Holland took sightings on coastal landmarks.[27] The corrections made by Cook and Holland were incorporated on a revised edition of the St. Lawrence map, dedicated to Saunders, that was engraved and published by Jefferys in 1760.

After the successful navigation of the St. Lawrence River, Holland resumed his role as acting engineer for Wolfe. Major Mackellar, who had been interned in Quebec and had prepared a report on the city's defenses, served as chief engineer. Wolfe's first moves were to land on Île d'Orléans and the south shore. Capturing high ground at Point Lévis, opposite Quebec, provided an ideal position from which British batteries could fire on the town and also cover British ships moving upriver (figure 1.5). Nevertheless, Quebec itself appeared a formidable military proposition. High cliffs pro-

1.5 · *A general view of Quebec from Point Levy*, by Richard Short, 1 September 1761. British troops are shown moving upstream in preparation for the assault on Quebec. Library and Archives Canada, Acc. No. 1989-283-1.

tected the city's western flank, while the St. Charles River and French fortifications protected the eastern side. To assess the situation, Wolfe reconnoitred the south shore above Lévis in the company of Holland and other officers. At the Etchemin River, opposite Anse au Foulon, a cove at the bottom of the cliffs, he asked the engineer for his telescope and observed "Indians and Canadians … playing in their canoes and then bathing in the river."[28] Wolfe ordered Major Goreham to establish a post at the river and instructed Holland to make frequent observations of French activities, suggesting to Holland that Wolfe's "design of landing at the Foulon [was] to be predetermined." After Holland reported that the cove was used by the French and Indians to wash clothes and that there were few "in number or in any shape on their guard," Wolfe took the engineer into his confidence: "Voilà mon cher Holland, ce sera ma derniere rescource mais il faut avant que mes autre projects travailent, et manquent. Je vous parle en confiance; en attendant, il faut deguiser mon intention à qui que ce soit et tachez de faire croire l'impossibilité de montez."[29]

According to Holland, Wolfe disguised the "grand object" by launching a series of attacks up and down the river. The first attack, which must have been planned well in advance of Wolfe's reconnoitre up river, was directed at French forces defending the Beauport Shore on the east side of Quebec. The British had landed at Montmorency on the north shore in early July but were separated from the French by the Montmorency River. On 31 July, Wolfe launched a frontal attack from the St. Lawrence on the Beauport defences, which was beaten off with some loss, including Captain Otherlong of the 60th Regiment. Holland served in Otherlong's company and after his death was promoted from captain-lieutenant to captain.[30] Foiled in his attack, Wolfe struck his camp at Montmorency, established his headquarters on Île d'Orléans, and sent expeditions up and down the St. Lawrence to lay waste the countryside. Brigadier Murray was dispatched upriver to establish a camp opposite Pointe-aux-Trembles in order to draw French troops away from Quebec. Holland served as Murray's engineer and took part in raids on Pointe-aux-Trembles and Deschambault. According to Holland, these attacks were all part of Wolfe's grand plan. "By deserters and our own observations," Holland recounted, "we learned that the French were thoroughly convinced we had relinquished all further serious design. They in consequence allowed the Canadian Militia to repair home to gather in their harvests, and the Indians and others on the Heights above Foulon likewise disappeared. General Wolfe's well conceived scheme was now ripe for execution and as masterly put in practise."[31]

In early September, Wolfe concentrated British forces near the Etchemin River in preparation for crossing the St. Lawrence and the assault on Anse au Foulon. Although Holland "begged the favor" of the general to participate in the main assault, he was ordered to launch a diversionary attack farther upriver at Sillery. During the early hours of 13 September, British troops crossed the St. Lawrence and landed at Foulon; Holland, meanwhile, took two gun boats to bombard Sillery, only to have his vessel run down by a schooner. The engineer swam to safety and joined the main force at Foulon. There, he met Captain Vergor, commander of the French guard post at the top of the cliffs and now a British prisoner, "who railed bitterly at *Les Diables D'Anglois* for attacking his Rear, which he had conceived impossible." Holland scrambled up the cliffs to join Wolfe on the Plains of Abraham where British troops were assembling for battle. The first attacks from the French came from militia units on the British left flank, and Holland was ordered to throw up a redoubt as protection. But as "fire was becoming brisk" and the task impossible to accomplish, Holland reported back to Wolfe, only to find him mortally wounded and being carried off the field. Holland was one of a handful of soldiers with Wolfe, claiming to have held "his wounded hand at the time he expired." Years later, Holland would complain bitterly about Benjamin West's famous painting of the *Death of Wolfe*. "For reasons (best known to Mr West the painter) your mem[oriali]st was not admitted amongst the group represented by that artist as being attendant on the General in his glorious exit, but others are exhibited in that painting who never were in battle."[32]

The French surrendered Quebec five days after the battle, and the British moved into a ruined city (figure 1.6). As strong French forces still threatened from the west, the British troops, now under the command of Mur-

1.6 · *A view of the Bishop's House with the ruins as they appear in going down the hill from the upper to the lower town*, by Richard Short, 1 September 1761. Library and Archives Canada, Acc. No. 1970-188-16 W.H. Coverdale Collection of Canadiana.

ray, hurriedly strengthened the town's defenses. Chief engineer Mackellar directed operations, with Holland in charge of the outer guard posts.[33] The following April, French forces attacked the outskirts of Quebec and drove the British back to the safety of the walled town. During the engagement, Mackellar was "dangerously wounded," and Holland took over as acting chief engineer, organizing the defense of the town until it was relieved by the arrival of British warships in early May. In September, French forces in Montreal, faced by three British armies, capitulated, ending offensive military operations on the continent.

For Holland, the conflict had provided enormous opportunities. He had demonstrated his cartographic ability in New York and his engineering skill and bravery at Louisbourg and Quebec, the two principal sieges of the war; he had also helped defend Quebec during the critical days of the French siege. His abilities had been recognized by commanding officers from Loudoun to Wolfe and Murray, and he had been rewarded by promotion from lieutenant to captain, as well as by appointment as acting chief engineer. Although he could scarcely have imagined it as he endured the bitter Canadian winter of 1759–60, he had begun a personal and professional relationship with Quebec that would last until his death more than forty years later.

Among Holland's assistant engineers in the Royal American Regiment was Joseph Frederick Wallet Des Barres. Born in Montbéliard, France, close to the Swiss frontier, in 1729, Des Barres matriculated from the University of Basle in 1750, where he received a good education in drafting, mathematics, and sciences. With few prospects in Catholic France, he moved to England and entered the Royal Military Academy in Woolwich in 1753, where he was trained in military engineering, surveying, and mapmaking.[34] Like other "foreign Protestants," Des Barres joined the newly-formed Royal American Regiment in 1756 and embarked for North America (figure 1.7). He served first in the Hudson Valley; in spring 1757, he was at the "frontier town, and vicinity of Schenectady, being much harassed by depredations and barbarities of Indians" and engaged in local skirmishes, capturing an Indian chief.[35] He then moved to Nova Scotia to take part in the campaign against Louisbourg in 1758. He was in the vanguard of the assault vessels that came ashore at Gabarus Bay, and later served as an engineer during the siege. The following year, he participated in the expedition to Quebec, serving, according to his own testimony, as an aide-de-camp to Wolfe, and was at his commanding officer's side when he died. Des Barres also claimed to have prepared "a chart with soundings and nautical remarks, from a collection of surveys and numerous French documents … of the river St. Lawrence … which gained him much honourable notice from the numerous eminent commanders who had experienced its utility." Perhaps

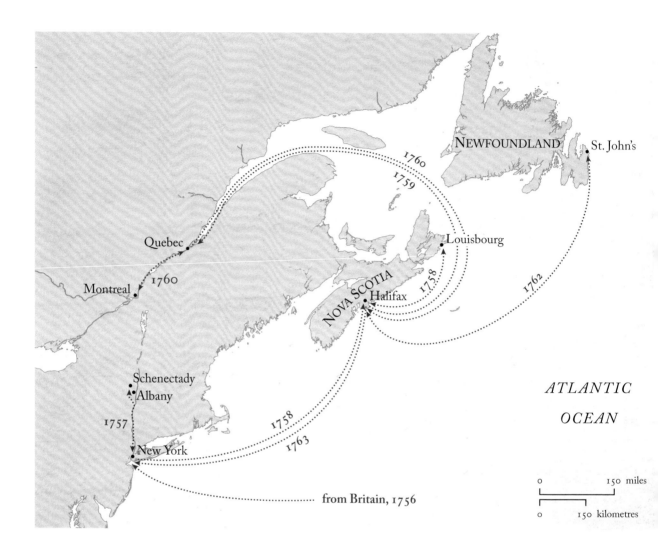

Des Barres assisted Holland and Cook in preparing their chart of the river. After the capture of Quebec, Des Barres surveyed the fortifications and the battlefield on the Plains of Abraham and took soundings in the harbour, again, most likely under the supervision of Holland. After taking part in the defense of the town in April 1760, he participated in the British campaign against Montreal, assisting in the capture of Fort Jacques Cartier and other strong points along the St. Lawrence. In fall, he was posted to Nova Scotia to assist resident engineer Lieutenant-Colonel John Henry Bastide in fortifying Halifax.[36] After the surprise French attack on southeastern Newfoundland in June 1762 and the capture of St. John's, British forces were hurriedly collected from Halifax and Louisbourg to regain the island. In the temporary absence of Bastide, Des Barres was ordered on the ex-

1.7 · Movements of J.F.W. Des Barres, 1756–1762

24 · SURVEYORS OF EMPIRE

pedition as assistant engineer and quarter-master general, and took part in the recapture of St. John's in September.[37] He was then ordered "round the coast of [the] island to take surveys of its principal harbours and ports, and to give in designs for their defence."[38] Exactly how far Des Barres got in this endeavour before winter set in is difficult to fathom, but thereafter he was posted to army headquarters in New York to "proceed on reconnoitering excursions, and report observations on the expediency of a project to establish a chain of military posts throughout the British colonies."[39] After the end of the war in February 1763, he was invited by Commodore Spry, commander-in-chief on the North American station, to undertake a survey of the coasts and harbours of Nova Scotia for the Admiralty.[40]

Survey of Canada, 1760–1763

After the capture of Quebec in September 1759, the British garrison under the command of Brigadier-General Murray was faced with considerable challenges. The town's defences had to be surveyed and repaired, communication had to be kept open with army headquarters in New York, and the surrounding countryside, with its large French population scattered along both banks of the river, had to be kept under surveillance. Town defences could be taken in hand, but French settlements and an overland route to New England were largely unknown. Information had to be gathered and army engineers were assigned the task.

The most pressing need was conveying dispatches overland to Amherst in New York in order to coordinate the British attack on Montreal planned for later in the year. Lieutenant John Montrésor of the Corps of Engineers volunteered for the hazardous mission (figure 1.8).[41] Accompanied by a detachment of thirteen men, including ten army rangers, Montrésor set out from Quebec in the depths of a Canadian winter on 26 January 1760.[42] Crossing the St. Lawrence by canoe and then skirting French villages on the South Shore, Montrésor's party dropped down into the Etchemin and Chaudière river valleys, which eventually took them to the height of land between the St. Lawrence and the Gulf of Maine. After crossing the watershed, the party worked its way across the frozen Rangeley Lakes and picked up a tributary of the Androscoggin River. Intense cold and deep snow hampered the party; provisions ran low. Men were forced to eat their moccasins and snow-shoe strings, as well as alder berries and raw woodpeckers. After an exhausting twenty-five day march, Montrésor and his party struggled into Topsham, Maine, having lost one man "frosted past recovery."[43] Throughout the ordeal, Montrésor kept notes of his journey and made a field sketch of the route, which he later worked up and dispatched to the

Board of Ordnance in London.⁴⁴ This was the first intelligence that the British had gathered about the route across the highlands between the St. Lawrence Valley and northern New England.

With the capitulation of the French in September 1760, the army found itself responsible for Canada, an enormous area that stretched from the Gulf of St. Lawrence, through the St. Lawrence Valley to the Great Lakes.⁴⁵ Within this vast arc, the greatest concentration of French settlement lay along the lower St. Lawrence. Apart from the principal towns of Quebec, Trois-Rivières, and Montreal, French settlers were spread along the north and south banks of the St. Lawrence, as well as along part of the Richelieu River south of Montreal. Some 70,000 French *habitants* occupied a thin strip of land approximately two hundred miles long. Whether Canada would remain in British hands or be traded back to the French in the ensuing peace negotiations remained to be seen in fall 1760, but the army needed accurate intelligence of the settled areas and their populations. Maps of Canada were completely lacking; in 1762, Murray reported that "no chart or map whatever [fell] into our hands" at the capture of Quebec.⁴⁶ The army needed accurate maps not only to administer the province but in case it was ever called upon again to invade the country. As Murray reported to William Pitt, British Prime Minister and great advocate of capturing Canada, "Happen what will, we never again can be at a loss how to attack and conquer this country in one campaign."⁴⁷

At some point in fall 1760, Amherst ordered Murray to conduct a survey of Canada (figure 1.9).⁴⁸ This enormous task would occupy the engineers for the next two years. Montrésor, who had made a name for himself by getting through to New York earlier in the year, and Captain Lieutenant William Spry, also from the Corps of Engineers, were placed in charge of the survey. Montrésor surveyed part of the north shore of the St. Lawrence from Montreal to Lac St. Pierre and also the Chaudière River, the height of land between the St. Lawrence and the Gulf of Maine, and the major rivers in southern Maine. Spry surveyed a few small islands in the St. Lawrence near Montreal and Île d'Orléans. The bulk of the work, however, fell to Samuel Holland and his deputy surveyors. Holland surveyed the South Shore from Montreal to Trois-Rivières, the Trois-Rivières area, and the north and south shores in the vicinity of Quebec. The rest of the surveying was undertaken by lieutenants Lewis Fusier, Joseph Peach, and Frederick Haldimand, Jr., and Ensign Philip Pittman. The surveyors followed standard field surveying techniques, using plane tables to draw upon, compasses and circumferentors to establish angles, and chains for measuring distance.⁴⁹ The surveyors also undertook a census of each parish, gathering information on population and the number of men able to bear arms. Such information would be vitally important in any future war with the

1.8 · *Colonel John Montrésor*, by John Singleton Copley, circa 1771. Tough, irascible, and proud of the Corps of Engineers, Montrésor, who is shown holding a copy of *Field Engineering*, was contemptuous of Holland and Des Barres, who served in a regular regiment. Founders Society Purchase, Gibbs-Williams Fund (41.37), The Detroit Institute of Arts.

1.9 · Survey of Canada, 1760–1762. After figure 4-1 in James Gordon Shields, *The Murray Map Cartographically Considered*, M.A. thesis, Queen's University, 1980.

26 · SURVEYORS OF EMPIRE

French for Canada. After two summers' hard work, the survey and census were largely complete. In November 1761, the surveyors gathered in Quebec to begin the laborious task of turning their field notes into maps.

Neither the French nor the British had previously attempted to produce a large-scale map of such an extensive area of North America.[50] Both nations had produced large-scale maps of individual towns, as well as cadastral divisions in particular counties and parishes, but none of these maps encompassed an area that stretched 200 miles. Moreover, the maps produced by the Survey of Canada were at the remarkably large scale of one inch to 2,000 feet or 1:24,000, a far larger scale than that used by the British Ordnance Survey in their mapping of Britain in the nineteenth century, which was at one inch to one mile or 1:63,360. In drafting the map, the surveyors followed standard cartographic conventions (figure 1.10).[51] Across many parts of the map, terrain was generalized, but prominent landmarks, such as Cap Tourmente on the north shore of the St. Lawrence and Mont-Royal on Montreal Island, were more accurately portrayed. Watercolour washes

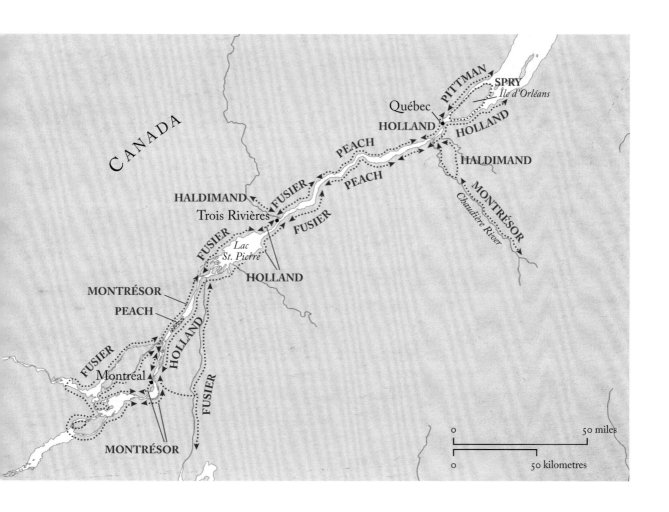

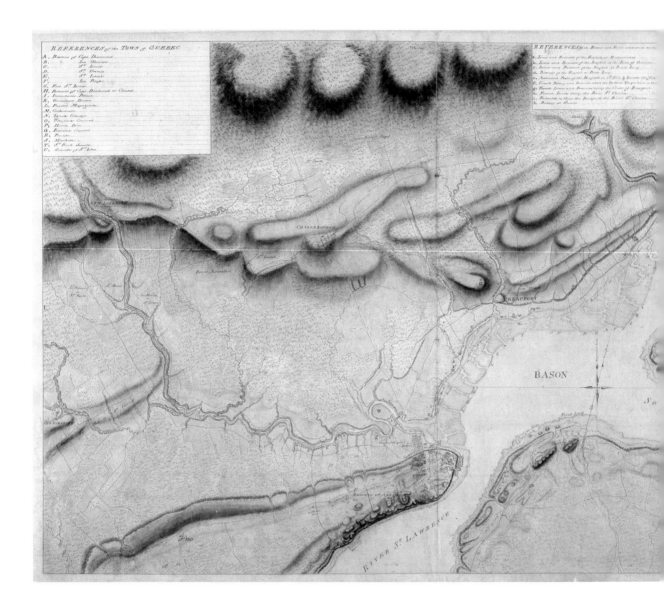

were used to sketch topography. Vegetation, too, was generalized, although the boundary between settled and forested lands was carefully delineated. Settlements were clearly shown, with important buildings, such as farmhouses, churches, and mills, picked out. Roads were also shown, although not always accurately. Each parish included census information.

Drafting the requisite number of maps took nearly two years, being finished in the summer of 1763. Two types of map emerged.[52] The first, taking the form of four large sheets, was intended as a presentation map for George III, William Pitt, the Board of Ordnance, the Secretary of State for

1.10 · Town of Quebec, 1761. A typical map from Murray's copy of the Survey of Canada, showing shaded relief, forest, cleared land, marsh, settlements, and roads. Library and Archives Canada, NMC 135067.

the Southern Department (which included North America), and General Amherst. The second type, comprising a "folio" of more than forty small maps that could be arranged to form a large map of the St. Lawrence, was made for the active military commanders, General Murray in Quebec, and General Gage, who replaced Amherst as commander-in-chief in North America in November 1763. Each type of map varied in its coverage of the St. Lawrence, and some included a detailed plan of Wolfe's battle on the Plains of Abraham.

The lengthy process of drafting the seven maps led to friction and conflict between Murray and his senior surveyors and also among the surveyors themselves. Murray was under pressure from his military superior, Amherst, to produce the maps, and so he, in turn, put pressure on Spry and Montrésor. As early as November 1761, Murray had lost patience with Spry, declaring that he could "neither draw nor survey to my knowledge."[53] A year later, near the end of the project, Murray gave Spry a dressing down that was "the most severe I ever received from any person in my life."[54] "A more ignorant gentleman is hardly to be met with in any profession" was Murray's verdict on his chief surveyor.[55] Montrésor scarcely fared better. After he was found guilty of secretly removing Holland's name from the map that the two men were preparing for presentation to William Pitt, Murray admitted to Amherst that "Montresor is not held in the greatest esteem here."[56] The feeling was mutual. In his journal, Montrésor referred to Murray as tyrannical and a "madman."[57] At least one cause of the dissent was the "inveterate animosity" that Spry and Montrésor, both from the Corps of Engineers, had for "foreign Protestants," like Holland, who were in regular regiments.[58] Montrésor barely concealed his contempt. "*Mem*[orandum]*s of British Folly*, not having any good subjects of their own," he wrote in his journal after the North American surveys began in the mid-1760s, "Mynheer Samuel Jan Van Hollandt Surveyor General at 2000 £ per annum *Sterling*. A.B.C.D.E.F.——Wallet des Barres Surveyor General Nova Scotia – 21 Shillings per diem Sterling. Van de Brahm Surveyor General to the Southward … *Quelles folies*."[59]

As relations between Murray and his two senior engineers disintegrated, the way was clear for Holland to establish himself as the most reliable and competent engineer in Quebec. By June 1762, when Murray submitted his "Report of the State of the Government of Quebec in Canada" to the Board of Trade in London, it was clear that Holland's star was in the ascendant. "I cannot slip the opportunity of recommending this gentleman to Your Lordship's [Lord Egremont] notice," wrote Murray, "he came to this country in 1756, and ever since the siege of Louisbourg I have been myself a witness of his unwearied endeavors for the King's service, in a word, he is an industrious brave officer, and an intelligent engineer, in which capacity he would be desirous, and deservedly merits to be advanced."[60] In October 1762, as the final maps were being prepared, Murray wrote to Amherst declaring that "sure I am if it had not been for the assistance of Holland and the other battalion officers, the survey of Canada would never have been taken."[61] A month earlier, Holland had embarked from Quebec on a ship bound for London, armed with maps from the Survey of Canada for presentation to the King, William Pitt, Lord Egremont, and the Board of Ordnance.[62]

Reorganizing British America

The Treaty of Paris signed on 10 February 1763 concluded the Seven Years War and confirmed Britain's massive new American empire. In taking over territories formerly claimed by France, Britain had acquired an enormous area stretching from the Gulf of St. Lawrence through the Great Lakes to the mouth of the Mississippi. The British also acquired Florida from Spain, France's ally in the war. In effect, Britain had taken over responsibility for all land in North America from the Arctic to the Gulf of Mexico and from the eastern seaboard to the Mississippi. The only area not included was a small enclave comprising New Orleans at the mouth of the Mississippi, which France had ceded to Spain. Within this enlarged British America lay several million people, mostly English and French settlers and African slaves, but also including some tens of thousands of Native people. Britain's challenge was to bring order to this continental space and to integrate it more firmly within the British Empire.

Yet in the early 1760s the British government scarcely had the institutional framework to develop a coordinated policy for such a massive empire.[63] Gathering information, crafting policy, and making decisions about British America were divided among many parts of government. As head of state, George III appointed ministers but could only proffer advice. Nevertheless, he had considerable interest in the empire, science, and mapping, and helped promote them in any way he could.[64] The Privy Council, a large body that included all male members of the royal family, prominent lords and bishops, and leading politicians, advised the king and played an important administrative role in the colonies. The Council oversaw colonial legislation, adjudicated appeals from colonial law courts, considered private memorials, and authorized a wide range of government actions affecting the colonies. Even so, the Council had no real decision-making authority. Major decisions concerning the colonies were made in the Cabinet, a committee of less than a dozen leading ministers, including those with direct responsibility for North America: the Secretary of State for the Southern

Department and the First Lord of the Admiralty. The Secretary of State for the Southern Department had responsibility for home affairs and Southern Europe, as well as North America, and usually delegated policy making for the colonies to the Board of Trade and Plantations. Consisting of a president and seven politicians in office, supported by a tiny civil service, the Board of Trade advised the secretary of state and served a vital administrative role overseeing the colonies. Leading politician Henry Fox called the presidency "the most important place in England," no doubt because of the position's influence over American policy.[65] Nevertheless, the Board had no executive authority to make decisions. Other major departments that had significant interests in North America included the Admiralty, which was responsible for protecting Britain from invasion and defending the nation's far-flung maritime interests; and the War Office, which was responsible for the army and its detachments overseas. Although the Cabinet served as the centre of executive decision-making, not all department ministers were members of Cabinet and many minor policy issues never reached Cabinet committee. Inevitably, a good deal of policy-making took place at department level. In social theorist Bruno Latour's conception, there were multiple "centres of calculation."[66]

During the early 1760s, these various "centres" made numerous efforts to gather information and make decisions about the administration of British America. Most immediately, the navy and the army needed information about the seas and lands they had to govern and protect. The Admiralty required charts of the coasts and waters of northeastern North America and the Gulf of Mexico. In the north, the navy's principal dockyard was at Halifax in Nova Scotia, and the navy had responsibility for protecting the immensely important cod fishery in Newfoundland, Nova Scotia, and the Gulf of St. Lawrence. In the south, the navy had to protect Florida and the Gulf Coast from any Spanish incursion from Cuba or New Orleans. More generally, the navy was responsible for enforcing the Navigation Acts. The passing of the Sugar Act in 1764 required the navy to police shipping, particularly between the sugar islands in the Caribbean and the ports on the American eastern seaboard. To apprehend smugglers, good charts were needed of intricate coastal waters. The army also required detailed topographic maps. It was responsible for defending North America from French or Spanish attack and for keeping peace on the frontier and in Canada. Army commanders needed maps of the captured territories, particularly maps that showed terrain, settlements, and routes of communication.

Yet these military needs were overshadowed by the British government's desire to recast the settlement, trade, administration, and revenue of North America. Faced with major fiscal deficits arising from the war and on-going outlays associated with rebuilding the fleet and garrisoning Nova

Scotia, Quebec, Florida, and the Indian Country, the government moved to reduce expense and raise revenue in British America.[67] One way of limiting expenses was to avoid further conflict on the frontier; this became even more essential after the outbreak of Pontiac's Rebellion in summer 1763. As a major cause of the unrest had been expansion of American settlers onto Indian land, the British government moved to set aside land solely for the Indians.[68] The great westward surge of settlers had to be halted, at least temporarily, and redirected to the northern and southern margins of British America.

The Earl of Egremont, Secretary of State for the Southern Department, began the process of devising a coherent policy for the settlement and government of the newly acquired territories. As early as March 1763, Egremont informed Amherst that a "general plan for the future regulation of that country [the trans-Appalachian west]" was under consideration.[69] Egremont worked closely with William Petty, 2nd Earl of Shelburne, who was appointed president of the Board of Trade in April. Both men were advised by knowledgeable civil servants. Egremont relied on Henry Ellis, former governor of Georgia, who appears to have drafted much of the new policy, while Shelburne drew on the expertise of John Pownall, secretary to the Board of Trade.[70] In May 1763, Egremont submitted to the Board a series of papers related to the new territories, including "Hints Relative to the Division and Government of the Conquered and Newly Acquired Countries in America."[71] Among its several proposals was the division of eastern North America:

> It might also be necessary to fix upon some line for a western boundary to our ancient provinces, beyond which our people should not at present be permitted to settle, hence as their numbers increased, they would emigrate to Nova Scotia, or to the provinces on the southern frontier, where they would be usefull to their mother country, instead of planting themselves in the heart of America, out of the reach of government, and where, from the great difficulty of procuring European commodities, they would be compelled to commence manufactures to the infinite prejudice of Britain.[72]

Essentially, Egremont was arguing for the diversion of settlers – "foreign Protestants or the King's natural born subjects"[73] – to Nova Scotia and the newly acquired territories of St. John's Island, Cape Breton Island, and Florida. He further proposed that "the island of St. John in the Gulph of St. Lawrence, and Isle Royal [Cape Breton Island], which are so near to Nova Scotia, should be united to it forthwith, and make a part of that government."[74]

In a lengthy response submitted the following month, Shelburne and the other Lords of Trade agreed that Nova Scotia should be extended northward to the highlands south of the St. Lawrence River. This meant that the "tract of land from Cape Roziere [Cap-des-Rosiers on the Gaspé peninsula] along the Gulph of St. Lawrence with the whole coast of the Bay of Fundy to the River Penobscot, or to the River St. Croix will be attended with this particular advantage, of leaving so extensive a line of sea coast to be settled by British subjects … and it will likewise be necessary to reannex the Islands of Cape Breton and St. John's to the government of Nova Scotia." The Board further offered its "humble opinion that the utmost attention should immediately be given to the speedy settlement of this tract of country and that instructions be prepared for your Majesty's Governor for that purpose, with particular regard to such officers & soldiers who have served so faithfully & bravely during the late war and who may now be willing to undertake such new settlements under proper conditions."[75] A newly enlarged Nova Scotia would not only absorb the overflow of settlers from the older colonies, but also accommodate thousands of demobilized soldiers and sailors.

Even though Egremont died suddenly in August and Shelburne resigned in September, the momentum for the reorganization of British America continued. George Dunk, Earl of Halifax, Secretary of State for the Northern Department, took over the Southern Department, while Wills Hill, Earl of Hillsborough, became president of the Board of Trade.[76] By then, of course, policy for reorganizing British America had been largely worked out; on 7 October 1763, the government's deliberations became enshrined in the Royal Proclamation. Among the Proclamation's major proposals were to define the boundaries of the Province of Quebec, divide Florida into the provinces of East and West Florida, annex St. John's Island and Cape Breton Island to Nova Scotia, set aside much of the trans-Appalachian west for the Indians, regulate the fur trade, and provide land grants for disbanded soldiers.[77] In just a few pages of text, the Proclamation had radically reorganized the settlement and government of British America. But putting these proposals into effect meant inscribing them across a vast continental landmass. That would need surveyors and map makers, backed by the resources and authority of the imperial government.

SURVEYING BRITISH AMERICA

Even as the Secretary of State for the Southern Department and the Lords of Trade were drawing up their policy for the government of the newly acquired territories, the Admiralty and the War Office were beginning surveys of North America. With so much of Britain's military success in the

Seven Years War dependent on the navy and its command of the sea, the Admiralty was acutely aware of the need for geographic information and the importance of surveying and mapping. The Admiralty recognized that naval operations had "suffered very much during this war for want of information and knowledge of harbours, roads, and accessible places on the coast of *France*" and set about to remedy the situation. On 26 October 1761, the Admiralty issued a set of instructions to flag officers, captains, and commanders of ships and vessels requiring them to gather:

> all useful information of the state of foreign coasts, ports, and roads … and to exert your utmost diligence in making the most accurate observations you possibly can of the latitude and longitude; of the variation of the compass; concerning the sands, shoals, soundings, sea-marks, dangerous winds to ships at anchor, roads, bays, harbours, time of high water, and setting of the tides; together with the depths of water at the time of the spring and neep tides, with the best directions for sailing into ports, or roads, and for avoiding dangers, and whether they require pilots, and where to be had; also the best anchoring-places, landing-places for troops, and watering-places, describing the best methods for watering, and wooding, and whether they are in such plenty, as to supply fleets; noting likewise what provisions of wine and refreshment may be had for the sick and ships.[78]

The instructions went on to observe that "the islands, keys, &c. in the *East* and *West Indies*, the harbours, &c. in our own plantations and colonies, and those of other nations, furnish a large field for such observations." In addition, navigation to "those colonies, the coasts of *Africa* and *India* afford occasions to remark the currents, trade winds, and variation of the compass, &c." The state of foreign fortifications also had to be noted. To expedite the process, the Admiralty supplied printed forms, with a series of questions, to all ships and vessels. The sailing master, under the watchful eye of the commanding officer, was responsible for entering the information and then sending it "in as expeditious a manner as may be" to the Admiralty in London. In essence, the Admiralty had begun a massive global gathering of geographic and hydrographic information.

In North America, the acquisition of information was tightly focused. In the Gulf of Mexico, the Admiralty needed surveys and charts of the waters around Florida and along the Gulf Coast (figure 1.11). In January 1764, naval surveyor George Gauld was assigned the task of charting the Gulf Coast from Cape Florida to the mouth of the Mississippi.[79] Gauld had originally served in the navy as a school master, but was discharged

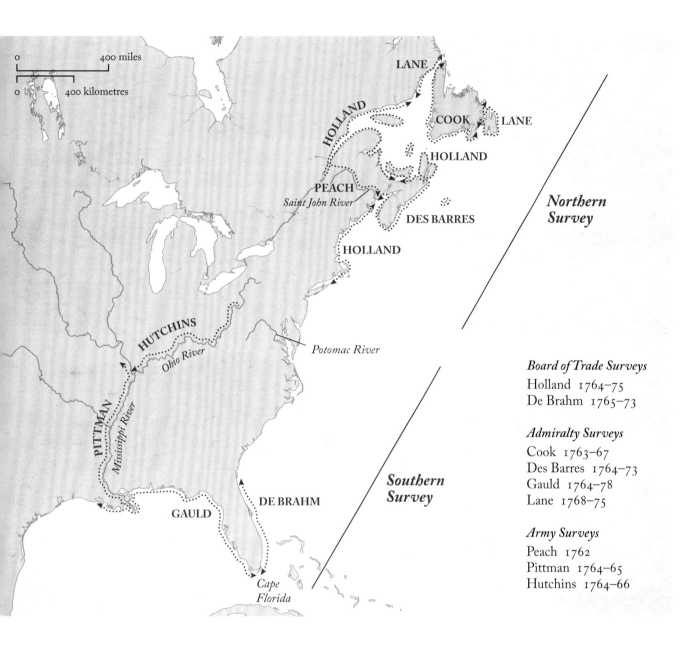

1.11 · Surveys of North America, 1762–1775

in 1759 and in the early 1760s received training in surveying. Desperately short of marine surveyors, the Admiralty recruited him back into service for the Gulf Coast survey. Gauld sailed from England on the *Tartar*, the same vessel that carried clockmaker John Harrison's H4 chronometer on its first sea trial to Barbados.[80] The success of the trial confirmed Harrison's chronometer as a practical means of ascertaining longitude at sea, which would soon revolutionize navigation, surveying, and map-making. While

Surveyors at War · 35

H4 returned to Britain, Gauld continued on to Kingston, Jamaica, and then to Pensacola, West Florida, where he began his Gulf Coast surveys, a task that would last sixteen years.

Far to the north, the Admiralty was interested in the waters around Newfoundland and Nova Scotia (figure 1.11). Newfoundland was significant for several reasons. First and foremost, the island had long been the centre of the English cod fishery. Since the 1570s, English ships had been visiting the inshore waters of the island to fish for cod; by the 1760s, the island was producing an annual average of £453,000 worth of dried cod, the fourth largest export from British America after sugar and rum from the West Indies and tobacco from the Chesapeake.[81] Withdrawal of the French from much of the Gulf of St. Lawrence and the acquisition of Labrador in 1763 allowed this immense English cod fishery to expand along the west coast of Newfoundland and along the southern shores of Labrador.[82] Second, the fishery was important as "a nursery of seamen." In the early 1760s, at least 10,000 English fishermen were employed seasonally in Newfoundland, a vast pool of maritime labour that could be impressed by the navy in time of war.[83] Third, the island was vitally important geo-politically. Although Britain gained title to Newfoundland under the terms of the Treaty of Utrecht (1713), the French secured rights to fish during the summer along an enormous stretch of coast, reaching from Cape Bonavista on the east side to Point Riche on the west.[84] The Treaty of Paris confirmed this access and also gave the small islands of St. Pierre and Miquelon, off the south coast of Newfoundland, to the French as safe havens for their bank fishing vessels. The navy's principal duty was to protect the English fishery from French incursion. Finally, the Admiralty was in charge of civil government on the island.[85] The naval commander on the Newfoundland station served as civil governor responsible for law and order.

Given Newfoundland's importance and the navy's responsibilities, Captain Thomas Graves, the naval commander in Newfoundland, quickly recognized the need for accurate charts of the island and lobbied the Admiralty early in 1763 for a marine survey. The survey was soon ordered and James Cook given the command.[86] After his work in the St. Lawrence in 1759, Cook was one of the few naval masters with any marine surveying experience and was already familiar with Newfoundland waters. In 1762, while serving on the flagship of the naval squadron that helped recapture St. John's from the French, Cook surveyed several harbours in southeastern Newfoundland.[87] Much of Cook's survey in Newfoundland was concerned with French possessions and the treaty shore. Upon arriving in Newfoundland in summer 1763, his first task was to survey the tiny islands of St. Pierre and Miquelon off the south coast before they were handed back to the French. After completing this survey, Cook moved to the northern peninsula, the *Petit Nord*, where the French had fishing rights, and surveyed that coast. He then surveyed the south coast, an area that lay strategically close to St. Pierre and Miquelon. Cook not only surveyed and mapped these coasts in considerable detail, but also gathered information on the potential for the cod fishery. Such work occupied him each summer between 1763 and 1767. By fall 1767, Cook was arguably the most experienced marine surveyor in the navy and the only one able to command a ship. Of the other marine surveyors then employed by the Admiralty, Gauld was neither a master nor a commander, while Des Barres was a commissioned army officer. Cook was thus an obvious candidate to command the great voyage to the Pacific then being planned by the Admiralty and the Royal Society.[88] After Cook's departure for the Pacific in 1768, Michael Lane, Cook's assistant, continued the Newfoundland and Labrador surveys.

The Admiralty was also concerned with Nova Scotia. Halifax served as the principal naval base in North America and commanded the sea lanes from the continent to the British Isles. Amidst these busy shipping lanes lay Sable Island, a shifting sand bank approximately one hundred miles southeast of Nova Scotia and a major navigational hazard. Moreover, the coasts and offshore banks of Nova Scotia offered plentiful opportunity for further development of the province's nascent cod fishery. As in Newfoundland, the impetus for surveying Nova Scotia came from the local commanding officer. Commodore Richard Spry, who had temporarily replaced Colvill as commander-in-chief of naval forces in North America, took up the issue with the Admiralty sometime in late 1762 or early 1763 and repeated his suggestion in a letter dated 4 June 1763. Spry thought that a survey would be "most useful and necessary … to the safety of ships coming on this coast, where there a number of good & safe harbours, known at present to a few fishermen only."[89] Spry also recommended J.F.W. Des Barres for the task. He "is in every respect qualified for such an undertaking, and who offer'd his service, to be employ'd on this useful business."

Colvill resumed command later in the year and also supported the need for surveys. Following the Admiralty order of 1761 to ship captains to make surveys and charts, as well as instructions issued by the Board of Trade in 1763 to colonial governors to create maps of their provinces, Colvill ordered his captains in October 1763 "to co-operate with and assist the persons employed by the said governors therin, as far as may be consistent with the other services you are employed on. You are also to cause accurate charts and surveys to be made of the said coast, with exact plans of the harbours noting the soundings, currents, tides &c conformably to the printed instructions you have received on that head."[90] By January 1764, Colvill had received instructions from the Admiralty to employ Des Barres on the Nova Scotia survey, and wrote to General Gage, commander-in-chief in North

America, to obtain leave of absence for the army surveyor.[91] In early December, the Admiralty agreed to employ Des Barres "in making surveys and charts of the coast and harbours of Nova Scotia" at a salary of 20 shillings per day, plus reasonable expenses for instruments, paper, and other incidentals.[92] At the end of 1764, the Admiralty had major surveys of the Gulf Coast, Newfoundland, and Nova Scotia underway.

The army also had interests in surveying parts of North America. After the Treaty of Paris, the army was responsible for maintaining garrisons in Newfoundland and Nova Scotia, as well as in the newly conquered territories in Canada, the Great Lakes region, the Illinois country, and Florida.[93] Lacking accurate maps, Amherst and his successor, Thomas Gage, had only the haziest notion of the location of some of the army outposts and the principal routes of communication. As a result, military engineers were ordered to survey fortifications and towns, surrounding districts, and routes of communication. After the Survey of Canada was completed in 1762, the most important surveys were the route surveys of the Saint John, Mississippi, and Ohio rivers (figure 1.11).

The Saint John survey was a by-product of the Survey of Canada; indeed, James Murray wanted to conduct the survey himself but was dissuaded by Amherst.[94] Instead, Lieutenant Joseph Peach, who had worked on the St. Lawrence survey, was detailed to survey the Saint John. He most likely began at the St. Lawrence, followed the old Indian and French canoe route from Rivière-du-Loup through Lake Témiscouata and the Madawaska River to the upper reaches of the Saint John, which could then be followed to its mouth on the Bay of Fundy.[95] This route provided the only overland communication between Quebec and Nova Scotia, and was particularly useful in winter when the St. Lawrence was frozen. The route maps were drawn by Montrésor in Quebec as punishment for his behaviour towards Holland, the cartographic equivalent of writing a thousand lines.[96]

Far to the south and west, the Ohio and Mississippi proved even greater challenges to survey (figure 1.11). Both were vital conduits into the heart of Indian territory and the French settlements in the Illinois country.[97] At the Treaty of Paris, the British army had only the slightest information about the two rivers and French settlements, but needed to assert its authority over the region, particularly after the outbreak of Pontiac's Rebellion in the trans-Appalachian west in summer 1763. Stunned by the withdrawal of their French allies and fearful of losing more land to already encroaching settlers, Pontiac's Indian uprising took the British by surprise and a string of forts were captured. Only Detroit, Niagara, and Pitt held out. Relief forces were hastily assembled. For the Ohio campaign, Henry Bouquet marched a small force over the Alleghenies to relieve Fort Pitt. After inflicting a defeat on the Indians at Bushy Run, Bouquet reinforced Fort Pitt, but it was not until the following fall that his force was strong enough to move down river to the forks of the Muskingum in the heart of Indian territory.[98] During the expedition, Ensign Thomas Hutchins of the Royal Americans surveyed the river, creating the first detailed map of the upper Ohio.[99] Two years later, Hutchins and his superior officer, Captain Harry Gordon, extended the survey to the junction with the Mississippi.[100]

Attempts by British army detachments to reach the Illinois country by going up the Mississippi from New Orleans also took time. A lack of coordination and Indian hostility doomed expeditions in 1763 and 1764. Even so, Lieutenant Philip Pittman, who had worked on the Survey of Canada in 1761, managed to survey the Iberville River.[101] With the Spanish holding New Orleans, the British thought that the Iberville would provide an alternative route to the main trunk of the Mississippi, but Pittman's survey proved otherwise. Pittman and an army detachment eventually reached the Illinois country in 1765. During the ascent of the Mississippi, Pittman surveyed its course, producing the first British map of the river. The map was supplemented by surveys of French settlements in the Illinois country. Pittman collected his various reports and maps and published them as *The Present State of the European Settlements on the Mississippi* in 1770, one of the earliest accounts of the region published in Britain.

Extensive as they were, these surveys by the navy and army were dwarfed by those of the Board of Trade. As the Lords of Trade began responding to Egremont's "hints" on the settlement and government of the newly acquired territories, Shelburne realized that a general survey of North America would be needed. According to Holland, Shelburne was "one of the first movers & patrons of this important business."[102] The need for a general survey, particularly of the northeastern and southern flanks of the continent, was essential. French and Spanish printed maps of these areas were available only at a small scale and were not based on systematic surveying. The Board needed to know the true extent of the coastline, location and extent of harbours, the nature of exploitable resources, and areas suitable for colonial settlement.

Holland appears to have played a significant role in Shelburne's deliberations. The army engineer had arrived in London in the fall of 1762, intending to serve in a new military campaign against the Spanish in Portugal, but with peace in the offing, his services were not needed.[103] Instead, he spent the winter of 1762–63 in London, perhaps staying with his patron, the Duke of Richmond, at Richmond House in the heart of Whitehall (figure 1.12).[104] During that time, he presented the Survey of Canada maps to George III, a keen geographer and map collector, who no doubt recognized Holland's cartographic abilities. Moreover, "Genl. Wolfe's favorable Representations of [Holland] while serving under him at the siege of

1.12 · *The Thames and the City of London from Richmond House*, by Canaletto, 1747. As guest of the Duke of Richmond, Samuel Holland no doubt enjoyed this view of the heart of empire before embarking on his survey of the wilderness of northeastern North America. By permission of the Trustees of the Goodwood Collection.

Louisbourg as well as the Duke of Richmond and General Murray's reports … to … Mr Secretary Pitt" had made Holland well known in government circles.[105]

At some point during summer 1763, Shelburne summoned Holland to the Board of Trade to discuss a survey of North America. Holland was given two tasks. First, he had to inventory all existing maps of North America that were in the Board of Ordnance's Drawing Room in the Tower of London and have relevant maps copied for use by the Board of Trade.[106] The Drawing Room supplied a draftsman to help Holland in the copying. Among the maps sent to Holland were surveys done in Nova Scotia of the basin and part of the river at Annapolis Royal, the islands and harbour at Canso, and the harbour at Halifax; as well as maps of the West Indies, including English Harbour in Antigua, the harbour of Port Antonio in Jamaica, and the harbour at Providence in the Bahamas. The Board of Ordnance's cartographic coverage of North America and the Caribbean appeared extremely spotty. Much, obviously, remained to be done.

The second and more important task was to prepare a proposal for a general survey of North America.[107] As with the Survey of Canada, the general survey was to be a large-scale, topographic survey, with detailed maps of important places. But there were also important differences. No doubt reflecting Holland's association with Cook and Simcoe, the survey was to include hydrographic information; it was also to be conducted according to the latest scientific principles. According to Holland: "The General Survey, shall be made by a scale of one mile to an inch, the places of note channels and harbours by a scale of four inches to a mile, being four times larger. The soundings of all harbours and channels shall be taken; with a natural and historical description of the countries, the rivers and lakes and whatever other remarks shall be thought necessary. The latitudes and longitudes of all capes, head-lands &c shall be determined by astronomical observations." Provincial surveyors were expected to provide copies of existing surveys. Holland reckoned the survey needed an armed cutter or other small armed vessel equipped with two whale boats and one large long-boat for transporting surveying parties along the shore. Armed vessels could also be used against smugglers, particularly trading from St. Pierre and Miquelon to the continent. In order that he could take leave from his surveying position in Quebec, Holland requested a deputy surveyor to carry on his work there. For the general survey, he required two assistant surveyors, a draftsman, and two parties of soldiers each consisting of a sergeant, a corporal, and twelve privates "to serve as camp, colour, and chain men, & to make signals alongshore, & on the tops of mountains" (figure 1.1). A variety of instruments were also needed. Altogether, Holland estimated the annual cost of the survey at £700, with an additional one-time expense of £208 for instruments. Holland reckoned the survey of the coasts, rivers, and bays of the Northern District would take five years.

Holland's "Proposals for carrying on the General Survey of the American Colonies" were received and read by the Board of Trade on 16 December 1763, and then recommended to his Majesty's consideration.[108] In their recommendation, the Lords observed:

> in consideration of measures proper to be pursued for the dividing, laying out and settling such parts of your Majesty's American dominions as it is expedient … to grant as soon as possible, in order that your subjects may avail themselves of the advantages which such settlement will produce to the trade, navigation and manufactures of this kingdom, we find ourselves under the greatest difficulties arising from the want of exact surveys of those countries, many parts of which have never been surveyed at all,

and others so imperfectly that the charts and maps thereof are not to be depended upon, and in this situation we are reduced to the necessity of making representations ... founded upon little or no information, or of delaying the important service of settling these parts of your Majesty's dominions.[109]

The Lords further recommended the division into Northern and Southern districts with a surveyor general of lands appointed to each. The Potomac River would serve as the dividing line between the two districts. A similar division had already occurred in the administration of the Indians. As Holland was already in the process of being appointed Provincial Surveyor of Lands in Quebec, he offered to serve as Surveyor General of the Northern District for no extra compensation. The Lords attached Holland's proposals, observing that "Captain Samuel Holland who has great knowledge of the northern parts of America, and who has not only distinguished himself as a brave and active officer and able engineer in your Majesty's service, but also is a skillfull surveyor in the accurate map he has made of the settled parts of your Majesty's colony of Quebeck."[110] The Privy Council approved the recommendation on 10 February 1764.[111] Two weeks later, the Board of Trade presented its "ESTIMATE of the Expence attending general Surveys of His Majesty's Dominions in *North America*, for the Year 1764" for perusal by members of the House of Commons.[112] The total budget was now £1,818.9s., reflecting an additional £700.17s. for the Survey of the Southern District, "the Instruments of Survey excepted." Holland received his commission appointing him Surveyor General of Quebec on 6 March 1764, and the commission appointing him Surveyor General of the Northern District of North America seventeen days later.[113]

During the first months of 1764, a surveyor general for the Southern District was also appointed. William Gerard De Brahm, a German emigrant who had been a long-time resident of Georgia, was selected.[114] He had arrived in Georgia in 1751 and during that decade had established himself as one of the region's leading surveyors and cartographers. He had worked in both Georgia and South Carolina, producing numerous maps of Savannah, Charles Town, and frontier forts, and was particularly known for his "A Map Of South Carolina And A Part Of Georgia," which was dedicated "To the Right Honourable George Dunk, Earl of Halifax First Lord Commissioner and to the rest of the Right Honourable the Lords Commissioners of Trade and Plantations." Thomas Jefferys published the map in 1757. Henry Ellis, who had been governor of Georgia between 1758 and 1760, knew De Brahm and no doubt recommended him to Halifax after he assumed the presidency of the Board of Trade in September 1763.[115] Although De Brahm had technical responsibility for all surveying and mapping in the

Southern District, he was instructed to focus his attention on surveying the coast south of St. Augustine to Cape Florida.[116]

In addition to the general surveys of the Northern and Southern districts, the Board of Trade instructed the governors of Quebec, Nova Scotia, and East and West Florida "as soon as conveniently may be to cause an accurate survey to be made" of their respective provinces by appointing an "able and skillful person," who will report "not only the nature and quality of the soil and climate, the rivers, bays, and harbors, and every other circumstance attending the natural state" but also how the provinces could be "most conveniently laid out in COUNTIES."[117] The surveyor was required to submit a written report and a map of the survey to the governor, which would then be forwarded to the Board of Trade. The navy was expected to provide support to these surveys, as Colvill explained to captains under his command in his instructions of 25 October 1763.[118]

In the twelve months after the Treaty of Paris, the British government had moved quickly to develop and institute policies for the settlement and government of the new territories acquired in North America. Although much scholarly focus has been on British management of the Indian country in the trans-Appalachian west, a crucial part of the new strategy was surveying and mapping territories in the northeast and south.[119] These territories were to be opened for settlement as rapidly as possible. The Board of Trade took on most responsibility, instituting major surveys of East Florida and the islands annexed to Nova Scotia. But the great military departments also played an important role. The Admiralty launched surveys of the Gulf Coast, Nova Scotia, and Newfoundland; the army began surveys of the Ohio and Mississippi rivers. Through its various departments, the British government was making a major investment in surveying the continent. By the summer of 1764, the navy had five survey vessels operating around the coasts of eastern North America, while the army had supplied surveyors, soldiers, and equipment. The scale of the proposed surveys was unprecedented. Thousands of miles of largely uncharted river and coast were to be surveyed, by far the largest surveying operation ever conducted by a European power overseas. All the surveys were to be done with the greatest scientific precision, using the latest surveying techniques and equipment. Well before Cook sailed for the Pacific, British imperialism and science were becoming entwined in these great North American surveys.

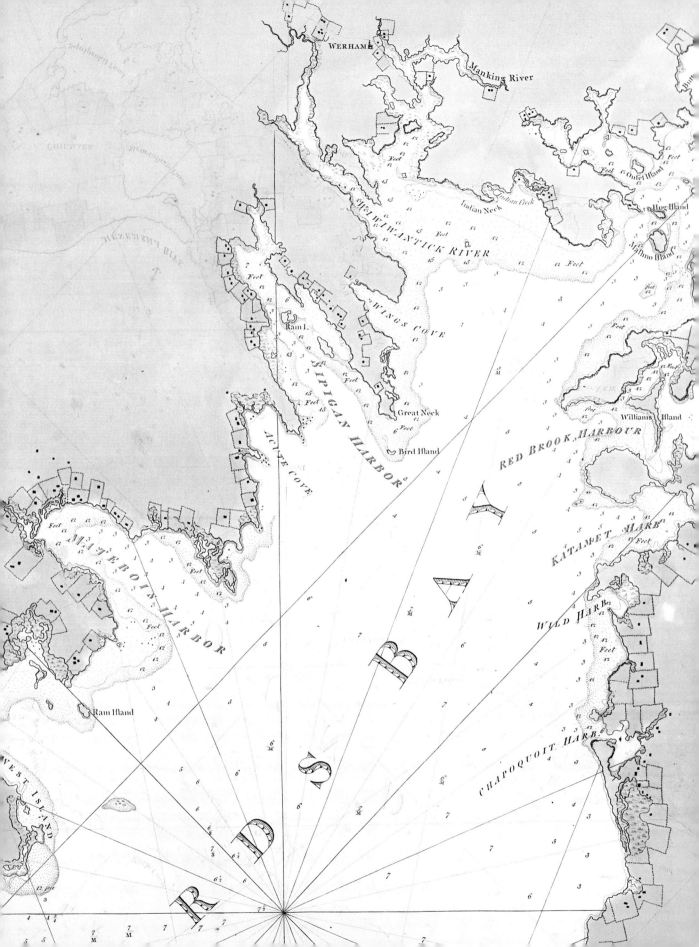

As Shelburne and Holland deliberated in London during summer 1763, they must have swiftly realized the immense task that lay ahead. Although the Survey of Canada had mapped the lower St. Lawrence between Montreal island in the west and Quebec in the east, a vast area encompassing the lowest reach of the St. Lawrence River, the Gulf of St. Lawrence, the coast of Nova Scotia, the Gulf of Maine, and the eastern seaboard from Cape Cod to the Chesapeake Bay still had to be systematically surveyed. The Board of Trade had the small-scale Mitchell Map of North America, published in 1755, to hand but it was based on prior surveys and maps, and hardly provided the large-scale detailed topographic information that the Board now needed.[1] Moreover, the map was poorly keyed to latitude and longitude. The Board's new survey had to provide both detailed geographic information and correct latitude and longitude. It also had to carry out cadastral surveys of the newly acquired islands of St. John and Cape Breton in the Gulf of St. Lawrence, which appeared to offer some potential for British settlement. In addition, the Admiralty required hydrographic information on coasts, harbours, and seabed. Des Barres would provide this material for Nova Scotia, and Holland had agreed to supply similar information for other parts of the seaboard. From the perspective of the government in

London, the two surveys were expected to bring northeastern North America fully into British view.

To fulfill these expectations, Holland and Des Barres faced enormous challenges. The General Survey of the Northern District was to begin surveying in the Gulf of St. Lawrence but little European settlement existed between Halifax and Quebec. Much of the area was a spruce-covered wilderness in summer and locked up in ice and snow in winter. Holland and his men would have to be supplied for months on end, and rely entirely on the navy for transport, supplies, and provisions. Even Des Barres, who was based in Halifax and had much less territory to cover, was surveying poorly known and thinly settled coasts. Between the imperial idea that created the surveys and the maps and reports that the surveys produced lay years of hard, dangerous, and exhausting work in the wilds of northeastern North America.

Holland's Surveys of the Gulf of St. Lawrence, 1764–1770

After the Privy Council authorized the General Survey of North America on 10 February 1764, the two departments of state responsible for the survey moved quickly to implement the great undertaking. The Lords of Trade had already selected their surveyor and made the formal appointment of Holland as Surveyor General for the Northern District on 23rd March. The Admiralty, however, did not have a vessel suitable for the survey. The Navy Board soon found one in the River Thames, which was then entered on the Navy List as the Armed Vessel *Canceaux*.[2] The senior Lords of the Admiralty also instructed the Navy Board "to cause the said ship to be cleaned, sheathed, graved & fitted in a merchants yard accordingly, and to cause her to be victualled & stored for a voyage to North America with all possible dispatch." The Lords appointed Lieutenant Henry Mowat as commander of the *Canceaux* on 14 March. Mowat immediately took command of the ship then moored at the King's Yard in Deptford (figure 2.1).[4] Born into a naval family in Scotland in 1734, Mowat had served in the navy during the Seven Years War and received his commission in 1759. At the age of 30, he took command of the *Canceaux* "under a promise from the Earl of Egmont, then first Lord of the Admiralty, of being soon promoted to the rank of master & commander." The promise remained unfulfilled at Egmont's death in 1770, leaving Mowat an embittered man.[5]

During late March and early April 1764, the *Canceaux* was fitted out at Deptford, a process that included painting and caulking the hull, stowing ballast and stores, getting the top masts and yards up, and setting the

1 · *View of Deptford*, by Joseph Farrington, circa 1774. © National Maritime Museum, Greenwich, London, BHC 1874.

rigging. On 18 April, the Admiralty "resolved that orders be given to Lieut. Mouat of the Canceau[x] Armed Ship to receive onboard Capt. Samuel Holland, Surveyor General of His Majesty's Lands for the Northern district of America, together with his assistants and servants, and proceed with them to River St. Laurence, in order to make surveys, and to proceed, from time to time, to such parts, and in such manner, for the executing the said service as Captain Holland shall desire." Instructions were also sent to Rear Admiral Colvill in Halifax to give Mowat further orders with respect "to the punctual and due performance of the service he is to be employed up[o]n."[6] Given the importance of the General Survey, the Admiralty was determined to ensure the navy's complete co-operation in North America.

Soon after the Admiralty gave its orders to Mowat, Holland and his surveyors and all their baggage and equipment came on board the *Canceaux* at Deptford. Holland's party consisted of deputy surveyors lieutenants Robinson, Haldimand, and Carleton, and civilian Thomas Wright.[7] Of these,

Surveys · 47

information survives for Haldimand and Wright. Haldimand had served as one of Holland's deputies on the Survey of Canada. He was a nephew of Frederick Haldimand, a distinguished senior officer in the Royal Americans, acting civil governor of Trois-Rivières in 1763–64, and a close friend of the surveyor general.[8] Wright had been educated at Christ's Hospital in London, where he had received a good grounding in drawing and mathematics.[9] He had moved to the American colonies in 1759 and spent most of his time surveying; in 1763, he drafted a manuscript map of Georgia and Florida.[10] Henry Ellis, former governor of Georgia and adviser to the Secretary of State for the Southern Department, most likely recommended Wright to Holland.[11] Aside from the surveying party, Holland carried the Board of Trade's dispatches for General Murray at Quebec. These included the commission appointing Murray captain-general and governor in chief of Canada, the seals of office, and instructions for governing the province.[12] These instructions reflected the government's new policy outlined in the Proclamation of October 1763, and were of considerable importance for the future development of Canada.

On 21 April, the *Canceaux* cast off, came to sail, and worked its way downriver.[13] The ship stopped to pick up guns from the arsenal at Woolwich, gunpowder from stores at Gravesend, and two months' pay for the ship's company from the navy pay officer at Chatham. Although Mowat and Holland were no doubt keen to make a fast trans-Atlantic crossing, the *Canceaux* had to take shelter at Spithead in mid-May to ride out gales and take on fresh water and beer from the dockyard at Portsmouth. On 22 May, the ship left its anchorage and made its way down the English Channel, passing Guernsey on the 30, the last time the commander and many of the crew would see English shores for seven years. The *Canceaux* took nearly a month to cross the Atlantic, reaching the Newfoundland banks on 27 June "where the fogg and contrary winds made us very unhappy."[14] On 11 July, the lookout caught sight of the low coastline of southeastern Cape Breton Island but the fog rolled in just as a strong current drew the ship towards land. The *Canceaux* came perilously close to foundering on rocks off the Island of Scatarie, saved only by an inshore fisherman from Lorembec who was "fishing amongst the rocks," saw the danger, and fired a musket to warn the ship. "We perceived the breakers," Holland recounted later, "within a cables length [600 feet] right ahead of us." The surveying expedition was almost lost before it had begun.

On 13 July, the *Canceaux* entered the Gulf of St. Lawrence and three days later made its way into the river. Encountering a northwest gale, the ship was forced to take shelter in the harbour at Gaspé, where Holland met French and English fishermen who informed him that he "had been long expecd at Quebec, & General Murray much in want of his Majesty's dispatches." As the westerlies continued, Holland, anxious to deliver the dispatches, left the ship on the 19 July and took a six-oared, open rowing boat and "a Canadian pilot who was well acquainted with every place along the South Shore as far as Quebec." Accompanied by assistant surveyor Lieutenant Robinson and two passengers, the little party worked its way up river. By the time they had reached Matane, the *Canceaux* had caught up with them and could resupply the party, but Holland did not tarry. By end of the month, the party had reached Rioux, near Trois-Pistoles, the first place that horses could be hired. From there, Holland and his party rode through the marshes along the edge of the river to L'Isle-Verte, where they transferred to an Indian canoe paddled by two Canadians who took them to the Island of Portage (Notre-Dame-du-Portage). From that point, roads were sufficiently good to allow Holland to hire a single horse chaise or "caleshes" for the remainder of the trip. The party arrived in Quebec on 2 August. During the journey, "the country people [i.e. the Canadians] were much rejoyced … when they heard that General Murray was to continue their governeur"; Murray's sympathy for the *habitants* was already well known.[15] Holland's fifteen-day trip up the St. Lawrence, covering some 400 miles, revealed not only the fragility of communications between London and Quebec, but also the immense distances that the General Survey would have to survey and map. The *Canceaux* did not reach Quebec until 12 August, twenty-eight days after entering the river, by which time it was "much in want of necessary reparations."[16]

During his short stay in Quebec, Holland organized the surveyor's office and prepared for the General Survey. Although he had wanted Des Barres to serve as interim deputy surveyor general, his army colleague had already accepted the Admiralty's appointment to lead the Survey of Nova Scotia. So Holland turned to John Collins, "a gentleman, qualified for the business, he having been imployed for many years as a deputy surveyor in the southern colonys, and was recommended to me, by Governor Murray, and several other gentlemen."[17] Until Holland took over the position in 1779, Collins was responsible for all survey work in Quebec, including fixing the boundary between the province and New York eastward from the St. Lawrence to Lake Champlain along the 45th parallel.[18] Holland also gathered supplies and men for the survey; he signed for tents and camp kettles from the Royal American Regimental stores, purchased stoves, and drafted sixteen soldiers from the garrison to serve as flag and chain men.[19]

The *Canceaux* left Quebec on 20 September and reached Port-la-Joye, St. John's Island, on 7 October. The most fertile island in the Gulf of St. Lawrence and already the object of speculative interest back in London, St. John's Island was to be the first campaign of the General Survey of the Northern District. At the time, the island was largely uninhabited. After its capture in 1758, the island had been cleared of most of its Acadian population; in 1764, probably fewer than 300 people remained.[20] A small army detachment garrisoned Fort Amherst, which commanded the entrance to Port-la-Joye, present-day Charlottetown Harbour. Remnant clearings lay scattered along rivers and parts of the coast, but much of the island remained thickly forested. Holland had a virtual *tabula rasa* on which to begin the survey.

With no good maps and only local guides available, Holland's first attempt at surveying nearly ended in tragedy. As the *Canceaux* reached the island's northwest coast, Holland dropped off Haldimand and a party to survey along shore to Fort Amherst. "An Accadian guide, which we had on board assured me," wrote Holland, "that the distance was only eighteen leagues [54 miles] and that at low water, it could be marched with ease along the sea coast, but on my arrival at the Fort … I found myself deceived by the Accadian, as the distance, was much more than what he told me."[21] Haldimand's party had only one week's provisions and no boat, which meant that all food, tents, and surveying equipment had to be carried. Sizing up the situation, Holland sent Robinson and Wright in a small boat with provisions for Haldimand's party. But after realizing that their progress would be slow because of the "many bays and inlets," Holland requested the fort's schooner to search along the coast for the two parties. Further calamity ensued. The schooner was wrecked during a fierce gale, although crew and some provisions were saved. The shipwrecked sailors soon encountered the two surveying parties, by which time Haldimand's party was "in very great distress, haveing been for three days without provisions." After this series of disasters, Holland recognized that surveying by land, at least during summer, would not be feasible and that much of the survey would have to be done by water in small boats.

Although the *Canceaux* was to overwinter at Port-la-Joye, Holland disembarked most of his party and set up his headquarters on land, a pattern that would be repeated throughout the course of the General Survey. Port-la-Joye was a good site, having a sheltered harbour, a central location, and the lone British outpost on the island. Even so, Fort Amherst was "only a poor stockaided redoubt, with scarce barracks sufficient to lodge the garrison." As all the houses in the vicinity had been pulled down for materials to build the fort, Holland and his men had to build their own quarters, as well as an observatory to house their telescopes.[22] Holland chose a sheltered spot in the woods near the shore and there raised the frame of a collapsed barn and covered it with salvaged boards. He called the site Observation Cove (figure 2.2).[23]

By early December, Holland and his party were settling into their winter quarters, warmed by the stoves purchased in Quebec and a plenti-

2.2 · Observation Cove and Fort Amherst, St. John's Island. Detail from *The south east coast of the Island of St. John* from *The Atlantic Neptune*. © National Maritime Museum, Greenwich, London, HNS 134.

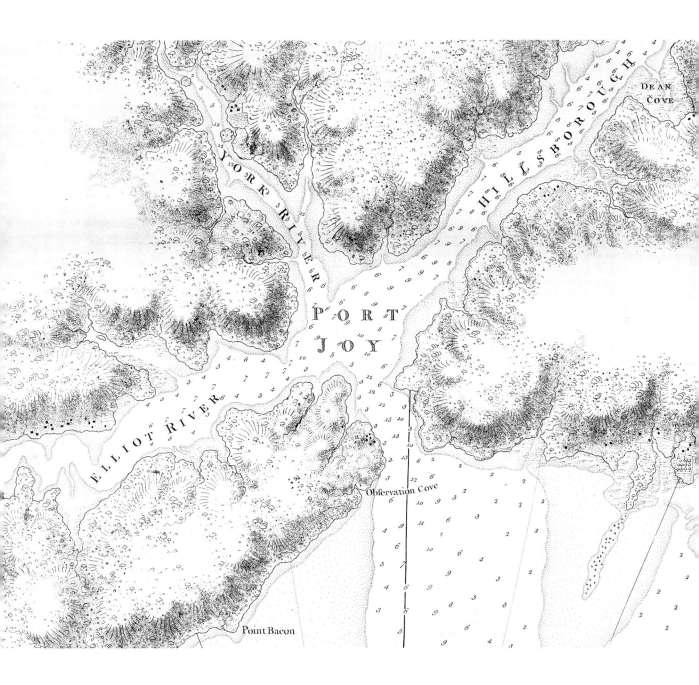

ful supply of firewood. The *Canceaux* had been moved farther up the bay into a protected cove in the hope of avoiding damage from ice and been unrigged. With the fort's schooner wrecked and the *Canceaux* immobilized, all "communication with the continent [had] been entirely cut off."[24] In this wintry isolation, Holland kept his men busy. As the Board of Trade had not received a copy of the Survey of Canada, Holland brought all the original maps from the surveyor's office in Quebec and set his draftsmen to work producing a complete set of finished plans at a large scale for the Board's use.[25] The finished map comprised sixty-two sheets. Holland also made regular astronomical observations, although the cold was frequently "so severe as to oblige us often to quit the telliscopes, to prevent the ill consequences of being struck by the frost."[26] But the main occupation was surveying.

Despite the cold, surveying in winter had its advantages. All waterways on the island as well as the surrounding sea were frozen, which allowed rapid movement by snowshoe and dogsled. Frequent high pressure systems brought brilliant, clear air, making winter ideal for taking sightings. Moreover, Holland had been informed that "in summer, the mosquitos are so very plenty, and troublesome, that neither man or beast can withstand them, in the woods."[27] In mid-February 1765, when day-time temperatures reached the mid-20s Fahrenheit and the waterways were well frozen, Holland sent out four parties, each consisting of a surveyor, three soldiers, a seaman, and an Acadian guide.[28] Each party was equipped with a single-dog sled, purchased from Acadian settlers, which carried the men's beaver skin coats, a buffalo or bear skin blanket, canteens, and about eight days' provisions. Holland and Robinson set out to survey Hillsborough River, Savage Harbour, and St. Peters Bay (figure 2.3). At St. Peters, Robinson left to survey along Fortune River to Fortune Bay, and then to Three Rivers (Cardigan, Brudenell, and Montague rivers), returning along the coast to Port-la-Joye, "not without deficulty being in distance about twenty leagues [60 miles] and the greatest part of the way covered with ice." Meanwhile, Holland, with new assistant surveyor Lieutenant John Pringle, surveyed along the north coast west from St. Peters Bay to Savage Harbour, Tracadie Bay, Rustico Bay, and Orby Head. Holland had hoped to survey all the way to Malpeque Bay, but "the rainey weather beginning rendered the ice unsafe to depend upon so that the scarcity of provisions, and the impractibillity of getting an additional supply, should the rivers break, suddenly determined me to return." The party retraced their steps to Tracadie and then took the portage over to Hillsborough River and thence back to Port-la-Joye. Wright was sent to survey St. Peters Island and Governors Island in Hillsborough Bay, and then along West River and through the woods to Malpeque. There, he surveyed the east side of the bay and then took the portage across to the east side of Bedeque Bay where he joined the survey to what had been done in the fall. Haldimand was sent to survey North River and the east side of Hillsborough Bay as far as Point Prim, but was unable to complete the survey having been "affected by the frost," which left him "incapable of service" for several weeks. After the experience of these surveys, Holland was convinced, despite the cold and frostbite, that it was "absolutely neces-

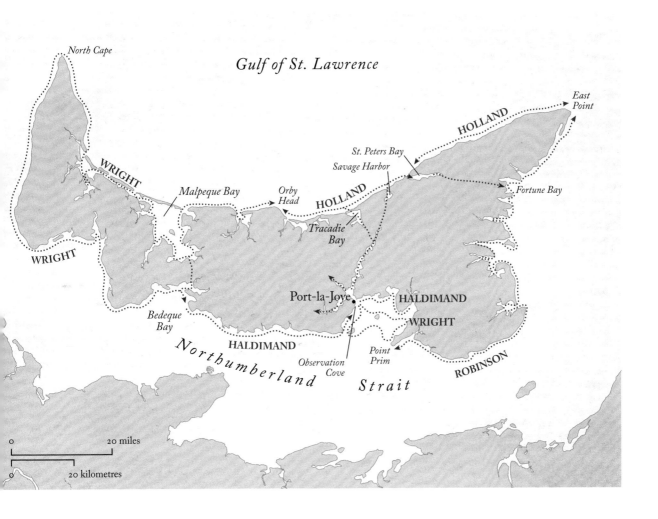

2.3 · Survey of St. John's Island, 1764–1765

sary to attempt doing as much as possible during the time which the rivers and lakes are froze."

In early April, the ice started breaking up. A piece about two miles long and one mile wide came adrift in Hillsborough River and carried the *Canceaux* from its cove as far as Governors Island in Hillsborough Bay, a distance of six miles. With jury masts and studding sails, southerly winds, a turn of the tide, and the assistance of a boat, the *Canceaux* managed to get back to its mooring, but for the next several days the crew had to place spars around the ship to prevent loose ice from ripping off the hull's copper sheathing.[29] By mid-month, Northumberland Strait was sufficiently clear of ice that one of the ship's boats crossed to the mainland and delivered letters to Fort Cumberland on the Chignecto Isthmus, the first communication with the outside world since late November.[30] In the first week of May, the

Surveys · 53

rivers were also clear. Later in the month, Holland began a new round of surveys.[31] Haldimand was sent to survey the Magdalen Islands in the Gulf, while Wright went westward to finish surveying that end of the island. A month later, Holland set off to finish the eastern end. The survey parties were now transported by ship's boats commanded by midshipmen from the *Canceaux* who also carried out inshore soundings.[32] By late summer, the surveys and soundings were complete, and Holland and the draftsmen were putting the finishing touches to the large-scale map of the island.[33] As their quarters in the barn were cramped, the team was "obliged to draw in a tent where we was much tormented by the misquoto's and sand flys."[34] The finished map revealed the surveyors' accumulated knowledge. Coastal and river-bound surveys had delimited the shape of the island and its main waterways, but, apart from two portage routes, the interior of the island had not been surveyed and was left blank.

Holland's next task was to survey Cape Breton Island. The island had served as the main base for the French cod fishery during the early eighteenth century, and coal deposits had been discovered outcropping along the eastern shore. The island thus had potential economic value. Moreover, the British knew virtually nothing about the interior of the island. Anxious not to waste time, Holland dispatched Wright and Pringle in a sloop at the beginning of September to survey the island's west coast.[35] A month later, the surveyor general and his party embarked on the *Canceaux* for the

voyage to Louisbourg.[36] Although heavily damaged during the siege in 1758, Louisbourg remained the only sizeable settlement (figure 2.4). In 1766, the island had about 700 inhabitants, at least 500 of them living in the ruins of the French fortress town.[37] Louisbourg served as a base for the island's meagre cod fishery, provisioned passing ships, and had a small garrison. Colonel Pringle, the commanding officer, made Holland welcome, providing the surveyor and his team with the best quarters in town, which included a "large drawing room with all conveniences for [the] whole party."[38] Apart from providing Holland with a convenient base, Louisbourg must have stirred powerful memories. The surveyor had directed the building of Wolfe's batteries and siege trenches, and he had mapped the whole battlefield. To commemorate his battalion commander and the British victory, Holland employed soldiers from the survey team and garrison to construct a memorial on the ruins of the French citadel. Made from stones from the old fortifications, the monument was inevitably in "rustick taste" but, as Holland explained to his old comrade-in-arms, Frederick Haldimand, "the injurys of time can make but little impression on it."[39]

Surveying Cape Breton proved much more challenging than St. John's Island. As the *Canceaux* made its way through the Gut of Canso that separates the island from the mainland, Holland could not have failed to notice the mountains running down to the sea along the northwest coast. "The dangers will much encrease in the survey of this island," he wrote later, "as it is almost inteirely surrounded by inaccessable rocks and clifts, exposed to the surf (or breaking of the oceans) & to which may be added the inconveneancy of the frequent foggs."[40] As Mowat dared not venture the *Canceaux* "on an unknown coast without much danger," the survey parties would again have to use whale boats, although one used by Wright and Pringle had already been smashed. Moreover, the enormous scale of the survey became clearer when Holland realized that the island was virtually bisected by the Bras d'Or Lake, which added hundreds of miles more shore line to survey. In addition to these challenges, Holland was faced with maintaining the strength of his survey parties. Robinson had gone back to England after the survey of St. John's Island finished in October, and Haldimand had died when he fell through ice on a pond near Louisbourg two months later.[41] Short of deputies, Holland recruited George Sproule from the garrison at Louisbourg, having found him "very fit in knowledge and constitution for this business."[42]

As on St. John's Island, Holland sent out survey parties during February when the "snow storms and severe frost … are a little abated."[43] But the bulk of the work was done after the ice broke up. In early summer, boats were sent out to complete the survey. Wright surveyed from Seal Island (Margaree Island) around Cape North to Cape Fumée (Cape Smokey);

2.4 · *A north west view of the town and harbour of Louisbourg on the Island of Cape Breton in North America*, by Thomas Wright, 1766. The *Venus* surveying sloop is shown under sail far left. Library and Archives Canada, Acc. No. R9266-432 Peter Winkworth Collection of Canadiana.

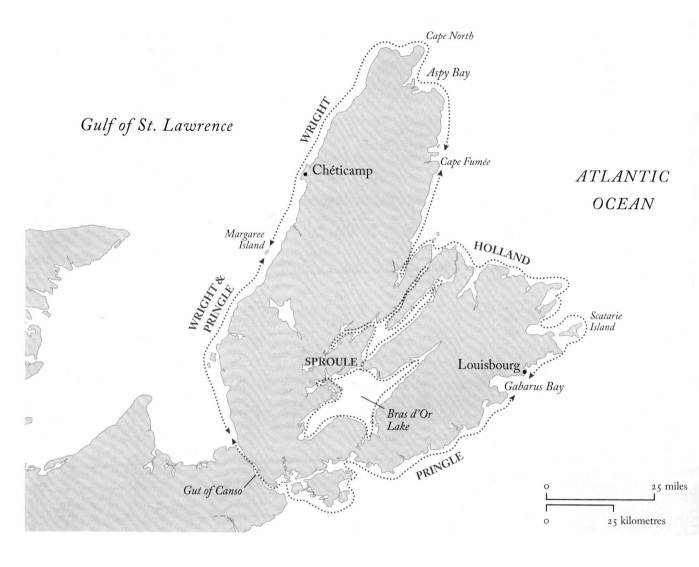

Sproule surveyed the Bras d'Or Lake; Pringle worked along the south shore from the Gut of Canso to Gabarus Bay; and Holland surveyed from Gabarus Bay to Cape Fumée (figure 2.5).[44] In addition, the *Canceaux* and naval crews in small boats carried out soundings. By the end of July 1766, the survey was largely complete, save for part of Great Bras d'Or Lake and Scatarie Island, the place where the *Canceaux* had nearly been wrecked two years earlier.[45] Surveying Cape Breton had not been easy. Wright had considerable difficulty surveying from Chéticamp around Cape North to Aspy Bay. The coast was virtually all rocks and cliffs, which meant the surveyors could not land to chain along the foreshore; instead, they measured using a rope strung between two boats. Although a schooner with supplies and

2.5 · Survey of Cape Breton Island, 1765–1766

56 · SURVEYORS OF EMPIRE

provisions supported the boats, the skipper found it difficult to keep station; prevailing westerlies frequently threatened to blow the vessel onshore, and easterlies off the island blew the vessel into the Gulf. As a consequence, Wright's surveying party was often "brought to a starving condition."[46] Meanwhile, Holland had his own difficulties surveying Scatarie. The fogs were "so frequent" that he was "twice obliged to return without performing [the survey]."[47]

Nevertheless, enough work had been accomplished during the summer that Holland divided the survey teams in August, sending Wright to survey and overwinter at Anticosti, a large uninhabited island in the northern part of the Gulf, and Pringle to survey and overwinter at Pabos, a small Acadian fishing settlement on the south shore of the Gaspé peninsula.[48] Holland and Sproule remained at Louisbourg to complete the survey and draft the maps. Although the winter was "very severe," the teams finished surveying several inland lakes and small streams, probably Millward Lake (Mira River).[49] As with the survey of St. John's Island, much of the interior of Cape Breton Island remained unsurveyed.

Holland also prepared his report on the island for the Board of Trade. Among his several recommendations was to erect a town at Dartmouth Harbour (modern Sydney), a deepwater, well-sheltered harbour on the island's east coast. He could not understand why the French had selected "cold, barren" Louisbourg, rather than Dartmouth Harbour, as their principal base.[50] He further recommended the political separation of Cape Breton from Nova Scotia, in order to encourage the island's economic development.[51] Both suggestions would be acted upon after the end of the War of American Independence. Convinced that "no part of North America can boast a more advantageous situation for commerce & fishing," Holland outlined the island's chief resources of coal and fish. He was sure that once the Lords of Trade had seen his plan and description, they would find Cape Breton Island "a greater acquesition to the British Empire in America, than was expected."[52]

During the "extraordinary severe" winter at Louisbourg, Holland turned his mind to more exciting ventures than writing reports and drafting maps. In January 1767, he sent Hillsborough an intriguing proposal for an expedition to find the fabled Northwest Passage.[53] "What I have to offer," Holland explained, "is only the result, of what I have collected, at my leisure hours, from sundry authors personally concerned." After outlining the progress of the Russians through Siberia to Kamchatka and thence across the Pacific to the northwest coast of North America, Holland suggested an expedition from Nelson River on Hudson Bay to the Pacific. Holland reasoned that the "rivers must run easterly into the Bay, or westerly into the Pacific Ocean" and that he could "transport my necessaries from thence thro' the several lakes, to the heads of the rivers that run west, to the Pacific Ocean, & there building two small vessels, seek the passage, & examine those discoveries laid down in the maps of other nations, whether they are real or feigned." Holland admitted that he could only outline the project but when "I arrive in Canada, I shall be indefatigable in gaining what intelligence I possibly can." Although his proposal revealed European ignorance of the western cordillera, it did suggest a strategy for finding the passage from the Pacific Ocean. By January 1768, Holland had heard nothing from the Board of Trade and considered the proposal a dead letter.[54] Nevertheless, Holland's idea reflected a gathering interest in a renewed search for the Northwest Passage.[55] In 1770, the Hudson's Bay Company, stung by criticism of its inactivity beside the "frozen sea," sent Samuel Hearne into the interior in search of a viable route. The young explorer descended the Coppermine River only to find that it disgorged into the Arctic Ocean rather than the Pacific. The continent was beginning to appear much wider than many supposed. Six years later, James Cook, on his third voyage to the Pacific, sailed to the northwest coast in search of the passage, confirmation that a maritime approach was worth pursuing.

Although Holland had planned to join Pringle at Pabos on the south shore of Gaspé in early spring 1767, the surveyor general spent a good part of spring and summer in Cape Breton. In late April, the *Canceaux* had to sail to Halifax for a refit in the naval dockyard and to take on provisions, and did not return until the end of June.[56] In early July, Holland and his surveyors and their baggage embarked on the ship and reached Gaspé at the end of the month.[57] There, they found Wright and his party from Anticosti, and within a week Pringle had arrived from surveying Chaleur Bay. But their reunion was short lived. Convinced of the utility of dispersing his parties around the Gulf, Holland divided them again. Sproule was left at Gaspé in order to fix its longitude through astronomical observation and to carry Pringle's survey around the peninsula into the St. Lawrence River as far as Matane. Meanwhile, Holland was conveyed on the *Canceaux* to Matane, from where he spent the rest of the summer surveying the South Shore as far as Bic. By the beginning of October, Holland was back in Quebec taking stock (figure 2.6).[58] Only thirty leagues (ninety miles) remained to be surveyed along the South Shore between Bic and the easternmost point reached by the Survey of Canada, but much of the North Shore still needed to be surveyed (figure 2.7). The Survey of Canada had got as far as Île aux Coudres, which still left an enormous amount to be done. Holland planned to start by sending a party to overwinter on the Saguenay River from where they could survey St. John's Lake (Lac Saint-Jean). Of the surveyors remaining in Quebec, Wright and Pringle were granted leave to return to England, having "had so great a share of our fatigues" and

2.6 · *A view of Quebec from the south east* from *The Atlantic Neptune*. © National Maritime Museum, Greenwich, London, HNS 125.

"compleated every thing agreeable to [Holland's] directions."[59] The other surveyors were to stay in Quebec, perhaps accommodated in the Intendant's Residence or the Jesuit College.[60] These members of what Holland called his "travelling academy" were to draw plans of Chaleur Bay, Anticosti, and the South Shore.[61] The *Canceaux* was despatched to Gaspé to overwinter and support Sproule's party.

Four years into the General Survey, Holland was undoubtedly feeling the strain. In January 1768, he complained to Haldimand about being "tired of surveying." He had not heard from the Board of Trade for more than a year and was out of pocket for survey expenses. "I had last year but little satisfaction for my troubles, things being in such confusion at home [London], that our services abroad are but little thought on." Holland's application for more money to pay for an extra surveyor and six soldiers had not reached the Board in time to be included in the army estimates submitted to Parliament, leaving him to cover the substantial cost himself. Moreover, his proposal for an expedition to the Pacific had been ignored, "so I think no more about it, & as I find so little notice taken of what is already don[e], I am determined not to entangle in any more bussiness than what I have now on my hands."[62]

Yet if the Board of Trade in London had lost track of the General Survey, General Gage in New York was realizing its importance. In the wake of the riots sparked by the Stamp Act in 1765, Gage had become much

Surveys · 59

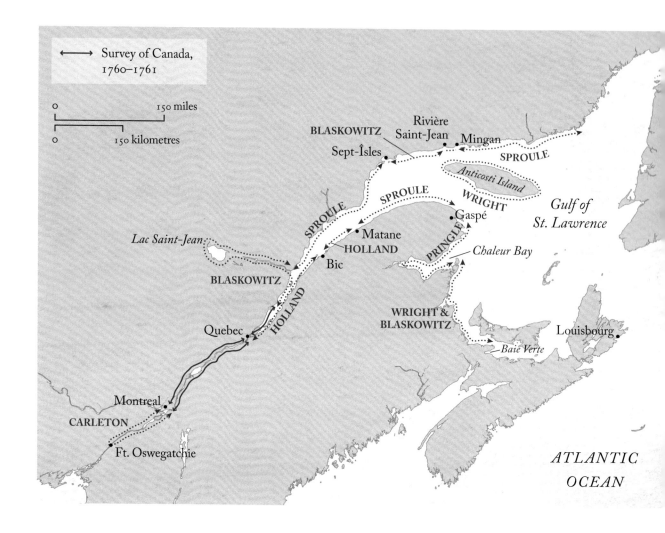

more aware of the difficulties of maintaining authority in the continental colonies. Good geographic information on towns and countryside was much needed. In mid-December 1765, Gage ordered Montrésor to survey New York and the countryside to the north and north east of the town; in September 1766, Montrésor also surveyed various islands in the harbour with a view to their fortification. Even more importantly, he prepared, in great secrecy, a military plan for moving "a considerable body of troops ... through an interior inhabited country in North America."[63] At some point in late 1766 or early 1767, Montrésor must have informed Gage about the lack of maps of the interior. Gage, who had been kept informed by Holland about the progress of the General Survey, expressed his concerns to the surveyor general. "When you shall have finished the coast," the general wrote in July 1767, "it is to be hoped you will be employed in surveying the interior

2.7 · Survey of the Gulf of St. Lawrence and St. Lawrence River, 1766–1770

60 · SURVEYORS OF EMPIRE

parts of this country; there is no map of the inhabited provinces of any use, for there is none correct, even the roads are not marked. A good military sketch of these provinces, was it no more, might be of great use."[64] Circumstances were changing in British America and government officials were beginning to respond. Even so, Gage never requested the Board of Trade to shift the focus of the General Survey from the Gulf of St. Lawrence to the continental colonies. Holland was left alone to continue tidying up the geography captured in the last war.

As the ice left the St. Lawrence in spring 1768 and communication was restored between Quebec and the Gulf, Holland caught up with the fortunes of Sproule and Mowat and learnt that another disaster had only just been averted. Although Mowat had planned to overwinter at Gaspé in support of Sproule's party, the *Canceaux* had encountered a severe northwest gale off Cape Rosier (Cap des Rosiers) in late November. As arctic air spilled into the Gulf, "a very hard frost" turned the ship into "a body of ice." With the *Canceaux* in serious danger of foundering, Mowat had no choice but to run before the gale and seek shelter, first in the lee of the Magdalen Islands and then in the Gut of Canso.[65] Having lost sails, rigging, and top masts, the ship could not beat back to Gaspé and so Mowat anchored in Cape Canso Cove for the winter. Sproule and his party were now in a desperate situation, running low on provisions more than 300 miles from their depot ship. Mowat attempted to help by transferring supplies to a sloop that had taken shelter at Canso and intended to go whaling in the northern Gulf, but the vessel failed to reach Gaspé.[66] Eventually, a shallop managed to resupply the survey party sometime in early January.[67] By then, Sproule's party had been "reduced to very great distress, having only three days provisions left."[68] The *Canceaux* herself had to be refitted in the dockyard at Halifax during March and April, before returning to embark Sproule in the first week of June.[69]

During summer 1768, Holland's parties continued working in widely separate places.[70] Sproule was transferred from Gaspé to the North Shore in order to survey from the Bay of Seven Islands (Sept-Îles) towards Tadoussac (figure 2.8). The party that had surveyed St. John's Lake (Lac Saint-Jean) during the winter continued its work down the Saguenay to the St. Lawrence. Meanwhile, Holland and another party finished surveying the South Shore. Another party began surveying west of Montreal up the St. Lawrence River towards Lake Ontario, getting as far as Fort Oswegatchie (Ogdensburg, New York). By November, Holland reckoned that the General Survey, after four years of work, had completed 5,000 miles "all chained," by far the largest survey ever conducted by the British.[71]

Over winter, Holland remained in Quebec collating the surveys and bringing them to a common scale. He also planned to produce a map of

seigneuries in the province granted during the French regime.[72] General Guy Carleton, who had taken over from Murray as governor of Quebec in 1768, was keen to sort out the legal basis of landholding in the province and secure the support of the Canadian seigneurs for his new administration. Building on the Survey of Canada, the map of the seigneuries was another part of the government's administrative stocktaking. Holland was also anxious to press on with the General Survey and sent Charles Blaskowitz with a party to overwinter at Mingan on the North Shore of the Gulf of St. Lawrence, where they were to establish the longitude and survey St. John's River (Rivière Saint-Jean), the eastern limit of the province.[73] Holland himself ventured out to survey Montmorency River, site of Wolfe's encampment ten years earlier.[74] With soundings in the Gulf largely finished, the *Canceaux* returned to Quebec to be laid up for the winter.[75]

The surveyor general was determined to finish the surveys of the Gulf in 1769, although he would play little direct role in surveying. In spring, Sproule was sent in a schooner to finish the survey along the barren, glaciated North Shore, which included a section near Tadoussac, the Manicouagan Shoals, and eastward from Mingan "as far as will be sufficient … to compleat the North Channel or entrance of this extensive river."[76] Sproule took the survey as far as Cape Joly opposite the eastern end of Anticosti Island.[77] Meanwhile, the *Canceaux* and small boats were employed sounding along the St. Lawrence.[78] Most of Holland's time, however, was focused on two major scientific and surveying projects: the transit of Venus and establishing the boundary line between New York and New Jersey.

The transit of Venus on 3 June 1769 was one of the major scientific events of the eighteenth century.[79] The movement of the planet Venus across the face of the sun provided astronomers with an opportunity to calculate the distance between the earth and the sun. As transits of Venus usually

2.8 · *A view of Cape Torment, part of the Island of Orleans, Isle Madame & Isle au Rot [Ruaux]*, by James Peachey, 16 November 1784. This view gives some sense of the immensity of the St. Lawrence River and the landscape that Holland and his deputies had to survey. James Peachey joined the General Survey in the early 1770s. Library and Archives Canada, Acc. No. 1989-218-7.

2.9 · *A view of Island Coudre, bearing N.W. distance 2 leagues, taken in crossing the traverse*, by James Peachey, 17 November 1784. Wright observed the transit of Venus from the island on 3 June 1769. Library and Archives Canada, Acc. No. 1989-218-9.

occur in pairs eight years apart in a cycle that takes more than a century to repeat, the transits that occurred in the late eighteenth century were extremely important.[80] The first transit in 1761 had attracted considerable scientific attention, particularly in France and Britain. The second, expected in 1769, drew even more interest and led to an international effort to record the planetary movement from observatories strategically placed around the globe. The scale of the British effort was unparalleled, reflecting the nation's rapid emergence during the 1760s as the pre-eminent European scientific power.[81] Among the expeditions financed by the British government and the Royal Society were Cook's voyage to the Pacific to record the transit from the newly discovered island of Tahiti, Wales and Dymond's trip to Churchill on Hudson Bay, and Bayley and Dixon's expeditions to Hammerfest and North Cape in northern Norway.[82] The transit also generated interest among scientific communities in British America, particularly the American Philosophical Society in Philadelphia, which sponsored several observations.

As one of the leading astronomers in North America, Samuel Holland was keen to observe the transit. He had been in regular communication with Nevil Maskelyne, Astronomer Royal at Greenwich, about calculating the longitude of various places in the Gulf of St. Lawrence and had received a pamphlet from Maskelyne concerning the transit.[83] Thomas Wright was also involved in the observations. While in London, he had been briefed by Maskelyne about the transit, and on his way back to rejoin the survey at Quebec stopped at Île aux Coudres in the St. Lawrence to make the observation (figure 2.9). In the event, Holland saw the external contact of the planet with the sun but missed the complete transit because of clouds, while Wright took his eye off the telescope at the critical moment of external contact but saw all the internal movement across the face of the sun.[84] Both

men submitted papers on their observations to the Royal Society, which were read and later published in the society's *Philosophical Transactions*.[85] Maskelyne appeared to be pleased: "[Wright's] observations seem to me to be made with proper care & knowledge of the subject, & likely to be useful, in conjunction with those made by Captn. Holland & the many other observers dispersed in various places, in determining the grand problem of the Sun's distance from the Earth."[86]

Soon after the transit on 3 June, Holland left Quebec for New York to join the commission to settle the boundary between New York and New Jersey. He had received the invitation from Sir Henry Moore, governor of New York, the previous year and felt honoured to be included among the commissioners.[87] As he explained to Hillsborough, "settling the limits of the two provinces in dispute, requires an accurate set of instruments for surveying & astronomy, as well as a great knowledge of the country; I believe no person in those provinces can equal me in the first of these points, & perhaps, not a great many in the latter, as I was long in those provinces at the commencement of [the] last war."[88] The trip proved useful. Holland went overland via the Richelieu River, Lake Champlain, and the Hudson Valley, taking a survey party with him. The surveyors were left at Fort Ticonderoga to survey south to Albany, while Holland continued on to New York. He attended the first meeting of the boundary commission in mid-July, and then spent the rest of the summer and early fall establishing the latitude and longitude of the boundary line.[89] Holland appears to have taken a portable observatory with him, although the telescopes used were those employed by Charles Mason and Jeremiah Dixon in settling the boundary between Pennsylvania and Maryland.[90] Among the surveyors working with Holland was David Rittenhouse from Philadelphia, who was already becoming known for his astronomical and mathematical interests. Indeed, Rittenhouse had just observed the transit of Venus (he had fainted with excitement) and no doubt the two men had much to discuss. Certainly, Holland found him "a most surprising genius," who had acquired his skills and knowledge "without any other instructions but by books." The meeting of the two men represented the coming together of two scientific networks, one based in metropolitan Britain, the other in colonial America.[91] Rittenhouse was to play an important role in the development of American science after the end of the War of Independence. Holland also noted that his map of New York, created for Lord Loudon in 1756 and published by Jefferys in 1768, was much in use, although the surveyor recognized its deficiences. He was now sensible that the sources he had used to compile the map were "extreamly erroneous"; he had also been informed by "several gentlemen of the navy" that the sea coasts of Long Island and New England had "scarce any resemblance." Once the General Survey arrived in New York, Holland hoped to "reform these errors."[92]

By the end of the year, the surveyor general was back in Quebec and planning the next season's campaign. The surveys of the St. Lawrence River and much of the Gulf of St. Lawrence were now complete, but Holland was chagrined to learn that Des Barres had no intention of extending his survey of Nova Scotia as far as Chaleur Bay. This meant that a stretch of coast between the south shore of Chaleur Bay and Green Bay (Baie Verte) still needed to be surveyed. Holland planned to have a party complete that task in the spring.[93] Holland was also concerned about the failure of Mowat to complete the soundings and sailing directions in the Gulf and St. Lawrence River. Although Mowat had done some marine surveying, he had spent a good deal of time cruising for smugglers, a potentially lucrative endeavour as he stood to benefit financially from any vessels and cargoes that he intercepted. The end result was that several of the maps submitted to the Board of Trade lacked soundings. But as Holland explained to Hillsborough, "the commander of His Majesty's naval forces in Nova Scotia can at any time get [soundings] performed by some able officer of his squadron, whose ambition to distinguish himself may be superior to other motives."[94]

As the last piece of the Gulf was being surveyed in spring 1770, Holland could report to Lord Hillsborough that "the survey of these northern parts is finished."[95] It had been an immense undertaking. Even with three or four deputy surveyors and the support of a ship and several smaller craft, the survey had taken more than five years, the time that Holland had originally allocated for the entire General Survey of the Northern District from the St. Lawrence to the Potomac. More than five thousand miles of coast had been surveyed, much of it among the bleakest mid-latitude environments anywhere in the world. During their surveys, Holland and his men had encountered glaciated rock, endless spruce, freezing seas, bitter winters, and almost no settlement. As early as 1767, the surveyor general was referring to the Gulf survey as the "savage part of our duty."[96] He was no doubt glad to inform Hillsborough on 6 June 1770 that "On Sunday next I shall embark with all my assistants to begin the eastern parts of New England, to join the surveys already made by Mr. Desbarres under the direction of the Admiralty; & I hope to winter at Portsmouth in the Province of New Hampshire, from which place the surveys will be carried on with more comfort & less danger than in the inhospitable regions we have had the good fortune to see finished."[97]

Des Barres's Survey of Nova Scotia, 1764–1775

Des Barres's survey of Nova Scotia was much different than Holland's survey of the Gulf of St. Lawrence. Although extensive, the coasts of penin-

sular Nova Scotia and Sable Island were not on the same scale as the coastlines of the Gulf, which meant that Des Barres's survey was much smaller than Holland's and could be based in one location. Unlike Holland, who had to relocate his headquarters every few years, Des Barres remained based at Halifax and his country home at Castle Frederick in Falmouth throughout the duration of the Nova Scotia survey. This ensured a short chain of command between Des Barres and the commander-in-chief in Halifax, as well as close logistical support from the naval dockyard. Moreover, Des Barres was surveying a province that had some European settlement. Nova Scotia had approximately 13,000 inhabitants in 1764. Some 3,000 lived in Halifax, making it the largest town between Portsmouth, New Hampshire, and Quebec (figure 2.10).[98] Many of the settlers in the province had recently arrived from New England and taken up former Acadian farmland in the Annapolis and Avon river valleys.[99] Des Barres lived among these settlers in Falmouth, one of several new townships laid out in the Avon Valley.

After the Admiralty approved Des Barres's appointment to the Nova Scotia survey, Rear-Admiral Colvill, commander-in-chief in Halifax, made sure that the survey was ready to start at the beginning of summer 1764. On 20 May, Colvill recorded in his journal that he had "ordered the midshipman commanding the Dispatch tender to receive Lieutenant Des Barres and to proceed with him to such harbour or part of the coast as he shall from time to time desire for the purpose of surveying, and to give him all

2.10 · *Town and harbour of Halifax as they appear from the opposite shore called Dartmouth*, by Dominic Serres, circa 1762. Collection of the Art Gallery of Nova Scotia, 1982.44.

2.11 · Survey of Nova Scotia, 1764–1772

the assistance in his power."¹⁰⁰ The *Dispatch*, as its name implies, was a small craft used for ferrying the admiral's dispatches to vessels patrolling northern waters. The actual survey began on 6 June when the *Dispatch*, with Des Barres onboard, left Halifax Harbour.¹⁰¹ The surveyor's original intention was to start at Cape Musquodoboit, about 25 miles east of Halifax, and then work eastward to Cape Canso, but in the event he began farther east at Beaver Harbour and then worked westward to St. Margaret's Bay.¹⁰² This change in plan most likely reflected the obvious need to survey and map the coastline adjacent to Halifax at the earliest opportunity (figure 2.11). From the start, Des Barres was intent on producing a marine survey, combining coastal topography and seabed soundings. As he explained to Gage after the first season of surveying, he was going to show "in the clearest and exactest

manner ... the true dimensions and appearances of every inlett, harbour, bay, headland, islands, rocks, shoals &c. And the particular soundings in each of those bays, harbours, &c. which soundings I have carryed to the distance of ten and twelve miles, along the shore, in its offing; and to this I propose adding a few observations concerning the tides, &c., and written directions for to sail into each of those harbours, bays, &c."[103] The division of work between Holland and Mowat that so plagued the General Survey of the Northern District was not going to be repeated in the Survey of Nova Scotia.

In summer 1765, Des Barres completed the survey of the Eastern Shore as far as Canso.[104] Both seasons had been slow going, mainly because Des Barres had encountered an immensely complicated coast. As he explained to Colvill after the first season: "scarcely any known shore [is] so much intersected with bays, harbours and creeks ... and the offing of it is so full of islands, rocks and shoals as are almost innumerable."[105] The following year, Des Barres switched his attention to surveying Sable Island, which lay approximately 100 miles southeast of Halifax. An ever-changing sand bar, Sable Island lay in the track of vessels sailing from North America to Europe, and had become a graveyard of ships and men. Des Barres spent only a few weeks surveying the island before being recalled to Halifax. Colvill was returning to England and the crew of the survey vessel were needed to man the commander-in-chief's flagship.[106] With no crew available, Des Barres's vessel was discharged, bringing the season's surveying to an abrupt end.[107] The 1767 season was not much better. Six weeks were lost trying to charter a vessel and find a crew and then severe weather seriously hampered work on Sable Island. A small schooner and two boats employed on the survey were lost, and Des Barres himself narrowly escaped from drowning.[108] The following year, Des Barres completed the survey of Sable Island and surrounding shoals, and then moved back to the Eastern Shore in order to extend the survey beyond Canso. During this work, a sloop used on the survey was wrecked in Chedabucto Bay.[109] Nevertheless, Des Barres appears to have got as far as Tatamagouche Bay. He finished the eastern part of the survey at Cocagne River (near Shediac, New Brunswick) at the end of the following season.[110]

By late 1769, the lack of a dedicated vessel for the survey had become such a problem that Commodore Samuel Hood, who had replaced Colvill as commander-in-chief, convinced the Admiralty to authorize the purchase of a Marblehead schooner for Des Barres's use. The Lords named the vessel *Diligent*, and appointed Lieutenant John Knight as commander.[111] Knight had served as a midshipman on the *Romney*, the commander-in-chief's flagship in Halifax Harbour, and was transferred to the survey in 1765. Given Knight's four years experience in marine surveying and sounding, Des

Barres was keen to ensure his promotion to the new command and lobbied Hood and the Admiralty.[112] Knight would have a distinguished career as a marine surveyor.

The 1770 season was spent sounding along the Atlantic coast of Cape Breton Island and surveying the South Shore of Nova Scotia.[113] Knight and the *Diligent* sounded the waters off Louisbourg and Scatarie Island during June, presumably because Mowat had not done so in the *Canceaux*. The remainder of the summer was spent surveying and sounding the South Shore, with occasional visits to Halifax for supplies and provisions. Des Barres had got as far west as St. Margaret's Bay in 1764 and now needed to extend the survey to Cape Sable. As with the Eastern Shore, the coastline of the South Shore was deeply indented and took an immense amount of time to survey. Knight was also employed in the offing, sounding the outer banks of Nova Scotia.

In August 1770, Des Barres and Holland crossed paths in Liverpool, Nova Scotia, and discussed plans for publishing their work. Although the two surveyors had corresponded, this appears to have been the first time that they had met since before the surveys started. Holland was in transit from Quebec to Portsmouth, New Hampshire, onboard the *Canceaux* and joined Des Barres for three days. Simeon Perkins, a Connecticut merchant who had recently settled in the small Nova Scotia outport (figure 2.12), recorded in his diary the visit of the two surveyors:

2.12 · Simeon Perkins House, Liverpool, Nova Scotia. Leading Liverpool merchant Simeon Perkins entertained Holland, Des Barres, and Mowat here in August 1770. Photograph courtesy of Scott Baltjes.

Surveys · 69

> Sunday, July 29th, – Mr. DesBarres, employed by the King, to survey the coasts of this Province, and has been here the whole of last week, removes to Port Mutton.
> Tuesday, Aug. 7th, – I go to Herring Cove to ship fish for Boston. Meet DesBarres, the surveyor, returned from Port Jolly, awaiting a ship, Canso [*Canceaux*]. There is some talk of the captain of the Canso about to search vessels for unaccustomed goods.
> Thursday, Aug. 9th, – Ship Canso arrives. Capt. Moit [Mowat], the commander, and another officer, go to my house. They do not intend to search vessels.
> Saturday, Aug. 11th, – I go on board the Canso with John and Samuel Doggert, Capt. DeMill and Capt. Adams.
> Sunday Aug. 12th, – The officers of the Canso call to see me, with Capt. Holland and DesBarres, and took a glass of wine.
> Monday Aug. 13th, – The Canso sails this morning.[114]

During the visit, Holland proposed to Des Barres a scheme "for extending the public benefit" of the two surveys.[115] According to Des Barres, Holland acknowledged that his charts "cannot be so generally beneficial to the public for want of proper soundings" and the few which had been laid down were "not to be relyed upon." Holland had given all his attention to the survey of the land and considered soundings "without his sphere of a *surveyor of the lands*." Nevertheless, Holland proposed to furnish Des Barres "with his astronomic observations and materials ... by the means of which [Des Barres was] to protract proper sea charts and to improve and compleat the same with soundings and necessary observations and remarks &c. peculiarly adapted for the purpose of navigation." In return, Holland wanted Des Barres's surveys on a small scale to join to his "geographic map," which he would then send to Des Barres "to look over, & improve." In Des Barres's account, Holland proposed "to distinguish in his geographic map what part he has received of me, and I am likewise to acknowledge parts with which he may have supplyed me should I insert them when I come to publish my sea charts." Although admitting a reluctance to "join the work of others to mine," Des Barres realized the great opportunity "to compose a general Mercator chart, comprehending the River and Gulph of St. Lawrence[,] the Islands of Anticosty, St. John, Cape Breton[,] the Isle of Sable and this continent to the Bay of Fundy, &c, considering how very much such a chart is wanted at present; those hitherto published being full of errors and dangerous omissions." He also recognized that it would "still be necessary to survey again ... all the harbors, without which the soundings can never be laid down properly, nor directions given how to sail into them." In essence, the two surveyors had laid the foundations for *The Atlantic Neptune*. They had agreed to share their surveys and that Des Barres would publish them with full acknowledgment of Holland's role. They also agreed that further soundings would be necessary to complete Holland's charts of the Gulf of St. Lawrence.

Bruno Latour has argued that information from colonial peripheries was assembled in "centres of calculation" in Europe and then used by European colonial powers to control and manage the peripheries.[116] The meeting of Holland and Des Barres at Liverpool revealed the process by which imperial agents operating on the colonial periphery influenced the creation and dissemination of knowledge in the imperial core. Holland appears to have had the initial idea and realized that he needed Des Barres's cooperation to make a large chart of the northeast. Des Barres seized the opportunity, no doubt recognizing the importance of such a work for his career. Indeed, he set about the task immediately. After the two surveyors sailed out of Liverpool Harbour on the morning of Monday, 13 August, Des Barres made shore at Port Hebert, about twenty-five miles farther along the coast, and from his camp wrote to Hood in Halifax outlining the proposal. Des Barres had set in motion the idea for a nautical atlas that would work its way up the chain of command and in the process cross the Atlantic from Halifax to London. But the initiative for *The Atlantic Neptune* had come, appropriately enough, from two surveyors meeting in a remote outpost on Nova Scotia's Atlantic coast.

After surveying the South Shore, Des Barres moved steadily around the coast of western Nova Scotia and into the Bay of Fundy. In 1771, the *Diligent* was active sounding off Cape Sable, while Des Barres and his men surveyed and sounded in the bay. At the beginning of October, the schooner was caught by a westerly gale in the vicinity of Petit Passage between Digby Neck and Long Island and driven onshore, "much damaged."[117] Knight managed to get the *Diligent* off the rocks and sailed up the Bay of Fundy to Minas Basin and into the Avon River near Windsor, where the vessel was laid up for the winter. The other survey vessels were also laid up nearby, all within a mile or two of Des Barres's country residence at Castle Frederick. For the winter of 1771–72 the entire survey operation, through force of circumstance, had relocated from Halifax to the Fundy shore. The following spring, the *Diligent* was patched up and sailed around to Halifax for careening in the naval dockyard. After examination by the master shipwright, the schooner was considered in good enough shape to be put back into service.[118] The summer of 1772 was spent sounding off Cape Sable and in the Bay of Fundy, while smaller craft completed the survey along the north shore of the Bay of Fundy as far as the Saint John River, the terminus point of the survey. At the end of the season, Des Barres notified Rear-Admiral John Montagu, the new commander-in-chief, that he had completed the Survey of Nova Scotia.[119]

Des Barres was now keen to get to London and begin engraving and publishing his charts, but leaving America took more than a year. In early November 1772, the *Diligent*, with Des Barres onboard, was ordered to Boston, where Montagu had his flagship.[120] In passage, Des Barres encountered Holland onboard the *Canceaux* at Townshend Harbor (modern Boothbay Harbor, Maine), which provided the opportunity for Holland to hand over copies of the General Survey to Des Barres.[121] After repairs to the schooner's hull in Boston, Knight was ordered to take the *Diligent* to Halifax for the winter.[122] In March 1773, the Admiralty gave permission for Des Barres to leave for England "in order to publish his draughts" and to sail on the *Diligent*, but Montagu did not receive notification until the April packet arrived in June.[123] Even then, Montagu did not authorize the *Diligent*'s departure until the first week of November.[124] In the interim, Des Barres and the survey vessels continued to sound along the western shore of Nova Scotia.[125] Des Barres arrived at Spithead most likely in late November or early December 1773. After establishing himself in London, the surveyor was so dissatisfied with the lack of soundings off western Nova Scotia that he lobbied the Admiralty to send Knight and the *Diligent* back to the Gulf of Maine.[126] The *Diligent* returned to American waters in summer 1774 and continued sounding the shoals, reefs, and banks in the Gulf until the following summer when Vice-Admiral Samuel Graves, the new commander-in-chief, "directed Lieutenant Knight in the *Diligent* schooner to leave the business of sounding for the present and come to Boston, having more pressing service for him than what he is now employed on."[127] The outbreak of the American Revolution had finally brought the survey to a close.

Holland's Surveys of New England, 1770–1775

Holland's move from Quebec to Portsmouth, New Hampshire, in 1770 brought him, Mowat, and the *Canceaux* into a much different world than that in the Gulf of St. Lawrence. Portsmouth lay on the eastern edge of the New England colonies. With a population of 4,590 in 1775, the town had grown prosperous on the export of lumber, dried fish, and provisions to the sugar islands in the West Indies (figure 2.13). Leading merchants had built fine houses that were considerably more elegant than any in Halifax or Quebec.[128] The town served as the provincial capital and centre of the powerful Wentworth family; in 1770, John Wentworth served as royal gov-

2.13 · *A view of Portsmouth in New Hampshire taken from the east shore* from *The Atlantic Neptune*. Note the artist seated in the left foreground. © National Maritime Museum, Greenwich, London, NS 107.

ernor of New Hampshire. Portsmouth also lay within Boston's hinterland. Merchants and townspeople were well aware of the growing social and political unrest in the Massachusetts capital, and of attempts by the imperial government to impose its authority. The army and navy were an increasing presence in Boston and its environs. Troops were quartered in the town in 1768, and Commodore Hood shifted his flagship to Boston Harbor in 1770. From being on the outer periphery of the colonial world, Holland and Mowat were now extremely close to its centre.

Holland arrived in Portsmouth in August 1770 and immediately began organizing the coastal survey (figure 2.14). Although Wright and Blaskowitz had been left in the Gulf of St. Lawrence to finish the survey from Bay Chaleur to Baye Verte, Holland still had the faithful Sproule and dispatched him with a party and nine months' provisions to overwinter in Casco Bay. Meanwhile, Mowat took the *Canceaux* back to England for a major refit, the first time that the commander and crew had been home for seven years. Without the *Canceaux* to transport him along the coast, Holland was tied to the Piscataqua area and spent the summer surveying the river.[129] During the winter, he pushed farther inland. As in St. John's Island and Cape Breton, the surveyors used the frozen waterways to facilitate movement in the interior. A party was sent to survey the Connecticut River from its source high in the White Mountains to the Massachusetts border (the Connecticut served as New Hampshire's western boundary with New York), as well as Lake Winnipesaukee and the adjoining Smith

Surveys · 73

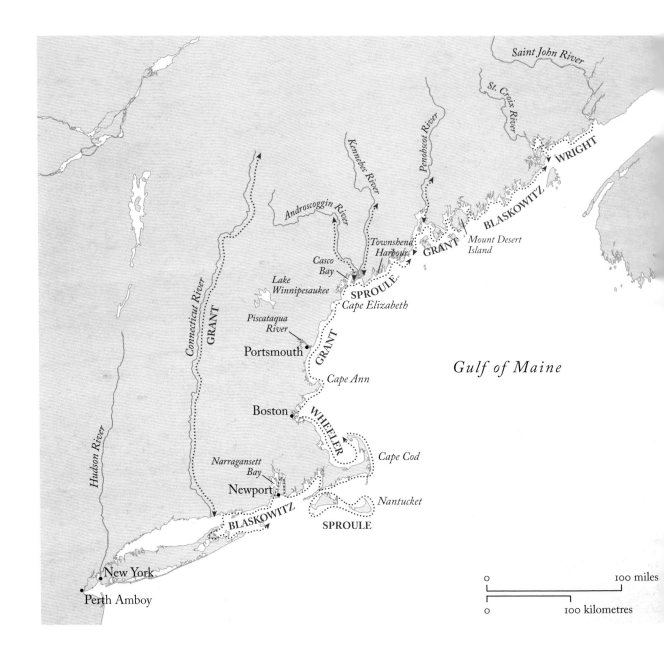

2.14 · Survey of New England, 1770–1775

Lakes (Lake Wentworth).¹³⁰ In the course of the work, Holland proposed to Governor Wentworth that the surveys should be extended to the entire province. With settlement expanding rapidly in the interior, particularly along the Connecticut River Valley, Wentworth realized the opportunity for a proper survey of the province and requested funding from the provincial House of Representatives, pointing out that "it is not probable the province will ever again have such an opportunity as now results from Capt. Holland's respectable offer of his services at an expense too inconsiderable to compare with the great & lasting advantages to ye publick which it comprehends."[131] But the provincial representatives, mindful of what they perceived as unnecessary expense, turned down the request.

Holland was also evaluating the coast between the Piscataqua and the Saint John River. In a letter to Hillsborough in December 1770, he recalled his visit to the mouth of the Saint John in 1758 in order to build a new fort.[132] During the course of the work, he took a sketch of the lower reach of the river and "could not help at the same time, admiring the amazing quantity of timber fit for every naval purpose."[133] Yet in conversations with a New England officer of the Rangers and carpenters working on the fort, Holland was informed that what he had seen "was nothing to compare between this river & that of the Kennebeck, especially in pine trees fit for masts." Ever after, Holland had made enquiries about the veracity of these statements and become convinced that "no tract in these northern parts is intrinsically of such value … to His Majesty, in supplying the Royal Navy in so essential a part of its construction."

Holland considered the region to be composed of three parts: from the Saint John to the St. Croix, a territory which belonged to Nova Scotia; from the St. Croix to the Kennebec, which belonged to the "Country of Sagadahook" (Sagadahoc); and from the Kennebec to the Piscataqua, which comprised the Province of Maine and belonged to Massachusetts. After conversations with Commodore Gambier in Boston and "some of the principal gentlemen of that place who wish well to government," Holland was convinced that Maine should be separated from Massachusetts. Indeed, he felt it was "hinted to me, that something of this kind was on the carpet at home, as the late conduct of the Province of Massachusetts Bay, was such as must surely affect their charter, & would give government an opportunity of seperating this territory of Main[e], politically, as it is seperated naturally from its present head." Fifty years ahead of his time, he recognized the advantages to the Crown of turning Maine into what he would later call "a grand magazine of masts."[134]

Holland also proposed new boundaries for the province, setting them out on a map that accompanied his letter (figure 2.15). Holland proposed the Saint John as Maine's easternmost boundary and the Saco River as the westernmost. "That part belonging to Nova Scotia," he explained, "would be no very inconsiderable loss to the Province; & for the small tract between Saco & Piscataqua Rivers now a part of Main[e], it may with propriety be annexed to the Royal Government of New Hampshire, that Province having so little sea coast & the River Saco would be a very natural limit as its eastern [boundary] line buts on part of it."[135] Appropriately, the new province would be called New Ireland for it lay between New England and Nova Scotia (New Scotland). Such a proposal to rearrange the political geography of northern New England reflected Holland's growing confi-

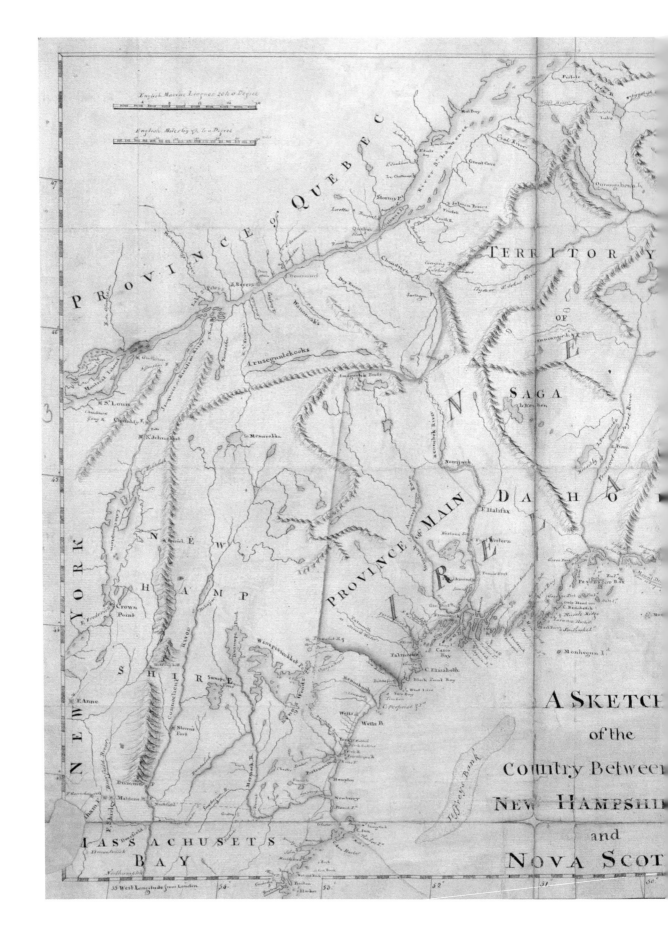

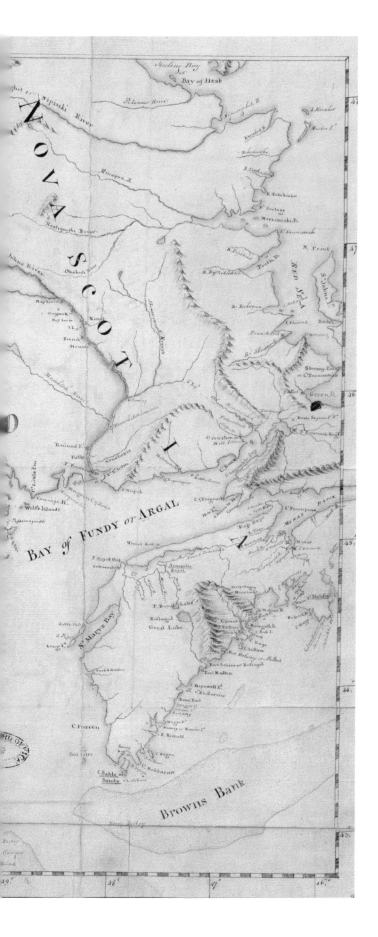

2.15 · *A sketch of the country between New Hampshire and Nova Scotia.* Holland's sketch map of the proposed province of New Ireland. The National Archives (UK), MPG 1/134.

dence as surveyor general and his utter dedication to advancing the interests of the Crown. By separating Maine from Massachusetts and transforming it into New Ireland, the Crown would have direct control over its government and economic development. Hillsborough found Holland's remarks "very sensible" and "useful," particularly as the government had instituted a survey of pine in the region.[136] Yet nothing was done until 1779, when the government briefly considered holding onto the area east of the Penobscot in order to accommodate Loyalists assembling at Majabigwaduce (modern Castine). The legal claim to the area was shaky, however, and the British dropped the idea during the peace negotiations that ended the War of American Independence.[137]

During summer 1771, Holland kept his surveyors working along the coast from Cape Ann in Massachusetts to the Kennebec River in Maine. James Grant surveyed from Cape Ann to Cape Elizabeth; Sproule from Cape Elizabeth, through Casco Bay, and up the Androscoggin and Kennebec rivers. Even more of the coast would have been done if the "multitude of islands in Casco Bay had not retarded the survey."[138] Holland and his deputies were now encountering an indented coastline much like the one that Des Barres had surveyed in Nova Scotia. Holland also continued working inland. The Board of Trade had instructed the governor of New Hampshire in 1761 to conduct a survey of the province and Wentworth used this decade-old directive to once again request funding from the legislature for a survey.[139] As the Board of Trade already paid the salaries of Holland and his surveyors, the only expense that needed to be covered was transportation, for which Holland had budgeted a hundred guineas. This time the House of Representatives felt more generous. In December 1771, the provincial government had finally been reimbursed for its expenses – more than six thousand pounds – incurred during the Seven Years War and the legislators felt flush. The house voted the funding for Holland's survey on 4 January 1772.[140] With this assistance, Holland dispatched three parties to survey the Merrimack, Salmon Falls, and Pemigewasset rivers, as well as fix the provincial boundaries using astronomical observations. Holland thought that if other provinces followed the example of New Hampshire in supporting surveys, "the geography of this country would soon be in a great degree of perfection."[141]

By November 1772, Holland was immensely appreciative of Wentworth, not only for his "kind endeavors" in securing the funding but also for "his constant compliance with every desire of mine, in expediting the business, as far as his assistance & authority were necessary."[142] During the year, Holland and his assistant Derbage had brought their survey plans and different private surveys of New Hampshire to a uniform scale, but progress on the provincial survey was painfully slow.[143] By April 1774, Holland was complaining to Lord Dartmouth, the new colonial secretary and president

2.16 · *A topographical map of the Province of New Hampshire* (detail). Holland's map provided detailed information on topography, settlement, roads, and civil boundaries. Library of Congress, Geography and Map Division, G3740 1784. H6 Vault.

of the Board of Trade, about the state of colonial surveying. "I find so little dependence can be put on the drafts or surveys made by those who pretend to practice surveying in this province," Holland wrote, "that I am obliged to examine every thing & make new surveys, as it were, only to bring the others to the exactness necessary in works of this kind. The unskillfulness I have mentioned of the practitioners in surveying of this Province (for I cannot call them surveyors) is such, as to have induced his Excellency Governor Wentworth to make application, to me, for Mr. Sproule to take on him the Office of Surveyor General of this Province."[144] The survey was completed in 1774 but it would be another decade before William Faden, Geographer to the King, would publish Holland's *A Topographical Map of the Province of New Hampshire*.[145] By then, of course, New Hampshire was no longer a province but a state.

Nevertheless, the map reveals, far more than any of the coastal surveys, the "exactness" of Holland's surveying (figure 2.16). No other map of colonial America shows in such detail the natural landscape of coast, marshes, waterways, and hills, and, even more impressively, the cultural landscape of roads, houses, churches, meeting houses, and mills. All these features were overlain with a political framework of townships and provincial boundaries. Gage had asked Holland for such a map of the interior parts of the country in 1767, but the commander-in-chief had not requested the Board of Trade to redeploy the surveyor to New York and the New England colonies. The New Hampshire survey, unique in its comprehensiveness, had come too late. A year after the survey had been completed, the British would begin

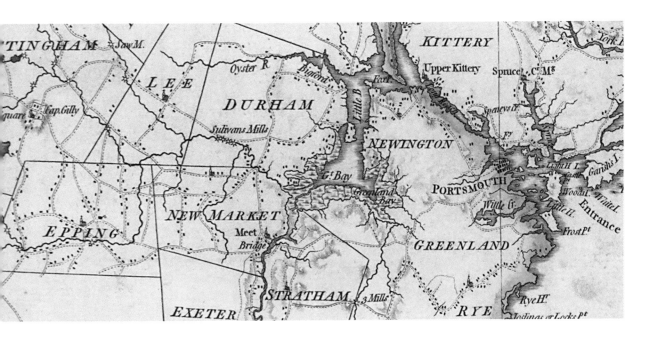

fighting the Americans without any detailed large-scale maps of the rebellious colonies.

Holland's survey of the coast of Maine was a major operation, as challenging as surveying any part of the Gulf of St. Lawrence. At the start of the summer surveying season in 1772, Holland pushed the survey as far as the mouth of the Kennebec River. He had the *Canceaux* back from its refit in England and two schooners which could land parties farther along the coast. While Sproule was left to continue working eastward from the Kennebec, schooners took Blaskowitz and Grant to Mount Desert Island and Wright to the Saint John River, the end point of Des Barres's survey. Blaskowitz was to survey eastward until he met Wright, while Grant surveyed westward until he met Sproule. Holland was under no illusions about the task at hand. As he explained to Pownall, "this extent [of coast] will be a laborious performance, as it is prodigiously intersected by harbors, bays, rivers, & inlets, as well as covered by a multitude of islands."[146] The surveyors were encountering the most complex coastline on the eastern seaboard. The mainland coast alone was approximately 3,500 miles in length; with the islands included, the total coastline was more than 5,000 miles. After a summer of "most severe and laborious service," the surveyor general had to apologize to Dartmouth for the lack of plans of the "Eastern Country." Another summer of work was needed to finish surveying the "infinite number of islands, & numberless harbors & bays."[147]

For the 1773 season, Holland sent Sproule, Blaskowitz, and Grant to finish the eastern surveys, while Wright was despatched to the Gulf of St. Lawrence to take more astronomical observations.[148] For his part, Holland took a cruise along the Maine coast onboard the *Canceaux* to fix locations through astronomical observations and to get a good view of the coast. He later wrote to Haldimand observing that the Maine coast consisted of a "multitude of islands, harbours, bays & inlets, in which no parts of America can equal in numbers in such a small extent."[149] Holland was accompanied by Wentworth, who, as Surveyor of the King's Woods, wanted to see the depredations being made on the pine trees of eastern Maine.[150] The three-week cruise took in Townshend Harbour (Boothbay Harbor), Goldsborough Harbour (Gouldsboro Harbor), and Cranberry Island Harbor.[151] Wentworth had hoped to get as far as Machias, the centre of lumbering in eastern Maine, but contrary winds slowed the *Canceaux*'s progress and Mowat was "apprehensive that the surveying partys … cou'd not be longer without the ship."[152] Nevertheless, Wentworth got some sense of the developing lumber trade and the extent of illegal cutting along the coast. He was extremely grateful to Mowat. Without a vessel of his own, the governor was completely dependent on the navy for transport, and made his appreciation known to the Treasury in London: "It is my duty here to observe," the governor wrote in November 1773, "the ready & most useful aid I have received from Mr. Mowat, in accomodating me with passages to these places, which was absolutely necessary, & his unwearied assiduity at all times in assisting me, by every means in his power to carry the laws for preserving pine timber into execution, which, but for this gentleman's good disposition & zeal, must have been utterly impossible for me to have so well accomplish'd."[153] The navy was crucial in lending logistical support to the civil power.

By October 1773, the surveys of the Maine coast were finished and Holland could dispatch to Dartmouth a small-scale plan of the sea coast from Cape Elizabeth to the Saint John River. He hoped that Dartmouth would "be pleased at seeing in one view, so great & valuable an extent of country … & prove I did not exaggerate … concerning its maritime advantages." Unfortunately, the "maritime advantages" could not be properly assessed as Mowat had done little sounding.[154] Holland now turned his attentions south and west towards Massachusetts, Rhode Island, Connecticut, and New York. He wanted to keep five deputies at work in order to finish the General Survey "as soon as possible, consistent with accuracy" and to relieve him of the physical stress of field surveying. At 45, he was "feeling the effects of seventeen years spent in the campaigns of [the] last war, & this employment."[155] Fortunately, the surveying would get easier. Cape Ann marked a significant break in the coastal geomorphology of the eastern seaboard. To the north, the coastline, as the surveyors had experienced, was rocky and broken, but to the south it was sandy and relatively uniform. Holland could now expect to make much faster progress.

Before the winter set in, Holland sent a party to extend the survey south from Cape Ann; he also despatched Grant to survey the upper course of the Connecticut River "through the woods to Canada" to connect the survey of the river to that of the St. Lawrence and to see if it was practicable to lay out a road "independent of water carriage."[156] No doubt Holland hoped to find an alternative route to the Hudson-Champlain valley and the Kennebec-Chaudiere rivers. As spring 1774 arrived, parties were sent south of Cape Ann. Wheeler was instructed to finish Boston Harbor (much of it already having been surveyed by George Callendar, master of the *Romney* in 1769) and then work towards Cape Cod.[157] Blaskowitz was sent to survey around the Cape to Rhode Island, and Sproule to survey Nantucket, Martha's Vineyard, and the Elizabeth Islands. The surveyor general prepared to move his headquarters from Portsmouth to Perth Amboy in New Jersey, a place close to New York and suited to "retirement, conveniency & neighbourhood."[158]

While Holland was redeploying his survey teams, the political situation in New England was deteriorating. The "Tea Party" had taken place

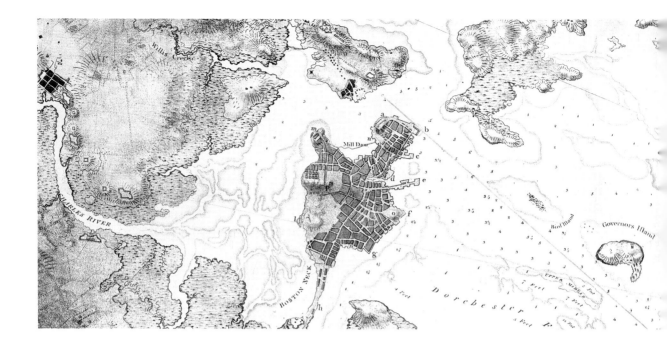

2.17 · *A chart of the harbour of Boston* (detail) from *The Atlantic Neptune*. The *Canceaux* took up station between Boston and Charles Town in May 1774. © National Maritime Museum, Greenwich, London, HNS 96.

in Boston in December 1773, and the British government was determined to punish the city. In May 1774, Rear-Admiral Montagu, after receiving orders to blockade the port effective the beginning of June, ordered all his ships to Boston.[159] The *Canceaux* sailed from Portsmouth on 21 May and took up station between Boston and Charles Town (figure 2.17).[160] Gage arrived the same month to take over from civilian governor Thomas Hutchinson; troops were disembarked in early June.[161] Even as these developments were happening in Boston, the General Survey continued along the coast. The *Canceaux* still acted as a depot ship, supplying stores and provisions to parties using the smaller surveying craft.[162] The survey parties in the field managed to avoid trouble. As Holland reported to Haldimand in June, "my parties are surveying in that province [Massachusetts] without any molestation, altho' Cape Codd & towards Road Island is the field of their labor; parts more exposed to an ignorant populace than the capital, as most persons of sence know the utility of our business."[163] Yet there were mishaps. In late August, one of the survey boats was wrecked to the west of Cape Cod, with the loss of two seamen, provisions, and supplies. Without a boat to reach the *Canceaux*, the survey parties were "greatly distressed before they could obtain provisions from Boston."[164] Meanwhile, Holland left for Boston in order to lobby Rear-Admiral Graves, the new commander-in-chief, for transport to move his servants, baggage, and equipment to New Jersey.[165] Unable to release the *Canceaux*, Graves procured a vessel leaving

Boston for New York and arranged the shipment of the surveyor general and his party from Piscataqua to Perth Amboy. Late in the year, Holland reached his new and, as it turned out, final headquarters.[166]

By December 1774, the General Survey had been brought round to Newport, Rhode Island, and Holland was busy settling another provincial boundary dispute. In late November and early December, he collaborated with David Rittenhouse to fix the 42nd degree of latitude on the Delaware River in order to run the boundary westward between New York and Pennsylvania.[167] Holland was also planning the next season's surveying, intending to extend the survey from Narragansett Bay to the Hudson River, including Long Island. After the clash of arms at Lexington and Concord in April 1775, he was much less sanguine. "I am affraid this summer's surveys will not equal those of former years," he wrote to Dartmouth in May, "as the present situation of public affairs is such as to make the continuance of our ship under the Admiral's immediate command, still necessary … but I will endeavor to do all that can be done."[168] Without the *Canceaux*, it was difficult to supply several survey parties operating along the coast; all that Holland managed to do was to equip one party to survey the environs of Perth Amboy (figure 2.18). Even so, the survey was cut short, perhaps by intimidation, and much of the summer was spent indoors drafting.[169] As rebel forces congregated around Boston, Holland was still thinking of surveying in New England. In September, he wrote to Dartmouth saying that he was ready "as soon as both, or either of the governors of New York & Massachusetts Bay shall signify the time most convenient for runing the boundary line between their provinces."[170] By then, of course, there could have been little realistic hope of any further surveying.

With British administration of the colonies breaking down, Graves was becoming increasingly concerned for Holland's safety. On 12 September, he wrote to the surveyor general expressing his "opinion that we are risquing the loss of all your drawings, plans, instruments &c, and perhaps the liberty if not the lives of all concerned by continuing to prosecute the business of the survey."[171] Graves ordered Captain Vendeput, commander of the *Asia* in New York Harbor, to receive onboard Holland and "such of the gentlemen and people employed on the survey as may prefer being secure in the Kings ship to remaining among the rebels," as well as all the equipment, charts, plans, and drawings "in order that the publick may not lose the benefit of [Holland's] labours purchased at so considerable an expence." Graves assured Holland that "affairs are growing worse, and that hostilities will not continue to be carried so partially as at present." In early October, Vandeput reported to Graves that Holland was still in Perth Amboy, where the "people are in general well affected."[172] A month later, Graves wrote to Vandeput in some desperation: "What's become of Captain Holland and

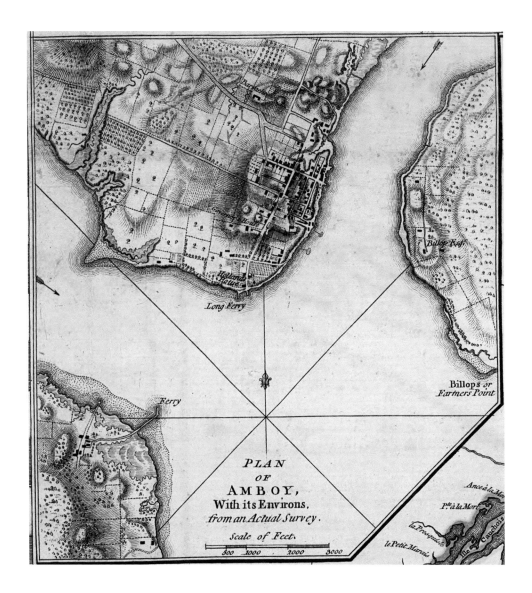

2.18 · *Plan of Amboy with its environs from an actual survey* inset on *The provinces of New York, and New Jersey, with part of Pensilvania and the province of Quebec*. Holland House is shown near the southern tip of Amboy. Library of Congress Geography and Map Division, G3800 1776.P6 Vault.

his officers on the survey? I observe you mention that his most valuable draughts are sent home, but he must notwithstanding have many articles remaining of too great consequence to fall into the hands of an ignorant, infatuated mob. If you can get a line to him, let him know that I would not have him put too much confidence in the Perth Amboyans, and desire him to place what he wishes to preserve on board you."[173] At some point, rebel leaders offered a military command to Holland, which he refused with "indignation and contempt." After that, his position in Perth Amboy was untenable. Leaving his family and all his effects "to the mercy of the rebels," Holland and his surveyors found safety onboard the *Asia*.[174] There, the party split: three surveyors were despatched to Boston with maps and charts for use by the commander-in-chief; Holland, two surveyors, and an assistant embarked on the packet *Earl of Halifax* for England.[175] The ship sailed on Wednesday, 15 November, a day of fresh breezes and clear skies, much different than the fog that had greeted Holland onboard the *Canceaux* off Scatarie Island eleven years earlier.[176] The General Survey of the Northern District of North America had come to an end.

Holland and his small party reached England by Christmas, and settled in London to complete the remaining drafts and "prepare the whole survey for publication."[177] For a man who had spent more than a decade surveying some of the bleakest coasts in North America, Holland must have been bitterly disappointed not to have completed the General Survey all the way to the Potomac. He reckoned that he would have completed the work the following season or "at most in one year more."[178] Nevertheless, he had enough drafting to occupy him, as well as contacts to make. Sometime between January and July 1776, Holland visited Cook in London, perhaps at his house in the Mile End Road. Cook had only recently come back from his second voyage to the Pacific – the search for the southern continent – and was now preparing for another voyage, this time to find the Northwest Passage from the Pacific side of North America. The two men were almost exactly the same age – Cook 47, Holland 48 – and both had achieved much since they had last crossed paths at the siege of Quebec in 1759. According to Holland, Cook was quite open about his debt to the army surveyor. "On my meeting him in London in the year 1776, after his several discoveries," Holland recounted, "he confessed most candidly that the several improvements and instructions he had received on board the *Pembroke* had been the sole foundation of the services he had been enabled to perform."[179] No doubt the two military men also reflected on the change in British fortunes in North America since their first encounter on the beach at Kennington Cove after the capture of Louisbourg.

The meeting of Holland and Cook in London marked the coming together of the two greatest surveyors of the age. Holland had just completed the largest systematic survey of any coastline in the world, Cook had explored and mapped more coastline than anyone ever before and helped open European eyes to the peoples and cultures of the south Pacific. Holland had demonstrated the scientific rigour and methodical practice of surveying by land, Cook had shown the possibilities of running surveys by sea. Both forms of surveying were essential for delimiting the geography of the world. The two men also represented different strands of British imperial endeavour. While Holland had spent nearly two decades helping to secure the British Empire in North America, Cook had extended it to the Antipodes. Although Cook and Holland could hardly have known it, their meeting at the time that colonial Americans were declaring independence marked a turning point in the future geographical expansion of empire.

In many ways, the two men personified successive stages of British imperial expansion. Cook represented the opening phase of exploration. He had made his name exploring the coasts of New Zealand and New South Wales and making the first small-scale charts of these lands and the great Southern Ocean. Holland, as well as Des Barres, represented the second phase of detailed surveying. Although the coasts of northeastern North America were well known by the late eighteenth century, they had not been surveyed and mapped at a large scale. Holland's achievement was to initiate this second phase of scientific and imperial endeavour. In this second phase, the surveys of Holland and Des Barres marked an important development in British and, more generally, European, knowledge of the world. Unlike Cook, who had sailed immense distances across the world's oceans, pinpointing islands and delineating continental coasts, Holland and Des Barres were based in particular parts of northeastern North America and systematically surveyed coasts and offshore waters. The two surveyors came to know these coasts and waters intimately, recording them in minute detail in field books, locating them precisely through astronomical observation, and delineating their intricate and varied topography and hydrography in manuscript charts and plans. Holland had also surveyed St. John's Island and Cape Breton Island in preparation for European settlement. At an abstract level, Cook had created the enormous spaces of the Pacific Ocean through exploration and mapping; Holland and Des Barres had created the smaller spaces of northeastern North America through triangulation and surveying.[180]

Yet there was a significant difference between the surveying methods of Holland and Des Barres. Holland was most comfortable surveying on land – coastlines, rivers, roads, settlements, terrain – and had little aptitude or interest in marine surveying, leaving that task almost entirely to Mowat and the navy. Des Barres, on the other hand, had been hired by the Admiralty to conduct a marine survey and worked closely with Knight in sounding the coasts of Nova Scotia. These different approaches would be reflected in the charts published in *The Atlantic Neptune*. The two approaches represented different ways of knowing land and sea, and would become increasingly familiar as British surveyors fanned out across the globe in the late eighteenth and early nineteenth centuries.

Surveying

3

To project scientific surveying expeditions from London across nearly 3,000 miles of ocean, support men and equipment along some of the harshest coasts in the world, and bring back information gathered over many years was a major undertaking. The British government had never attempted anything remotely like it before. Yet in 1764 the government launched the General Surveys of North America by Holland and De Brahm and the Admiralty surveys of Des Barres, Cook, and Gauld. Of these, Holland's General Survey of the Northern District was by far the largest and best equipped expedition to leave Britain before Cook's voyage to the Pacific in 1768. Although soon overshadowed by Cook's great venture, Holland's survey nevertheless represented a major commitment by the government to using science in the administration of empire. While much smaller than Holland's survey, Des Barres's Survey of Nova Scotia was also a rigorous scientific enterprise and stood comparison with Cook's survey of Newfoundland, De Brahm's of East Florida, and Gauld's of the Gulf of Mexico.

Even though British imperial and scientific ambition was great in the heady years of the early 1760s, the government scarcely had the specialized bureaucracy to handle large survey expeditions overseas. The Admiralty, which would be responsible for many of the logistics, did not possess a hydrographic office and treated marine surveying on an *ad hoc* basis. Surveys were commonly left to individual ship's captains and the making of charts to commercial map dealers. The practices of marine surveying were not well established, and astronomical observation, needed to establish latitude and longitude, was a specialized endeavour. Much had to be learned as the surveys progressed. Further complicating matters were confused chains of command, slow and tenuous trans-Atlantic communications, shortage of trained personnel, reliability of scientific equipment, and dependence on local knowledge. To produce the charts and reports needed in London, a great assemblage of men, vessels, equipment, and knowledge had to be put together and made to work in frequently trying and difficult circumstances over many years.

CHAINS OF COMMAND

At the centre of the British Empire stood the imperial government in London with its great departments of state arraigned along Whitehall. Several departments played important roles in the surveys, but the Board of Trade and the Admiralty had principal responsibility. Both departments served as heads of chains of command that stretched thousands of miles from the board rooms and offices of Whitehall to the survey teams and vessels operating along the coasts of eastern North America. The structure of these chains of command and the time it took for commands to cross the Atlantic were critically important to the successful operation of the surveys.

Established in 1696, the Board of Trade was intended to be the main government department overseeing England's plantations and colonies.[1] For much of the early eighteenth century, however, the Lords of Trade had not taken an active role in administering the colonies, content to leave them in a state of "salutary neglect." With the appointment of George Dunk, Earl of Halifax, as First Lord in 1748, the situation changed dramatically (figure 3.1).[2] Halifax quickly became a master of colonial affairs and was soon intervening directly in the colonies. His greatest achievement was the British settlement of Nova Scotia and the founding of its new capital, Halifax, in 1749.[3] Even after he left office in 1760, the Board continued to play an active role in colonial affairs. In fall 1763, the Board was probably at the height of its influence, advising the Secretary of State for the Southern Department on policy for the settlement and government of the newly acquired territories and helping draft the Royal Proclamation. At that critical juncture, the Earl of Hillsborough served as First Lord of the Board of Trade. An Anglo-Irish landlord, Hillsborough understood that expansion of interior settlements in North America had much to do with the great rage for emigration from the British Isles, particularly from northern Ireland, the centre of his estates.[4] He thus tried to restrict emigration and American settlement

3.1 · *Lord Halifax and his secretaries*, attributed to Daniel Gardner, after Hugh Douglas Hamilton, circa 1765–67. Halifax is shown with under-secretaries Edward Sedgwick and Lovell Stanhope. Holland named places on St. John's Island after Halifax and Sedgwick. National Portrait Gallery, London, NPG 3328.

while encouraging mercantile exploitation of the newly acquired territories, particularly their forests and fisheries. In 1768, Hillsborough was also appointed the first Secretary of State for the Colonies, a position created to deal with the growing political crisis in America. Other members of the Board of Trade during the 1760s were political placemen: Soames Jenyns, Edward Eliot, Edward Bacon, John Yorke, George Rice, Lord Orwell, and Bamber Gascoyne.[5] Some of these members had served on the Board for several years and accumulated considerable experience in colonial affairs; nearly all would be commemorated by the General Survey of the Northern District. The other significant figure at the Board of Trade was the secretary, John Pownall. Beginning as a junior clerk at the Board in 1741, Pownall rose through the civil service grades until appointed secretary in 1758, a position he held until retiring in 1776.[6] As secretary, Pownall prepared the agenda for Board meetings, reviewed all incoming and outgoing correspondence, oversaw the small clerical staff, and met all manner of people, from colonial agents to London merchants, interested in colonial affairs. Of immense assistance to Halifax during his term in office, Pownall also impressed Hillsborough, who testified to his "ability and extraordinary application."[7] During his work on the General Survey, Samuel Holland reported to both Pownall and the First Lord.

The Admiralty, the other department involved in the surveys, was almost an empire in itself (figure 3.2).[8] The Lords of the Admiralty over-

Surveying · 89

saw an immense military machine, comprising fleets, dockyards, victualling suppliers, and overseas establishments. After the naval victories in the Seven Years War, the navy's prestige could not have been higher. As a consequence, the Admiralty had immense political clout, reinforced by numerous naval officers and Admiralty officials who served as members of Parliament. But like all government departments, the Admiralty was subservient to the Cabinet. The First Lord of the Admiralty was a political appointee and Cabinet member. In September 1763, John Perceval, Earl of Egmont, took over as First Lord and remained in office for three years. Second in importance was the Senior Naval Lord. In April 1763, Richard Howe, Viscount Howe, was appointed to the position and stayed for two years.[9] Howe had been captain of the lead ship at the Battle of Quiberon Bay (1759), an engagement fought in treacherous inshore waters. He was thus well aware of the navy's need for accurate charts, particularly of foreign shores.[10] The secretary of the Admiralty, like the secretary of the Board of Trade, also played a pivotal role in the administrative system.[11] Philip Stephens was appointed secretary in June 1763 and remained in office for more than thirty years, a remarkable testament to his abilities. During that time, he accumulated vast experience, immense knowledge, and considerable influence, particularly over patronage. In general, the Lords of the Admiralty communicated with subordinate admirals and commodores who, in turn, instructed their captains and commanders.

3.2 · *Board room of the Admiralty*, published by Rudolph Ackermann after Thomas Rowlandson and Augustus Charles Pugin, 1808. Note the rolled wall maps, globes, and wind-dial. © National Maritime Museum, Greenwich, London, PAD 1358.

As both the Board of Trade and the Admiralty had responsibility for the General Survey, the chain of command was not always clear and this soon led to difficulties. Almost immediately after the *Canceaux* arrived at Port-la-Joye on St. John's Island in October 1764, Holland ran afoul of Mowat. To survey the island's coast, Holland needed small boats and crew, but Mowat refused to authorize either, insisting that the survey was under his direction.[12] To get Mowat's cooperation, Holland followed two courses of action. First, he reported the matter to the Board of Trade, which required sending a letter from St. John's Island across the Atlantic to London. After receiving and reviewing the request, the Lords of Trade instructed Pownall to communicate with Stephens, who in turn placed the matter on the agenda of the Board of Admiralty. After due consideration, the Lords of Admiralty sent instructions across the Atlantic to Colvill in Halifax, who in turn issued orders to Mowat. The Lords also wrote to Mowat directly, so "as may prevent any complaint of the like kind for the future."[13] From Holland submitting his request to the Board of Trade in November 1764 to Colvill issuing his order to Mowat in June 1765 took eight months.[14] Holland also approached Colvill, which proved far more expeditious. Upon receipt of Holland's letter and copy of his commission, Colvill took decisive action. Writing to Mowat in early November 1764, the admiral made clear that "the service [Holland] is employed upon is of too great importance to be neglected on any account whatsoever" and then severely reprimanded the naval lieutenant: "I take occasion to recommend to you in the strongest terms to maintain perfect harmony with Captain Holland, to look upon yourself as a joint labourer with him, and to suffer no such distinction as ships duty and surveyor's duty to exist but one and all, with heart and hand, to pursue a cause which is so much the object of his Majesty's attention and care."[15] Mowat had been put in his place. By January 1767, after two hard seasons of surveying, Holland had come to appreciate Mowat's cooperation. On hearing that commanders of armed vessels were about to be changed, Holland wrote to Hillsborough "to beg of your Lordship, to interfere should Lieut.t Mowat ... be under that order: notwithstanding, some misunderstandings between me & the former, I should still be glad to keep him, as those matters have been settled, & I may have the same disagreeable steps to undergo with his successor."[16] Mowat remained in command.

Although Colvill supported the surveys in any way he could, there were occasional lapses in communication between the Admiralty and newly appointed commodores, particularly as other matters pressed for attention. By the early 1770s, civil unrest in Boston had led commodores on the Halifax station to shift their flagship to Boston harbour. Distant from office files and pre-occupied with local affairs, Rear-Admiral John Montagu had no idea about the coastal surveys. "Having no article in my orders from their

Lordships that mentions either Captain Holland or Capt. DesBarre [sic] who are surveying the coast of Nova Scotia and North America, I am at a loss how to act," he reported to the Admiralty in September 1771.[17] During the winter of 1771–72, Holland directed Mowat to purchase boats, stores, and other material for the use of the survey but failed to inform Montagu. "As Capt. Holland is acting under the Board of Trade's orders," explained the admiral to the First Lord of the Admiralty, "he does not think he has any business to apply to me for what he may want. I would be glad Mr. Holland may be ordered to correspond with me and make me acquainted with his wants before he directs Lieut. Mowat to purchase any." Montagu further complained about supernumerary seamen that Holland had entered on the *Canceaux*'s books. "If he is allowed to enter men as he please there is no knowing where it will end. As it is, it appears to me a very expensive affair."[18] Chains of command were becoming entangled and became ever more so as demands on the navy increased.

Much of the navy's role in North American waters was to enforce British laws on trade and navigation.[19] During the 1760s, the British government, keen to raise more revenue, tightened the regulations. In 1764, the Sugar Act stipulated that no trade could be carried on between the French islands of St. Pierre and Miquelon and the North American continent, and that sugar and molasses could not be imported from French islands in the Caribbean to American or other colonial ports.[20] Cheap French molasses was in great demand by American rum distillers. The North American squadron played an important role in enforcing these laws, but the number of warships had been reduced after the end of the war, putting pressure on the few that were available. Nevertheless, there was considerable incentive to seize vessels engaged in illegal trade. Any vessel that was condemned in a vice-admiralty court or by a collector of customs was sold at public auction, with part of the proceeds given to the officers and crew of the seizing vessel. Given this financial inducement and the location of the General Survey in the Gulf of St. Lawrence, Mowat had considerable incentive to leave off sounding and cruise for smugglers. In late October 1767, Mowat intercepted a brig from Guernsey in the lower St. Lawrence carrying illegal goods; the vessel was escorted to Quebec where it was condemned and sold. Yet the seizure and sale took valuable time.[21] When the *Canceaux* headed back down the St. Lawrence in early November, the ship was caught in a gale that blew her beyond Sproule's party at Gaspé. The following May, Mowat seized another brig, this time at Petit-de-Grat in the southwest corner of Cape Breton. The brig was later condemned and sold in Halifax.[22] Holland was sufficiently fed up with Mowat's extra-survey activities that he reported to Hillsborough: "the naval part is much behind the land part of the survey; as the Canceaux Armed Ship, cannot compatibly attend the two businesses of sounding & watching for smugglers."[23] The pressure on the navy only increased as troubles grew in New England. By October 1771, Rear-Admiral Montagu, needing every vessel on the Halifax station to cruise for smugglers, ordered "Lieut Mowat of the Canceaux in addition to his former orders from the Lordships to endeavour to seize all vessels employ'd in illicit trade, but on no account to deviate from the service of surveying."[24] Mowat spent much of the survey serving two masters.

While the Board of Trade's General Survey had divergent, often conflicting, chains of command, the Admiralty's survey of Nova Scotia was much more straightforward. Although an army officer, Des Barres was employed by the Admiralty and reported directly to the commander-in-chief in Halifax. Commanders of the survey vessels reported to the same officer. As the survey was based in Halifax, Des Barres and the naval commanders were in frequent contact with the commander-in-chief; indeed, Des Barres became a personal friend of Samuel Hood.[25] There also seems to have been less pressure on Knight to cruise for smugglers. The commander of the *Diligent* intercepted the occasional schooner from Marblehead fishing on the Nova Scotia banks, but never made a seizure.[26]

Further complicating the administration of the General Survey was the slow speed of communication. Even though the Atlantic world was becoming more integrated in the eighteenth century through the introduction of packet boats that sailed monthly between New York and Falmouth, England, these vessels were a long way from Holland's surveys in the Gulf of St. Lawrence.[27] Holland had to send communications by whatever local vessel was sailing at the time and make duplicates of all his materials. This meant not only having a clerk make copies of all outgoing correspondence, but also having draftsmen make copies of all maps. In November 1766, Holland, then based at Louisbourg in Cape Breton, intended to send Pownall "by way of Halifax or New York … two small plans with a copy of the description [of Cape Breton]; these plans are intended by way of duplicates … should the large survey miscarry, but if the contrary; will if their Lordships have no objection to it, present one to the Admiralty for the public service & the other for my benefactor the Duke of Richmond."[28] In this case, the plans were not sent via Halifax or New York because a local vessel became available. "Mr. Urquhart is just arrived from St. John's Island," Holland wrote, "& informs me of a vessel there being ready to sail, at his return, for England, on account of Messieurs Spence & Co; & not to lose this opportunity, we have worked day & night, to have the large plan & description, ready to send."[29] But a fall gale undid all their work. "Since my writing the above," Holland continued, "Messrs Howard & Urquhart, with the large survey, on their passage for St. John's Island, have put back, & with great difficulty made this harbour [Louisbourg]. In the interim there has arrived a brig from Newfoundland, for Boston, who will sail the first fair wind, for that place; I shall send the plan therefore by this opportunity, & charge Mr. Howard to proceed in any vessel, shall be ready for England, & if none there to go by Road Island or New York, & there embrace the first occasion shall offer, either of the packett, a merchantman, or man of war; by which means I hope soon to have the pleasure, to hear of its being safe arrived at your office."[30]

Given vagaries of weather, particularly during winter, and infrequent shipping, considerable time could elapse between Holland sending material to the Board of Trade in London and receiving a reply. Holland sent his covering letter, maps, and description of Cape Breton on 10 November 1766 and they were read by the Board on 12 March 1767.[31] By 2 October, Holland had not received a reply, and wrote in some desperation requesting "an answer to … whether our labors, had obtained the approbation of their Lordships, I have not yet, had the honor to receive; an event that has given me the greatest uneasiness … as there is nothing I so much dread, as the displeasure of their Lordships, my patrons & protectors; so I hope … that this delay had been owing to the unavoidable multiplicity of business, which necessarily passes thro' your hands."[32] A further complicating factor was Holland's frequent movements. After surveying Cape Breton, Holland moved to Quebec and began receiving a backlog of correspondence, including letters from General Gage in New York. One letter arrived "at the proper time," having taken two months, but another "made the tour of Louisburg" and took nearly four months.[33] When Holland was based in Portsmouth, New Hampshire, a letter from Pownall sent on 7 July 1774 finally arrived on 14 December, "having made the tour of Canada; it being marked Montreal October 30."[34]

From their "centre of calculation" in Whitehall, the Lords of the Admiralty and Secretary Stephens must have quickly realized that the joint chain of command for the General Survey of the Northern District contrasted poorly with the single chain of command for the Survey of Nova Scotia. They must have also realized that having an army surveyor on a naval vessel was not ideal, and that the best solution was to have a naval surveyor, such as Gauld, serving under a naval officer or, even better, a naval surveyor, such as Cook, commanding a naval ship. Through trial and error, the Admiralty was working out the best practice for conducting hydrographic surveys far from Britain. By 1768, the Admiralty would have had a pretty good idea of who was needed to command the Pacific voyage then being proposed. For all the panegyrics about Cook's character and abilities, the influence of Commodore Palliser, Cook's patron, on the Earl of Sandwich, the First Lord of the Admiralty, and the shrewdness of Stephens as a judge of men, Cook was the only viable candidate in the navy who could both survey a coastline and command a ship.[35] The complex chains of command for the North American surveys had been resolved for the Pacific voyage, a solution that would serve the Admiralty well in the years ahead.

SURVEYORS, SOLDIERS, AND SEAMEN

At the start of the General Survey in 1764, Holland's team was relatively small, comprising fewer than two dozen men. Apart from Holland, there were four deputy surveyors: Lieutenants Haldimand, Robinson, and Carleton, and civilian surveyor Thomas Wright. The survey team also included a draftsman, a clerk, a clock and instrument maker, three volunteers (George Goldfrap, James Grant, Charles Blaskowitz), and twelve soldiers.[36] By the time the survey reached Louisbourg for the survey of Cape Breton Island in 1766, Holland had been forced to make several changes.[37] As Robinson had returned to England and both Carleton and Haldimand had died during the course of the survey, Holland recruited George Sproule and Lieutenant John Pringle as deputy surveyors. In addition, Holland engaged William Brown as another deputy surveyor, two assistant surveyors, and six privates from the 59th Regiment. A corporal, who had served as clerk, was replaced, having been "given to drunkeness, & other debaucheries." In his stead, Holland hired George Derbage "who is a gentleman of a good education, well acquainted with trade, & has several civil employments in this province [Nova Scotia]."[38] In November 1766, the survey team stood at thirty-one.

Six years later, when Holland was headquartered at Portsmouth, New Hampshire, for the survey of the northern coast of New England, the survey team numbered thirty-five.[39] The deputy surveyors and draftsmen were Thomas Wright, George Sproule, James Grant, Thomas Wheeler, and Charles Blaskowitz, with George Derbage serving as clerk. Holland had two servants, and indulged in some nepotism by employing two of his sons, John and Henry. Each deputy surveyor and clerk also had a servant. In addition, a sergeant, a corporal and twelve privates served as flag and chain men. A typical surveying party consisted of a deputy surveyor, his servant or batman, two chain bearers, two men for station colours, and one man for carrying instruments. The composition of the party appears to have been standard army practice; it was exactly the same as the survey teams on the Survey of Scotland some twenty-five years earlier.[40] For the General Survey, soldiers were seconded from the second battalion of the Royal American Regiment; they continued to receive their regimental pay plus sixpence per day from the Board of Trade, which, according to Holland, "keeps them scarce in shois and cloading." On learning that the Royal Americans were to be transferred to the West Indies in 1772, Holland feared that he would lose his men and protested to Pownall, Haldimand, and Gage, pointing out that the soldiers were "trained & disciplined to this business by a series of surveying campaigns – have enured themselves to the way of life incident to it, – & are seasoned to its hardships."[41] The accounts of the General Survey for 1773 and 1774 continue to list the soldiers as an expense, but no longer specify that the men were from the 60th Regiment.[42] Holland's accounts also list entries for two slaves, the "Negress Bet," purchased in Quebec in 1769 for £40, and "Jeanne, the Negress," purchased in Portsmouth in 1772 for £27.[43] These slaves may well have served as the survey team's washerwomen; they are a reminder that even in northeastern North America domestic slavery was part of everyday life.

An essential part of the surveying parties was the naval detachment. Small boat crews commanded by midshipmen were needed to ferry the surveying parties along the coast and carry out sounding along shore. Soon after taking command of the *Canceaux* at Deptford, Mowat recognized that he had insufficient crew to man the vessel and assist the survey, and got authorization from the Admiralty to increase the ship's company by a further ten men including a midshipman.[44] Even so, Mowat found himself short of men when the survey started. After an appeal to Colvill, he was instructed to "enter as many supernumeraries as may answer [Holland's] demand" and supernumerary seamen became essential to the survey.[45] As Mowat explained to Montagu in May 1772, supernumeraries were "absolutely necessary at present" for he could not "with safety spare twenty seamen out of the Canceaux's complement without endangering the safety of the ship, in particular upon a coast which I am entirely unacquainted with and where the nature of the service that she is employed upon requires her strict attendance; and consequently often obliged to beat off the shore or liable to anchor in very disagreeable situations which I have often experienced in the course of these eight years past." At the time, Mowat had to supply each of the five surveying parties with a boat crew consisting of four seamen and a midshipman. He also had six officers and servants, a clerk and gunner's mate, a cook for the officers, a cook for the ship's company, an apprentice from Christ's Hospital, and ten seamen who remained on board ship. Although officers were not allowed to have servants carried on the ship's books, Mowat made clear "the dignity of His Majesty's Navy never meant that [officers] should go and cook their victuals themselves, and do many other necessary matters which distinguish them from the foremast men besides that of walking the quarter deck."[46] In total, the ship's complement was forty-five. With the additional thirty-five surveying personnel, the total onboard the *Canceaux* was eighty. This included six seamen as supernumeraries, an additional expense for the navy.

The Survey of Nova Scotia had a much smaller team. In the first year, Des Barres had a personal servant and an assistant surveyor, one William Lloyd, a conductor of artillery from the garrison in Halifax.[47] Robert Borthwick, a midshipman, seconded from Colvill's flagship, *Romney*, moored in Halifax Harbour, commanded the *Dispatch* for the first two years of the survey, and then was replaced by a Mr. Evans, mate of the *Romney*.[48] Two other midshipmen, John Knight and James Luttrell, also from the *Romney*, were sent to assist, probably commanding smaller vessels. In 1769, Luttrell was posted to the West Indies, and Des Barres lobbied hard to retain Knight, who successfully passed his lieutenant's exam and took command of the *Diligent*.[49] Like the *Canceaux*, the *Diligent* carried a large complement of thirty men, enough to sail the vessel and crew the survey boats.[50] Many of these men were from the *Romney*. Throughout the duration of the survey, the North American squadron had a shortage of men; few wanted to serve given the conditions and low pay.[51] As a result, successive commanders-in-chief had to scratch around for crews. "Several of the ships are very short of their compliments," wrote Colvill to the Admiralty in September 1764, "and we are still in want of men for the sloops and schooners; I have therefore written to Captain Palliser commander in chief at Newfoundland for a supply of men ... from the Newfoundland Squadron."[52] In these circumstances, men from the *Romney* were frequently detailed to serve on the survey vessels.

In spring 1767, Des Barres could not get men released from warships in Halifax and "with much difficulty" had to hire local seamen. He was sufficiently dissatisfied with this arrangement that he remonstrated to Hood, the new commander-in-chief, about the manning problem. The survey suffered, Des Barres explained, "by having new setts of people to collect every spring, who as soon as they became acquainted with the nature of the service and their duty were to leave it again."[53] As a solution, Hood ordered that twenty-eight men be hired as supernumeraries; Des Barres recruited them from among Acadian settlers in Nova Scotia, some of them, perhaps, tenants on his estates. By August 1770, the surveyor was pleased with the new arrangement. "The work goes on quicker and with less labour in proportion than formerly," he explained to Hood, "and I begin to feel the advantage (tho' but late) of being attended by people who are at length become acquainted with the nature of my duty. These people are inhabitants of this country and more valuable from their knowledge of many parts of the coast."[54] At the end of the surveying season, the men were released back to their farms and trades. The Navy Board in London, however, considered this arrangement irregular and "made difficulties" releasing funds to pay the seamen. With the men threatening not to work unless they were paid in Halifax every three months, Gambier and Montagu continued to over-ride the Board until the survey was concluded.[55]

Without a department of government responsible for surveying and mapping, arrangements for staffing the General Survey of the Northern

District and the Survey of Nova Scotia were necessarily makeshift. Senior surveyors and deputies were drawn from the ranks of the army and the surveying profession, while survey parties were made up of army privates, naval seamen, and local settlers. Inevitably, such arrangements ran afoul of official regulations, particularly in the navy, and both Holland and Des Barres spent considerable time sorting out personnel issues. As with the chain of command, the Admiralty would eventually realize that having a naval officer commanding a naval vessel with a naval crew was the best solution for marine surveying.

Discipline and Punishment

Ensuring that men remained with the surveys over many years and obeyed orders required strict discipline backed up by the threat of harsh punishment. Such a disciplinary regime, of course, was standard in the army and the navy in the eighteenth century, and any man who volunteered or was impressed expected such treatment. Nevertheless, the conditions endured by the surveyors, soldiers, and seamen engaged in the General Survey and the Survey of Nova Scotia were extreme, even by military standards. In particular, the six years that Holland spent surveying the Gulf of St. Lawrence must have taxed all involved. The actual field surveys, carried out mostly by small boat during the summer and on foot during winter, were physically demanding and occasionally dangerous; Haldimand and several crew died during the course of the surveys. Moreover, winters were brutal. Apart from winters spent amid the ruins of Louisbourg and the more comfortable surroundings of Quebec, the survey teams over-wintered in primitive accommodation on St. John's Island and at Pabos, Gaspé, Anticosti, Mingan, and Lac Saint-Jean, enduring a frozen world of ice, blizzards, and sub-zero temperatures. The crew of the *Canceaux* also experienced two winters – at Port-la-Joye and Cape Canso – when the ship was frozen in ice.[56] Once the General Survey moved to Portsmouth, conditions improved. Winters were much less severe and the town offered grog shops and brothels. Even so, the General Survey was in its seventh year and by that stage officers and men onboard the *Canceaux* were long overdue their pay and leave. When Hood received instructions from the Admiralty in September 1770 to send the *Canceaux* back to England for much needed repairs, he "ordered her home, to the very great joy of the petty officers and seamen, who have complained to me, for want of their pay."[57]

    Maintenance of discipline began at the top of the chain of command. The Board of Trade and the Admiralty were both sources of patronage and Holland, Des Barres, Mowat, and Knight knew well enough to keep in their good books. Holland was especially keen to receive the Board's approbation for his work and the Lords obliged, although they took their time. After four years surveying, during which time Holland had submitted major surveys and descriptions of St. John's Island and Cape Breton Island, the surveyor general finally received praise. Writing in March 1769, Hillsborough informed Holland that his plan of Cape Breton, "having been laid before the King, His Majesty has graciously expressed his approbation of your diligent attention to your duty."[58] In his reply, Holland expressed his gratitude: "your Lordship's letter … had given me real happiness in the high approbation His Majesty & your Lordship express for the past performance of my duty; so it has also given me the greatest encouragement to use my utmost endeavors towards bringing a service so necessary for the public good to a happy conclusion."[59]

Holland, in turn, encouraged his deputies, soldiers, and sailors. At the start of the survey, Holland accepted three volunteers – Goldfrap, Grant, Blaskowitz – into the service, paying them a nominal salary.[60] No doubt keen to get a training and win advancement in the surveyor general's "travelling academy," the three volunteers gradually took on more field and drafting work over the course of the survey and were rewarded by Holland with increased salary.[61] He also rewarded seamen with premiums and bounties for "saving the whaleboat from being lost" and for taking care of Sproule's boats.[62] Small boats, of course, were essential to the survey and not easily replaced. Apart from financial rewards, which were limited, given the parliamentary grant to the survey, Holland tried to instill an *esprit de corps* by creating a special uniform: "As a mark of distinction & attachment to the Right Honorable Board," Holland wrote in November 1766, "I have agreed with the gentlemen & privates of my party to wear a green uniform, with leather caps, which last are much more convenient, & less troublesome than hats, in our excursions thro' woods, & by water. In the front of the caps we beg leave of their Lordships, to have embossed the emblem & motto of Trade & Plantations, this favor … will when we come into a more cultivated part of the continent, both show that we think ourselves honored by the service of the Right Honorable Board, & procure us respect & assistance, besides it will inspirit the men, with more vigor & emulation, when they see themselves seperated & distinguished from those of their own level."[63]

Without Holland's journals, it is impossible to assess the disciplinary measures the surveyor general used if orders were not obeyed, but a good record survives for the naval branch of the surveys. Discipline on naval ships was governed by the Articles of War.[64] These laid out standing regulations passed by Parliament and were read to the assembled ship's company every month. For the period 1765–70, when the *Canceaux* was closely involved in surveying the Gulf of St. Lawrence, an average of four men per year were punished, although there was great variation.[65] In 1766, eight men were disciplined; the following year, only one. Given that the ship's complement was forty, these were not high numbers. Most infractions were for drunkenness, neglect of duty, disobedience, disrespect of officers, and mutinous expressions. Before he sorted out the manning problem on the small boats, Des Barres complained of the "many disappointments (and I might well say distress) … proceeding from drunkness [sic], negligence or want of zeal" among the crews.[66] Nevertheless, it is unlikely that the incidence of indiscipline on the Survey of Nova Scotia was exceptionally high. For the only complete years of the *Diligent*'s log, four men were disciplined in 1771 and three the following year.[67] Punishments varied. A minor infringement such as unhelpfulness and disrespect to an officer could incur as few as four lashes; mutinous words and disobeying orders usually resulted in twenty-four lashes, sometimes as many as thirty-six.[68]

The greatest problem was desertion. Both the *Canceaux* and the *Dispatch* lost men. Soon after the *Canceaux* arrived in Quebec in 1764, eight men jumped ship, twenty percent of the crew; in Halifax and Louisbourg two years later, a further eight left, although three were later caught. In Quebec in 1767, another two went.[69] When men were brought back, punishments were usually severe. Two men captured in Halifax received fifty lashes each, a third a hundred.[70] But the worst case occurred in the first year of Des Barres's survey along the Eastern Shore of Nova Scotia. Six men from the *Romney* employed on the survey took one of the small boats and made for Cape Breton, where five were picked up by the army officer commanding the coal mines at Spanish River (Sydney).[71] They were brought back to Halifax and imprisoned in the brig on the *Romney*. Two managed to escape, however, using a ship's boat, but were soon recaptured. Faced with severe under-manning on his warships, Colvill was determined to set an example. The two repeat offenders were tried by court martial and suffered "death by being hanged by the neck at the yard arm."[72] The three other men were sentenced to "receive seven hundred lashes each upon their bare backs with a cat of nine tails."[73] Of these men, one was given clemency because he was "a sober orderly good seaman who was probably misled by others." The other two received 350 lashes "in the usual manner from ship to ship" and the "rest when they are able to bear it."[74] Unfortunately for the man let off, he was suspected later of "robbing the house of a German settler" in Halifax and was condemned to have his original sentence executed, and then delivered up to the civilian authorities for further punishment.[75] The desertion of these men from Des Barres's party was testament to the tough conditions on the survey. At their trial, the prisoners said nothing in their defense except that "the work on board the sloop [*Dispatch*] being too hard was the reason of their first deserting."[76] It must have been, given that their chances of evading capture were remote. Scarcely any settlement existed along the Eastern Shore of Nova Scotia, while the army garrisoned Louisbourg and the coal mines at Spanish River on Cape Breton Island. Men were deserting into a wilderness, where they would either die or be picked up by a naval patrol. The one man who evaded capture spoke French and no doubt slipped away among the Acadian settlers in Cape Breton.[77]

Ships, Schooners, and Boats

Ships used on expeditions to the Pacific during the late eighteenth century, such as Cook's *Endeavour* and la Pérouse's *L'Astrolabe*, have been regarded not just as a means of transport but as scientific instruments in their own right.[78] The ship allowed exploration to take place and also served as a mobile laboratory where specimens – natural and cultural – collected on the voyage could be examined in a controlled environment. In particular, the captain's cabin, because it was a large and secure space, became a centre of research and calculation. In this interpretation, the ship was "the Enlightenment's principal geographical instrument – the Enlightenment's 'vessel of modernity.'"[79] Yet this grand claim was much less true for the ships and vessels used on the General Survey of the Northern District and the Survey of Nova Scotia. Both the *Canceaux* and the *Diligent* had multiple uses, some of them quite prosaic, and much of the actual survey work was done from small boats.

In his memorial to the Board of Trade outlining his proposal and budget for the General Survey, Holland was quite specific about his shipping needs. He requested "an armed cutter or othe[r] small armed vessel … with two whale boats and one large long boat which will … serve to transport the party over bays and arms of the sea."[80] Holland must already have realized, no doubt from his experience with the Survey of Canada, that he needed a small vessel and boats for the survey, rather than a large ship. The Lords of the Admiralty accepted the surveyor's requirements and instructed the Navy Board to purchase exactly the kinds of craft he specified, the vessel to be about 120 tons.[81] In the event, the Navy Board could not find a suitable vessel "in the river," only a larger ship of about 180 tons.[82] The Lords approved the purchase.[83] For comparison, Cook's *Endeavour* was 366 tons, its larger size reflecting its original purpose as a collier.[84] The ship purchased for the General Survey cost £1,500 and was a general merchantman, originally called the *William*. The Admiralty had her registered on the Navy List as an Armed Ship called the *Canceaux*, and the Navy Board fitted her out with six three-pounders and eight swivels (figure 3.3). The ship's complement was set at thirty, enough to handle a ship of her size, but that was soon increased to forty in order to provide crews for the small boats.[85]

The shortcomings of having such a relatively large vessel quickly became apparent. Soon after the *Canceaux* arrived at St. John's Island in October 1764, Holland requested from Mowat "three large boats and eighteen

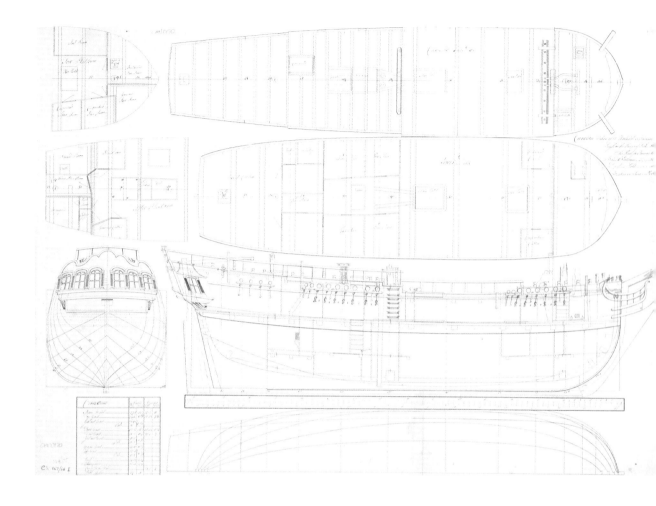

seamen in order to attend the different parties employed upon the survey where the ship [*Canceaux*] cannot go."[86] The following year, Holland reported to Pownall that the *Canceaux* could not provide support to the survey parties because Mowat was unwilling to "venture the Canceaux on an unknown coast without much danger."[87] During the entire survey of St. John's Island, the *Canceaux* remained moored at Port-la-Joye and only came to sail when Holland and his party removed to Louisbourg. In October 1767, Holland observed to Pownall that the *Canceaux* "however necessary, she may be for a total transportation of our party; our instruments; our stores … yet in the more immediate & frequent occasion we have for her, she is altogether unfit; such as the plying about an unknown coast; the serving as a magazine to all our different parties; circumstances it is impossible she can answer."[88] Essentially, the *Canceaux* had become little more than a depot ship, supplying provisions, supplies, and men to the survey parties in their small boats. The Armed Ship *Canceaux* never became the "Enlightenment's principal geographical instrument."

3.3 · Lines plan of the *Canceaux*, February 1771. © National Maritime Museum, Greenwich, London, J8539.

Of more practical use to Holland were the smaller craft and boats that he quickly assembled for the survey. After an appeal to Colvill in November 1764, Holland received a small schooner and two whale boats the following spring.[89] Holland named the schooner *Jupiter*, and a sloop, purchased in August 1765, *Venus*.[90] Classical gods were less on his mind than astronomical observations. A fishing schooner was also hired in August 1766 to convey Wright's party, provisions, boats, and instruments to Anticosti.[91] By the following year, Holland was clear in his own mind about the composition of the survey craft. "Instead of the Canceaux Armed Vessel, & her present complement of 40 seamen, boats, &c.," he explained to Pownall, "let there be two schooners allowed, of about 40 or 50 tons, with 4 whaleboats; these if purchased in America, could be got cheap; to man these at 5 seamen to each schooner, & boat, there would be in all only thirty: These to be under the absolute direction of me, & the gentlemen of my party. These vessels could attend me & my party any where, from their small draft of water; they would cost little at first, & with economy & care, would be but a triffling yearly expence, for the future."[92] Essentially, Holland was proposing to eliminate the principal naval contribution of the *Canceaux* and take complete charge of both the survey parties and the survey boats. Although nothing came of the proposal, the General Survey evolved close to this model in practice, particularly when Mowat took the *Canceaux* cruising for smugglers. Schooners, shallops, long boats, and whale boats became the ubiquitous craft of the survey, carrying surveyors, soldiers, instruments, tents, and provisions along the rivers and coasts of the northern district.

Unlike Holland's survey, which was fitted out in London, Des Barres had to make do with whatever vessels were available in Halifax. At the start of the survey in 1764, Colvill assigned the *Dispatch* to the survey; it was "a stout sloop" that served to carry the admiral's dispatches.[93] The vessel lasted for two years until it was beyond repair and discharged, at which point the schooner *Charlotte* was hired for the season.[94] In 1767, Des Barres convinced Hood to buy a proper vessel for the survey but authorization from the Admiralty took time.[95] Meanwhile, the surveyor continued to make do. During 1767 and 1768, he employed a sloop until it was wrecked in Chedabucto Bay; he also had a small schooner of about sixty tons and a decked shallop (figure 3.4). The shallop (a large heavy boat with one or two masts) was needed for taking soundings in shore because small boats were liable to capsize in an Atlantic "surge."[96] After further appeals from Des Barres and Hood, the Admiralty finally gave permission for the purchase of a dedicated schooner in March 1770; the vessel was to be "armed, fitted and manned, and called the *Diligent*."[97] The schooner was used to take soundings in the "great offing," while Des Barres continued to employ a small schooner, a shallop, and a boat "to carry on the survey and soundings of the sea coast and harbors."[98] Smaller craft were especially "fit to sound close

3.4 · Ship model of a cargo schooner, circa 1780. This type of vessel was used on the Survey of Nova Scotia. © National Maritime Museum, Greenwich, London, SLR0545.

in shore amongst unknown sunken rocks and hidden dangers." The small schooner was named the *Sable* after the island that Des Barres had spent three years surveying and had a complement of seven men; the shallop was named the *Tatamagouche* after one of Des Barres's Nova Scotia estates and also had a complement of seven. In addition, fourteen men crewed small boats and a canoe.[99] The *Sable* lasted until spring 1773, when it was found to be irreparable and replaced by a small schooner hired for the final season of the survey.[100]

For both the General Survey and the Survey of Nova Scotia, the geographical instrument of choice was not the ship but much smaller craft: schooners, sloops, shallops, boats, even a canoe. The ship was the appropriate vessel for the great swings across the Pacific by Cook and la Pérouse, but it was hardly suited to inshore surveying work. Indeed, Cook, like Mowat, was always reluctant to venture the *Endeavour* too close to the coast, and for that reason missed observing the true shape of the Banks Peninsula on the east coast of New Zealand in 1770.[101] La Pérouse's two ships got too close inshore and were wrecked on a coral atoll in the Solomon Islands.[102] The ship, then, was a blunt instrument for detailed surveying; much smaller craft were needed to dodge among the rocks, reefs, and islands of the coasts of St. John's Island, Cape Breton Island, Nova Scotia, and Maine. Small schooners, shallops, and boats were the fine scientific instruments used in the intricate coastal surveying of northeastern North America.

EQUIPMENT, INSTRUMENTS, AND TECHNIQUES

Although small craft were essential for ferrying surveyors and their baggage along shore and for sounding inshore waters, survey parties also needed a great range of equipment, from tents to telescopes, from dipsey lines to plane tables, and the knowledge of how to use them. The ultimate success of the General Survey and the Survey of Nova Scotia would depend on how accurately coasts were surveyed and sea bottoms sounded. Holland was determined that the General Survey be "done with the greatest exactitude imaginable."[103] Des Barres no doubt felt the same about the Survey of Nova Scotia.

A good sense of the typical equipment carried by a survey party comes from the *Canceaux*'s log. On 21 June 1766, George Sproule's survey party left Louisbourg in the *Jupiter* schooner to survey and sound Bras d'Or Lake on Cape Breton Island. The party comprised ten men: Sproule, a servant, five soldiers, a midshipman, and two seamen. They had provisions for three months and considerable equipment: sixty-five yards of old canvas tents, deep water lead and line, two hand lines, a lantern, spare cable, eighty-nine yards of bunting for colours (to mark sighting poles; see figure 1.1), three reels of twine, three sail needles, two padlocks, one grapnel, and thirty-five fathoms of rope.[104] Although not recorded in the log entry, the party would have had camp kettles for cooking and brewing spruce beer, a necessary ascorbate to prevent scurvy (figure 3.5). In addition, Sproule had surveying instruments: a plane table, an amplitude compass, a quadrant, and a notebook. A year earlier, Holland had written to Richard Cumberland, his agent in London, to procure three plane tables, one each for the deputy surveyors, "having found the great expedition and convenience of the plan[e] table in the survey of harbours and rivers preferable to the theodolite or circumferentor."[105] Holland had used a plane table in his survey of Louisbourg in 1758; the portable instrument – a large board mounted on a tripod – was well suited to topographic surveying and map-making.

The method of coastal surveying using a plane table was straightforward and comprised two stages.[106] First, the surveyor measured out a base line along shore and positioned the plane table at one extremity (figure 3.6). The base-line was scaled and marked on the field-sheet on top of the plane table. The surveyor then sighted on flags situated on prominent landmarks and recorded the angles on the field-sheet. This exercise would be

3.5 · *A view from the camp at the east end of the naked sand hills, on the south east shore of the Isle of Sable* from *The Atlantic Neptune*. Des Barres's survey party pitching camp. Des Barres is the figure wearing the red coat and tricorne hat in the left foreground. © National Maritime Museum, Greenwich, London, HNS 77.

repeated at the other extremity of the base-line, thereby creating a series of intersecting lines or triangles. Distances could then be calculated between points of the triangles. Such triangulation rapidly created a framework of points that fixed the main features of the landscape on the field-sheet (figure 3.7). Des Barres explained the process to Colvill in 1765:

> I measured a base of 350 fathoms [2,100 feet] along a flatt on the western side of Exeter Harbour, and from its extremities, having with a theodolite, taken the angles of visual rays to objects placed on the opposite shore which being calculated trigonometrically and protracted in their proper bearings on paper fixed upon a plain-table, I then repeated, with the plain table, the same operations over again, and intersected the same objects, from the same extremities of the base line, by which and other intersections, a series of triangles, I had the distance between an object placed on Point Bulkeley, and another on Newton Head, from whence, by

farther intersections performed in the same manner, I determined the true emplacement of Winter's, Roger's, and Barron's Islands, and of all the ledges, thence repeating the former operations from all these islands, I found all the angles and distances to agree with what I had layd down, from the above mentioned observations, before, from points as were most commodiously situated on those islands, and head lands, I observed the distant head lands, bays, islands, points, and other remarkable objects, as far as they could be distinguished.[107]

Some parts of the coast were so rocky that it was impossible to land and lay down a base line along the beach. In such cases, survey parties used a rope stretched between shore and boat or between two boats to establish a base line. Again, Des Barres described the process:

Soon after coming to reflect on the extremely irregular, winding and broken shape of this coast, which is mostly covered with spruce and thick underwood, and thence perceiving the difficulties I had to encounter with, in finding convenient places of sufficient levell and extent for to measure base lines upon; I tryed the following experiment: I soaked a dipsey line in salt water, till it

West shore of Richmond Isle, near the entrance of the Gut of Canso from *The Atlantic Neptune*. A survey party working along a rocky, spruce-covered shore typical of the Atlantic coasts of Nova Scotia and Maine. © National Maritime Museum, Greenwich, London, NS 64.

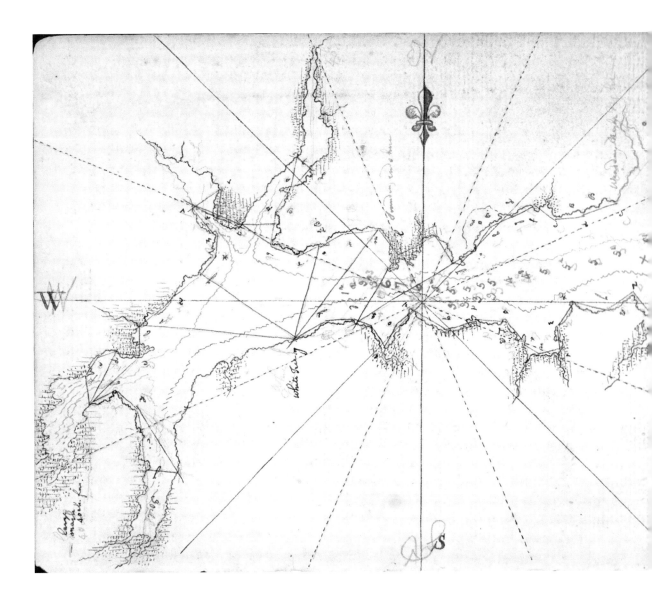

was fully imbibed, and then stretched and rubbed it tought, and with an iron chain, measured 100 fathoms [600 feet] of it, with marks at every 10 fathoms [60 feet]. Just before the change of the tide, on a calm day, I fixed the one end of this 100 fathom line to a station on Point Bulkeley and, with the other end, rowed right out for another station on Newton Head; (whose distance I already knew) when I got the line tought, I made its end fast to the grapnell, and let it run to the ground; after this, I caused another boat to take the first end (which was fixed to the station on shore) and hawl in the whole line till it came to be perpendicular with the

3.7 · Field survey of Grand River, St. John's Island, 1765, by Thomas Wright. Triangulation lines and soundings are clearly shown. Public Archives and Records Office of Prince Edward Island, Acc2330/sG-2/pg33.

grapnell let down by the first boat and thence to proceed rowing out again for the said station on Newton Head, till the line got to be tought, made fast to the grapnell, and let down as before, and so continued: By which measurement, I found the distance to be 510 fathoms [3,060 feet]; longer, by 19 feet than I had found it to be, by the method of triangles. Many subsequent tryalls and examinations of this new method have convinced me how surprisingly it coincides with measurations performed, by the means of an iron chain, on shore, and, from these considerations and other cogent reasons, I have been induced to apply it very advantageously during the course of my survey.[108]

Des Barres also took backsightings to ensure complete accuracy. "I went along shore, and reexamined the accuracy of every intersected object, delineated the true shape of every head land, island, point, bay, rock above water & etc and every winding and irregularity of the coast; and, with boats sent around the shoals, rocks and breakers, determined, from observations on shore, their position and extent, as perfectly as I could." The final stage was to sketch in the intervening land-forms between the observation points. "The irregular and hilly nature of the lands along shore was, everywhere, sketched off, upon drought, on the spot. The interior parts of the country were layd down from the accounts of bearings, distances and descriptions, observed, with a common pocket compass, by Acadians."[109] At the end of this process of triangulation and sketching, Des Barres had created a miniature landscape on the field-sheet, which could then be copied and scaled to make a complete map of the coast. The surveyor had transformed the coastline into a representation that was easily portable and reproducible.[110] Indeed, Des Barres was anxious to provide charts as quickly as possible to the boat crews sounding offshore. "When the map of any part of the coast was compleated in this manner, I provided immediately each craft with copys of it."[111]

While surveying parties worked along the coast, boats and larger vessels sounded offshore. The best practice was worked out by Des Barres. As he explained to Colvill: "The sloop was employed in beating off and on, upon the coast, to the distance of ten and twelve miles in the offing, laying down the soundings in their proper bearing, and distance, remarking every where the quality of the bottom. The shallop was, in the mean time, kept busy in sounding, and remarking around the head lands, islands, and rocks in the offing; and the boats within the indrought, upwards, to the heads of bays, harbrs. & etc."[112] In other words, different size of craft sounded different zones of the coast: small boats within headlands, shallops between headlands and outer islands, and schooners in deep water some miles off-

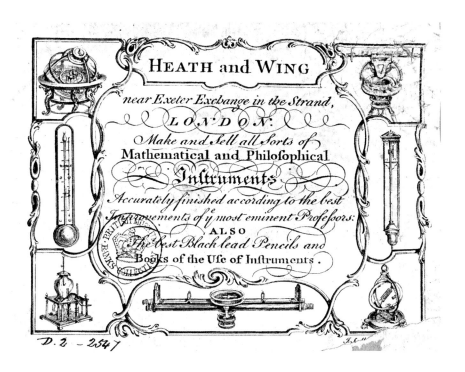

shore. This division of labour continued when Knight commanded the *Diligent*. Des Barres worked along the coast with the boats and shallops, and Knight and the schooner spent most of the time in the offing.[113] Indeed, Knight frequently sounded as far out as the edge of the continental shelf.[114] This efficient division of labour, which resulted in the systematic sounding of the coasts and banks of Nova Scotia, was never achieved by Holland and Mowat. Most commonly, midshipmen, who commanded Holland's survey boats, sounded inshore waters, while Mowat in the *Canceaux* failed to sound in the offing.

Even though triangulation, sketching, and sounding created a detailed record of the coast and seabed, the survey still needed to be fitted into a larger grid of latitude and longitude. Establishing latitude was relatively easy and could be done with a quadrant and astronomical tables. A quadrant fixed the altitude of the sun at noon, allowing a surveyor to calculate his distance north of the equator. Each surveying party was equipped with a quadrant to fix the latitude of the local survey. On St. John's Island and Cape Breton Island, Holland used "an astronomical quadrant, or equal altitude instrument of two feet radius, divided by Mr. Sisson, and improved, with an horizontal circle and stand, by Mess. Heath and Wing."[115] This was most likely a Hadley quadrant, which was standard issue for navigators during the mid-eighteenth century and made by several instrument makers,

3.8 · Heath & Wing trade card. Holland purchased many of his instruments from Heath & Wing. © Trustees of the British Museum

3.9 · Holland's astronomical clock made by George Graham, circa 1750. Canadian Museum of Civilization, D-5579.

such as the London firms of Sisson and Heath & Wing (figure 3.8).[116] For his survey of Anticosti Island in 1766, Wright took "a HADLEY's sextant of 18 inches radius, having a brass arch and index, and ivory vernier made by Messieurs HEATH and WING."[117] Hadley's quadrant was gradually replaced by the superior Bird quadrant. Holland received one of these instruments while he was at Louisbourg, sometime between 1765 and 1767, and used it to establish the latitude of Quebec in March 1769.[118] As Holland had used "an astronomical quadrant, of the old construction [Hadley's]; which I found to be very erroneous, when I compared it to BIRD's," Holland and Wright had to rectify their earlier observations taken at Port-la-Joye in 1765 and Île aux Coudres in 1769.[119] Apart from a quadrant, astronomical tables were needed for working out the declination of the sun or other stars during the year. Holland first used tables calculated by French-Italian astronomer and mathematician Giacomo Maraldi and then those by French astronomer Jérôme Lalande.[120]

Fixing longitude was more complicated. Before the widespread use of chronometers, which were not available to either Holland or Des Barres, astronomical observation was the only way of establishing longitude. In 1610, Galileo observed the orbit of Jupiter's four satellites and immediately realized that they acted as a celestial timepiece, which could be used to establish longitude.[121] During the seventeenth and eighteenth centuries, observation of the eclipses of Jupiter's satellites, particularly the immersion and emersion of the first satellite, became the standard means of establishing longitude on land; the method was used extensively by the Cassini in their survey of France. The Marquis de Chabert also used the method in his astronomical observations at Louisbourg in 1750 and 1751.[122] By the late 1760s, Nevil Maskelyne had developed an alternative method using lunar distances. The method required a set of annual tables, which Maskelyne began publishing as the *Nautical Almanac* in 1767.[123] By the end of his surveying, Holland had acquired a copy of the *Almanac*, suggesting that he was at least considering using the lunar distance method.[124]

To make an observation of the satellites of Jupiter, an observer needed a telescope, a clock, and a quadrant. Holland visited the leading instrument makers in London before he left in April 1764, and purchased a two-foot Gregorian reflecting telescope made by Short for £36.15s., a ten-foot refracting telescope made by Dollond for £21.16s, a regulating clock by Mudge & Dutton for £39.11s., and an astronomical quadrant from Heath & Wing.[125] Holland also brought "a monthly astronomical clock, or timepiece (with a compounded pendulum, and a spring to keep it going when the clock is wound up), made by the late Mr. George Graham" (figure 3.9). These were among the finest instruments of their type, and Holland took good care of them in the wilds of North America. As he explained to the

Surveying · 111

Board of Trade, who had paid for much of the equipment, "In building my winter habitation on St. John's Island, I constructed a strong stone chimney, to the back of which, I secured the clock, with the greatest precaution, & the room was kept temperate by an iron stove. In a few days the clock was regulated to mean or equal time; & always examined & compared by equal altitudes of the sun & stars, at or near the time, – when any immerssions or emerssions were to be observed. As the going of this clock is not inferior to any made by that renowned artist Mr. Graham, it will not be necessary to insert here a multitude of equal altitudes, & other observations to prove the exactness of this clock."[126] When the Surveyor General moved to Louisbourg and settled into spacious rooms in the former Intendant's palace, he took similar care: "At my arrival ... I brought the instruments in order; fixed the clock, (to a brick wall in a room kept warm by a stove,) & regulated it to equal or mean time. I also, put up in a different room, near the fire place, another monthly clock, sent me from London, by Messrs. Mudge & Dutton, with a simple or common pendulum: This clock with little trouble, was soon brought to keep time with the other clock, tho' when the fire was neglected it was soon perceived in its motion."[127] At some point during his stay at Louisbourg, Holland received a clock by leading-instrument maker John Shelton (probably in the same shipment as the Bird quadrant). The Royal Society had issued Shelton clocks to observers of the Transit of Venus in 1761 and issued them again in 1769, including to Cook's expedition to Tahiti.[128] With clocks made by George Graham, his former pupil Thomas Mudge, and John Shelton, Holland had some of the finest timepieces available. They were essential for recording the exact time of the immersions and emersions of Jupiter's satellites.[129]

Having several clocks also meant that observations could be made at different locations. Towards the end of the survey of Cape Breton, Holland sent Sproule and Watts to Gaspé, "where with a pendulum clock & telescope, they will determine the longitude of that place, to compare with my observations, which I shall make here at the same time."[130] Sproule took the Mudge & Dutton clock, leaving Holland with the Graham.[131] At the same time, Wright most likely took the Shelton clock for his observations on Anticosti Island, at the appropriately named Jupiter's Inlet.[132] Over the course of the survey, Holland also added to the telescopes that he brought from England, purchasing a four-foot and a twelve-foot telescope by Dollond.[133] After astronomical observations were made, Holland sent the results to Nevil Maskelyne so that they could be compared with observations made at the Royal Observatory in Greenwich. Only by comparing times of immersion and emersion of the satellites of Jupiter at two different locations could longitude be calculated.[134] For establishing longitude in northeastern North America, sites in the western wilderness and the Royal Observatory in southeastern England were vital centres of calculation.

3.10 · Latitudes and longitudes taken by Holland and his deputies, Des Barres, and Chabert.

Des Barres made do with much less instrumentation than Holland, a situation he much resented, as he made clear in a letter to Hood in August 1770: "To carry on the Survey of the Lands, the Lords of Trade have fitted out their surveyor [Holland] most compleatly with all manner of astronomical instruments &c. (and do allow him eight hundred pounds per annum for paying assistants, and extraordinary expenses, stationary, &c. independent of his salary and the fees of his office). How great the difference when I come to reflect on the public importance of my work to the latest posterity."[135] Des Barres purchased astronomical instruments in 1767, including a "large reflecting telescope wth some equatorial parts" for £12, and a "large brass astronomical quadrant of the compleatest sort" for

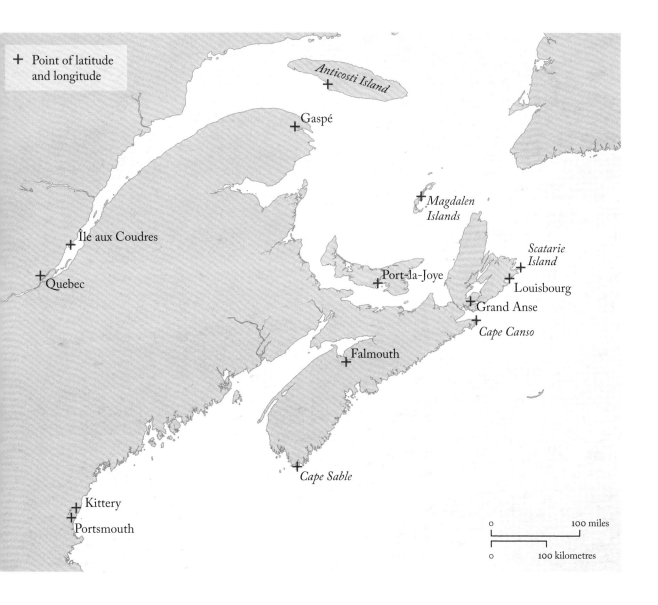

£60.[136] To make accurate timings of observations, Des Barres would also have needed a pendulum clock. The surveyor would certainly have made observations to establish latitude during his survey, but whether he also fixed longitude is not clear. He had an observatory at Castle Frederick and may have owned a pendulum clock, which would have allowed him to establish the longitude of Falmouth, but there is no record of him taking a portable observatory and clock on his survey. At their meeting at Liverpool in 1770, Holland offered Des Barres his astronomical observations, suggesting that the Admiralty surveyor had not established points of longitude.[137]

All this astronomical observing produced a particular geography of observation that provided the framework for constructing maps of northeastern North America (figure 3.10). As it was much easier to establish latitude, both the General Survey and the Survey of Nova Scotia produced numerous points of latitude, ensuring that many places would be correctly located on their meridian. But as fixing longitude required pendulum clocks, telescopes, and portable observatories, far fewer points of longitude were established. By 1772, these points included Holland House in Quebec, Port-la-Joye on St. John's Island, the Magdalen Islands, Louisbourg, Gaspé, Anticosti Island, Île aux Coudres, Kittery, and Portsmouth. Des Barres probably also fixed the longitude of Falmouth. In addition, Holland made use of Chabert's observations made in 1750–51, which included Louisbourg, Scatarie Island, Canso, Grand Anse on the north side of the Gut of Canso, and Cape Sable.[138] Essentially, Holland was creating an abstract framework of latitude and longitude into which the topographic field surveys could be fitted.[139] Fixing the graticule of parallels and meridians was absolutely essential, for it was the lattice upon which everything else was located. The creation of this abstract space, carefully calculated from observations of the planets, was necessary for fixing local points measured by triangulation. Local places were now embedded into a universal geometric framework. Maps produced by the survey could then be combined and scaled with other maps of the world. Such maps would not only allow the Board of Trade to organize its territories in North America but also allow ships of the Royal Navy to safely navigate their Atlantic realm. More importantly, colonial spaces in northeastern North America were bound through the Greenwich meridian to the cartographic space of Britain. An imperial cartography of the North Atlantic was beginning to take shape.

PILOTS AND GUIDES

Even with this great assemblage of skilled men, specialized vessels, and scientific equipment, the General Survey of the Northern District and the

Survey of Nova Scotia relied to a considerable extent on local knowledge provided by European settlers and Native peoples.[140] Canadians, Acadians, and Mi'kmaq had long familiarity with the coasts, rivers, and seabeds of the northeast, and their knowledge was soon incorporated into the surveys.

Before Holland and his surveyors left London, the Admiralty "resolved that Lieut. Mouat of the Canceau Armed Vessel be directed to bear Jean Baptiste Geynordlaine a pilot for the Gulph & River of St. Lawrence, as long as he finds it to be absolutely necessary."[141] Given the importance of the voyage, the Admiralty was taking no chances in the risky navigation of the St. Lawrence. Geynordlaine was most likely a Canadian, perhaps one of the river pilots captured by the British in 1759 and still awaiting repatriation to Canada. Mowat also took an Acadian pilot for the voyage down the St. Lawrence to St. John's Island in fall 1764. During the first summer of the survey, Holland spent £36.10s. for pilots.[142] For the winter of 1764–65, Holland hired an Acadian guide for each of the survey parties; these "guides during the winter on the ice" cost a further £19. The following summer, Haldimand hired "guides at the Magdelan Islands" for £12.6s., while Holland provided Acadian guides for St. John's Island, again costing £36.10s.[143] For 1766–67, the surveyor general spent £25.12s. for "guides & pilots for the parties," probably covering the expeditions by Pringle to Pabos and Wright to Anticosti.[144] Hiring of guides continued in Quebec; for 1767–68, Holland paid £2 for "guides, snow shoes &c. for Mr Sproule's party," and £9.16s. for "guides, snow shoes &c. for Mr Blaskowitz's [party] on the Saguenay survey."[145] In the last year of surveying in Quebec, Holland paid a further £8.12.3 for "snow shoes, slays, guides &c." for Blaskowitz's survey.[146] Even after the move to Portsmouth, New Hampshire, the General Survey continued to hire guides. From December 1769 to December 1773, accounts record that Holland paid Mayon and Fontaine, most likely Canadians, £25.5s. "as guides."[147] From the beginning of surveying in summer 1764 to leaving Portsmouth in fall 1774, the General Survey hired pilots and guides every year. Des Barres, too, made use of Acadians for the Survey of Nova Scotia. Between 1770 and 1773, the surveyor hired twenty-eight Acadian men each summer to crew inshore boats.[148] Given the demand for seamen elsewhere in the North American squadron, the Nova Scotia survey might well have taken much longer to complete without Acadian assistance.

Both Holland and Des Barres incorporated information from Native people into their surveys (figure 3.11). During the survey of Cape Breton Island, Holland relied heavily on local Mi'kmaq. In a letter to Pownall in November 1765, Holland hoped "to forward the business much during the winter ... by the intillenge [intelligence] of the inhabitants of this Island."[149] He went on: "I have sent to the Indians to get a new sett of snow shoes and dog slays made for the whole party, and to get some of them to serve me as guides." In 1765–66, Holland paid native guides £28.12s. Des Barres relied more on Acadian than Native guides, but he did encounter an Indian party during his survey of the South Shore of Nova Scotia and made use of their local knowledge. At LaHave River on 23 June 1770, the surveyor "met a number of Indians on the beach at New Dublin, who told me that there are no less than sixteen lakes and seventeen carrying places from the head of LaHave River to Annapolis River."[150] As Bruno Latour has argued, such a transfer of knowledge had considerable significance. On entering a new territory, European explorers and surveyors were dependent on Native people for local geographical knowledge. In this respect, Europeans were weak, Native peoples were strong. But once indigenous geographical knowledge had been transferred to Europeans and incorporated into reproducible manuscripts and texts, indigenous knowledge was no longer needed.[151] Local Native knowledge about the environment had been replaced by the universal knowledge of European maps and travel descriptions. The power relationship had been reversed.

Native peoples in northeastern North America were well aware of this shifting relationship. In 1761, Montrésor, while surveying the upper Penobscot and Kennebec rivers in Maine, observed that "Abenquis, jealous of the knowledge of their country, took care to leave but few vestiges of their route."[152] Three years later, civilian surveyor Joseph Chadwick, also survey-

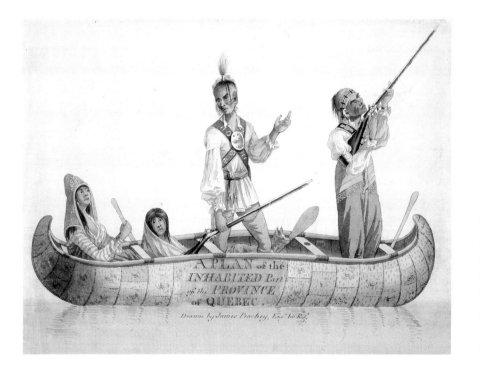

3.11 · *A plan of the inhabited part of the province of Quebec*, by James Peachey, circa 1785. An unused design for a cartouche, showing Mi'kmaq or Maliseet Indians. Library and Archives Canada, Acc No. R9266-334 Peter Winkworth Collection of Canadiana.

ing in eastern and northern Maine, noted that "the Indeins are so jealous of their countrey being exposed by this survey: as made it impractable for ous to preform the work with acqurice[y]." After some dispute with his Indian guides, Chadwick agreed "that I should take no draughts of any lands but only wrightings."[153] In summer 1766, Holland encountered a similar reaction from Mi'kmaq in Cape Breton encamped at Whycocomagh on the north shore of Bras d'Or Lake. "They behaved very peaceably," he noted, "but seemed dissatisfied at the lake being surveyed, saying we had discovered now all their private haunts, which the French never attempted to do."[154] During that summer's survey of the northwest coast of Cape Breton, the party tracing the Margaree River, one of the major rivers on the island, failed to find its source. As Holland later recorded, "this river has its source by the descriptions of the Indians from a lake at a great distance in the country … It was not possible to survey, this river further than what is laid down in the plan the stream being too rapid for a boat … and no Indians there at that time from whom to have bought, or hired, a birch cannoe."[155] Conceivably, the Mi'kmaq, having seen Bras d'Or Lake sounded and surveyed, were not prepared to reveal to the surveyors the upper waters of the Margaree and the large inland body of water now known as Lake Ainslie. When the General Survey moved to Quebec, Holland sent a party led by Lieutenant Carleton, nephew of governor Guy Carleton, to survey from Montreal to Lake Ontario. The party got as far as Fort Oswegatchie (Ogdensburg, New York) "where he could not with safety go further, as the Missasawga Indians were jealous of the operations he was carrying on, imagining it a preliminary to the taking of their lands, & no arguments were powerful enough to gain their consent for the party to survey or even proceed further."[156] Wherever possible, Holland and his deputy surveyors used Indian guides and incorporated indigenous knowledge into their surveys, but where field surveys ended abruptly it was often due to the absence or resistance of Native peoples.

Over more than a decade, the General Survey of the Northern District and the Survey of Nova Scotia assembled a great range of people and objects to observe, measure, and record the natural facts of coasts and seabeds in northeastern North America.[157] The total assemblage included more than a hundred surveyors, soldiers, and seamen; European settlers and Native peoples; vessels and boats; and scientific instruments and manuals – all mobilized to survey and sound the islands, peninsulas, coasts, rivers, banks, and reefs of the continental foreland. These actors and instruments were not detached from the greater Atlantic world. In various ways, they were linked to larger networks of science and empire in Britain. Observations and calculations flowed between Holland and his deputies in North America and Maskelyne in Greenwich. Survey results were presented at meetings of the Royal Society, and published in the society's *Philosophical Transactions*. Scientific instruments made in London were used in the wilderness of America. Communications, orders, and regulations flowed from the Board of Trade and Admiralty to surveyors and naval commanders. Such links tied the General Survey and the Survey of Nova Scotia back to the metropole; they formed a system of long-distance control exercised by the British government and scientific community over surveyors, soldiers, and seamen operating thousands of miles from home.

Major parts of this assemblage worked very well. Both surveys chained, triangulated, and sketched thousands of miles of coast and sounded thousands of square miles of inshore waters and ocean. Surveyors fixed latitudes and longitudes of significant landmarks and observed the transit of Venus. They also assessed the resources – both marine and terrestrial – of the newly acquired territories. All this information, contained in field-sheets and notebooks, was brought together and processed in the drafting rooms at Quebec, Observation Cove, Louisbourg, Portsmouth, Perth Amboy, and Castle Frederick. Some of it was sent almost immediately to the Board of Trade or the commodore on the Halifax station; much of it was retained to draft the maps and compose the descriptions that would eventually be sent to London. Yet not all the assemblage worked properly. Almost from the beginning of the General Survey, the division of command between the surveying branch and the naval branch created difficulties that ultimately compromised the maritime part of the survey. Employing men from different services and from the settler population also created problems, and large vessels were not always appropriate for the tasks at hand. In many ways, the Board of Trade and the Admiralty, through their agents, Holland and Des Barres, were learning as they went, trying to figure out the most expeditious way of surveying enormous areas under trying conditions thousands of miles from Britain. This was particularly a concern for the Admiralty, the department of government responsible for marine surveying. To project naval power around the globe, the Admiralty needed many things, but hydrographic charts of distant coasts were a necessity. The General Survey and the Survey of Nova Scotia were laying the foundations for later, even more extensive marine surveys of empire in the late eighteenth and nineteenth centuries.

4
Plans and Descriptions

In the early 1760s, the British government had only limited cartographic coverage of northeastern North America. The Board of Trade had commissioned John Mitchell's map of North America in 1755, but the great map was compiled from published sources and only showed the continent at a small scale.[1] For the newly acquired territories in the northeast, the British relied on maps published by the French Dépôt des Cartes et Plans de la Marine, hardly a long-term solution.[2] Moreover, the government had few large-scale maps of the region and those that it possessed had been generated for military purposes during the Seven Years War. The General Survey and the Survey of Nova Scotia were expected to produce scientifically accurate maps at small and large scales of the region. The surveys were also expected to produce descriptions or reports of the areas surveyed. These reports, usually quite lengthy, gathered together information that was not presented in cartographic form or explained information that was included on the maps. The reports described territory, inventoried resources, and proposed divisions of land. The reports accompanied large-scale, detailed maps and were cross-referenced to the maps. Consequently, maps and descriptions have to be seen as inter-related texts, a relationship often lost in scholarly discussion of British imperial surveys.[3] Both plans and descriptions were the major products of the surveys, and formed an important part of the growing imperial archive in London. Samuel Holland also submitted copies to General

Gage in New York for his use as commander-in-chief. On both sides of the Atlantic, plans and descriptions helped shape the British government's view of northeastern North America.

Plans

Manuscript plans produced by the two surveys varied considerably in scale, size, and information. In his proposal to the Board of Trade in 1764, Holland offered to make plans at two scales. The General Survey was to be at a scale of one inch to a mile, and the "places of note, channels and harbours" to be at a scale of four inches to one mile. For such an extensive survey, this was a remarkable level of resolution. The surveyor general also intended to include on the plans soundings of "all harbours and channels," and to determine "latitudes and longitudes of all capes, head-lands &c."[4] In the event, Holland did not adhere to his proposal, no doubt realizing the sheer impracticability of producing such detailed surveys for such an enormous area. Instead, he produced maps at a scale of one inch to 4,000 feet (i.e., a larger scale than one inch to one mile), which was "intended to shew as minutely as was necessary the situation of each place," and others at a scale of one inch to two miles which was "intended to connect these places, & shew an extended tract of country."[5] Holland also reduced large scale surveys to a smaller scale, usually one inch to four miles, allowing an entire geographic unit to be included on one sheet.[6] Des Barres, like Holland, also produced charts at two scales, although he eschewed feet for fathoms. His large scale maps were at a scale of one inch to 400 fathoms [2,400 feet], which allowed the surveyor to show "minutely the hills, valleys, barrs, shoals, soundings &c." His small-scale maps were at a scale of one inch to 1,600 fathoms [9,600 feet] or approximately 1.75 miles. This scale was used for showing large areas, such as the map of eastern Nova Scotia and Sable Island, "shewing the situation of the isle relative to the continent with the extent of the shoals, soundings, &c."[7]

Producing maps at these scales usually resulted in large sheets. Large scale maps at one inch to 4,000 feet were relatively manageable in size, because they covered only small areas. An unfinished topographical map of Penobscot Bay, Maine, at this scale covered a sheet 42.5 x 30.25 inches.[8] But small-scale maps at one inch to two miles could cover several sheets. A map of the coast from Saint John, Nova Scotia (New Brunswick), to Goldsborough Bay, Maine, was 32.5 x 54 inches; another, showing the coast of Maine from Falmouth (Portland) to Mount Desert Island, was 30.5 x 63 inches.[9] Even larger, Holland's map of the Maine and Nova Scotia (New Brunswick) coast from Cape Elizabeth to the Saint John River was 68 x

104 inches.[10] Des Barres's charts were of similar size; his chart of part of the coast of Nova Scotia from St. Margaret's Bay to Canso measured 27.5 x 96 inches.[11]

Des Barres justified such large maps in correspondence with Colvill in May 1765, when the first maps had been produced from the previous season's survey: "At the first glanze of the draught, it will, I am sensible, appear, inconvenient, by its great size; but when Your Lordship comes to reflect on the nature of the coast, surrounded with so many bays, harbours, inlets, & etc & small islands, rocks, shoals, & etc, between most of which there are passages for ships and vessels, you'll obviously conceive that the confused indistinctness unavoidably resulting from a contraction of the scale would have rendered it of little or no service upon approaching the land, where the necessity is greatest." Des Barres had hoped to reduce the survey to a scale of one inch to six miles, which would have resulted in a much more manageable sheet for Colvill, but as he explained to the admiral, "the short winter's days, notwithstanding the closest assiduity, scarcely afforded me sufficient time for compleating one single original draught from my plain table sheets, I was forced, for the present, to give up the thoughts of it."[12]

The information presented on the maps was relatively uniform. Both Holland and Des Barres gave greatest attention to charting the intricacies of coastlines and major rivers. Given the complexities of the coasts of Nova Scotia and Maine, maps had to be exceedingly detailed, showing capes, bays, islands, reefs, and rocks at low tide. Both surveyors also indicated terrain, particularly if it was visible along the coast and could give a bearing to mariners. Following contemporary cartographic convention, draftsmen used hill shading; terrain was picked out in green, with black or grey used to indicate steeper slopes and cliffs.[13] A generalized tree symbol was used to indicate forest or woodland. Sandbanks and mud flats were shown at low water, an immense task to map given the many estuaries along the coast and the enormous tidal range in some parts of the northeast. Holland's map of St. John's Island showed numerous mud flats (figure 4.1). Although tides around the island were not as great as those in the Bay of Fundy, the coastline had been submerged in post-glacial times, creating innumerable estuaries and creeks that all had to be surveyed and mapped.

In addition, offshore banks had to be sounded. As Des Barres explained to Colvill in December 1764: "I am present employed in completing a map of the [eastern] survey, which I shall lay before your Lordship as soon as it is finished wherein will be layd down the soundings of the bays, harbours, & etc and those along shore to the distance of ten or twelve miles in the offing, to which I shall add a few remarks on the tides and the necessary directions in writing for sailing into each bay, harbour, & etc."[14] With considerable assistance from Knight, Des Barres put great effort into

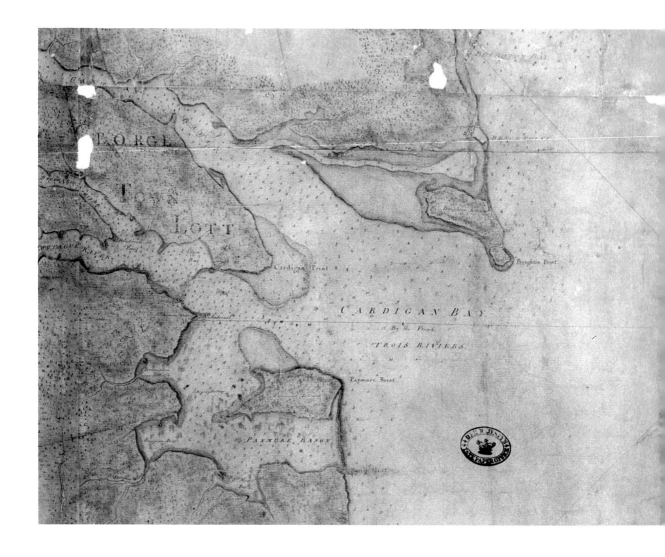

4.1 · Mud flats at low tide, St. John's Island. Detail from Holland's manuscript map of St. John's Island. The National Archives (UK), CO 700/Prince Edward Island 3.

sounding the coast and offshore banks of Nova Scotia and depicting them on manuscript charts and, later, successive editions of *The Atlantic Neptune*. The surveyor wanted to ensure that "ships, even when fog[g]y or dark weather deprives them of a observation … may readily find out where they are by the quality and depths of their soundings only."[15] Holland had far less cooperation from Mowat. In a letter written to Pownall in 1772, which accompanied plans of the west side of the St. Lawrence and a survey of Piscataqua Harbor, Holland admitted: "Tho these plans are as perfect in our branch as possible, the whole being exactly surveyed, & the position accurately ascertained by astronomical observations, yet they are far from being so, in what concerns the Naval Department; as except the few soundings done at Casco Bay by Mr. Sproule (while surveying) for his own satisfaction to embellish his work, there is not one remark or observation, either on the currents, the tides, sailing into harbors, or indeed once circumstance relative to the instruction of navigators yet done."[16] By the early 1770s, Holland was so fed up with Mowat's lack of attention to soundings that he added a *nota bene* to maps: "The omission of shoals, and soundings &c. in this plan has been occasioned by the want of Naval assistance."[17] The surveyor general was determined to let the Board of Trade know where responsibility for nautical omissions lay.

The authority of the maps and charts was created not only by their scientific accuracy but by their form. Lines and lettering were neat, accurate, and refined. A variety of lettering was used to emphasize the hierarchy of information. All lettering was in serif, which created a formal and dignified text, but there was also some variation in lettering between maps. On some maps, Holland lettered portions of titles, major political divisions, and large geographic features in upright capitals, sometimes with a larger point size for the first letter. On other maps, he lettered with slanted capitals. For minor political divisions and small geographic features, he used slanted capitals with lower case. Lettering of larger political units and geographic features were emphasized by using a greater point size. On Holland's maps of St. John's Island and Cape Breton Island, counties were lettered in a larger point than parishes, lakes larger than bays. Apart from lettering, maps contained elegantly drawn, geometric compass roses, and small-scale maps had a border delimiting latitude and longitude and a black frame.[18] Essentially, the surveyor general was producing relatively standardized maps that gave an impression of ordered, controlled landscapes.

Holland was careful to acknowledge the work of his deputy surveyors and draftsmen. On his first great manuscript map, "A Plan of the Island of St. John in the Province of Nova Scotia," created at Observation Cove and dated 19 September 1765, Holland listed who was responsible for what part of the survey. Holland, Haldimand, Robinson, and Wright were listed alongside each area that they had surveyed; similarly, Mowat, Brown, and Watt were listed alongside the waters they had sounded. As Holland explained to the Board of Trade: "The respective surveys of each gentleman and what I have surveyd myself, I have inserted in the plan so that their Lordships can judge of the industry and progress they have made."[19] The large scale survey of Cape Breton Island is, unfortunately, missing, but on other large-scale maps Holland continued to acknowledge each surveyor's contribution.

As Holland completed the St. Lawrence surveys, the production of a small-scale map of the northeast increasingly became an issue. In December 1769, the surveyor general wrote to Hillsborough, Secretary of State for the Colonies, that the draftsmen "are now busily employed in reducing the whole of our surveys to a small scale in order to form a general chart which shall give at one view an idea of these parts of America."[20] But such a map does not appear to have been produced, probably because of Holland's difficulties with Des Barres. Although the two surveyors met in August 1770 and agreed to share their maps, Des Barres did not uphold his end of the bargain. Without Des Barres's charts of Nova Scotia, Holland could not join his maps of the Gulf of St. Lawrence to those of the Gulf of Maine. In September 1771, Holland wrote to Pownall begging that "you will move the Right Honorable the Lords Commissioners for Trade & Plantations, to procure me an order to Mr. Desbarres (who is employed by the Admiralty in surveying the coast of Nova Scotia), to give me leave to take copies of his surveys for the use of their Lordships, that they may be joined to the general map I am preparing for them."[21] As Holland explained: "The general map I have so often mentioned & which I have greatly at heart, is at a stand, for want of copies of Mr. Desbarres' Nova Scotia Surveys … as I know this gentleman has little leisure to send me copies, I would be glad to have his plans sent me, to be returned as soon as I had obtained this reasonable request." By the following fall, Lord Dartmouth, the new Secretary of State for the Colonies and First Lord of the Board of Trade, clearly realized the importance of having an accurate general map of northeastern North America. "The surveys … appear to be far more correct and accurate than any that have been before made," he wrote to Holland, "and as they already comprehend a very large and important district of North America, I cannot but express my wish of seeing them collected together and united into one general plan: Such a plan could not fail of being of great public utility; and if it does not interfere with the other objects you have more immediately in view, you will take the proper steps for having such a general map prepared and transmitted to me."[22] By May 1773, Holland had received a copy of the Nova Scotia survey from Des Barres, but soon discovered "defects." The following April, Holland wrote to Pownall wishing "to have the most accurate drafts possible, as well as for the satisfaction of Ld. Dartmouth & the Honble. Board I serve, as that all the parts of my map may correspond as to care & exactness."[23] At the end of 1774, the surveyor general again wrote to Pownall, pointing out that "our surveys, each being connected with the other, & nothing left undone by us of our district fr[o]m Canada to Cape Anne, but Newfoundland & part of the Province of Nova Scotia, which tracts tho' within my instructions, I left to Mr. Cook & Mr. Desbarres who were surveying them for the Admiralty; as I would not put Government to an unnecessary expence (which was approved of), in repeating surveys that could be so easily communicated or obtained."[24] Yet the matter lay unresolved. With the Revolution breaking out, Holland confided to Haldimand in July 1775: "I have wrote frequently to Mr. Desbarres to send me a correct copy of his Nova Scotia surveys agreeably to the promise he made when he took copies of my surveys to make his projection complete: I wish he was advised to send me without more delay, those plans, as he knows I must be in absolute want of them & that it would be doing Government an essential service."[25] By then, Des Barres was back in London and busy preparing *The Atlantic Neptune*, a work that would include a general map at a small scale of northeastern North America.

Dividing St. John's Island

As part of its overall policy for the settlement and government of new territories in North America, the Board of Trade drew up precise instructions for the division and settlement of land. These directions were most clearly formulated in the "Instructions to General Murray," dated 7 December 1763 and delivered by Holland to Murray the following year. Among many instructions to the new governor of Canada, three paragraphs covered British land policy in the province.[26] The first requirement was for "an accurate survey" and a written report describing "not only the nature and quality of the soil and climate, the rivers, bays, and harbours, and every other circumstance attending the natural state of it; but also … in what manner it may be most conveniently laid out into counties." A map was to be attached to the report showing the proposed civil divisions. The second requirement was for laying out townships. The Board made clear that settling planters in townships had two advantages: first, settlers could assist each other in their civil concerns; second, they would provide security "against the insults and incursions of neighbouring indians, or other enemies." Townships were to be of "convenient size and extent," with the Board recommending that each township consist of about twenty thousand acres and having, as far as possible, natural boundaries and frontage on the St. Lawrence River. Finally, a town site was to be marked out in each township, "sufficient to contain such a number of families as you shall judge proper to settle there," with town and pasture lots convenient to each house lot. The town was to be "laid out upon, or, as near as conveniently may be, to some navigable river, or the sea coast."

The Board's policy for the division and settlement of St. John's Island was similar but exists in more abbreviated form. In a report presented to the king, dated 23 March 1764, the Board laid out its plan for the division and allotment of lands.[27]

> That the said Island should be forthwith surveyed by Your Majesty's Surveyor for the Northern District.

That it should be divided into counties of five hundred thousand acres each, so near as natural and proper boundaries would admit. That the said counties should be laid out in like manner into parishes of one hundred thousand acres each.

That each parish should be laid out in like manner into townships of twenty thousand acres each.

That each county, parish, and township, should be laid out in like manner as to partake, as much as possible, of the natural advantages of the country, especially those which arise from the sea coasts, and from the sides of navigable rivers.

That there should be laid out in each county a sufficient quantity of land for the scite [sic] and accommodation of a town, in the best and commodious part of the said county for the situation thereof; and that there should be reserved, in each parish, a proper scite [sic] for a church, and a proper number of acres near the same for a glebe for a minister.

In granting land "no one person should possess more, under such grants, than *twenty thousand acres*, and that *each should have a separate grant*."

Unlike Canada, mainland Nova Scotia, and the thirteen continental colonies, where European settlement expanded from existing frontiers, St. John's Island was a world in itself, an entire geographic unit that could be completely divided and arranged. The Board's instructions outlined a new hierarchy of space; Holland's challenge was to fit that hierarchy into the confines of the island (figure 4.2). The largest units were counties, which the Board recommended be 500,000 acres. The surveyor squeezed three onto his map: Kings County at 406,000 acres, Queen's County at 458,420 acres, and Prince County at 467,000 acres. The medium units were parishes, which were to be 100,000 acres. Holland divided the counties into fourteen parishes, ranging from 124,000 acres to 63,000 acres, with five at exactly 100,000 acres. The smallest units were townships, which were to be 20,000 acres. The surveyor created sixty-seven townships, the great majority of which were 20,000 acres. In addition, Holland provided three town lots or "royalties": George Town and Prince Town each had 4,000 acres, Charlotte Town had 7,300 acres. If Holland had a challenge creating relatively uniform civil divisions, he also had to ensure that as many townships as possible fronted water. Of sixty-seven townships, only two – townships sixty-six and sixty-seven – did not front river or sea. Bound by the directives of the Board of Trade, the surveyor general had managed to create a geometric grid of uniform divisions. But unlike many grids oriented to true north, Holland used magnetic north, 15½ degrees off true north, "being the most easy way of running the lines for the surveyors," laying the basis

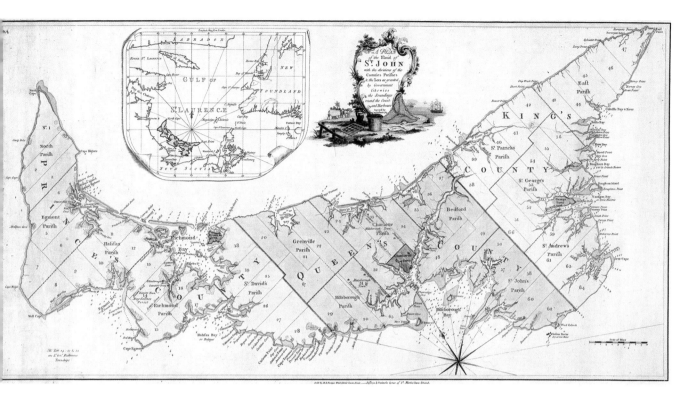

4.2 · *A plan of the Island of St. John ... Survey'd by Capt. Holland, 1775.* Library and Archives Canada, NMC-23350[1].

for one of the most distinctive grid patterns in North America.[28] When the map was used by the provincial surveyor of Nova Scotia as the basis for laying out townships and lots in the late 1760s, Holland's grid began to appear on the landscape.[29] As settlement expanded, roads and property boundaries tended to follow the grid, creating a mass of parallel lines that became inscribed on successive maps of the province during the nineteenth and twentieth centuries and which are clearly visible in the landscape today.

NAMING ST. JOHN'S ISLAND

If the Board of Trade instructed Holland in the hierarchy of civil divisions of St. John's Island, it appears to have given him *carte blanche* in assigning names. Almost like Gulliver towering over Lilliput, Holland gazed over his paper island and systematically allocated names to civil divisions, townsites, and geographical features (figure 4.3). What had been blank spaces on the map became identifiable places. Through this process of naming, Holland revealed, as historian E.P. Thompson succinctly put it in another context, how the "mind meets the world."[30]

Plans and Descriptions · 127

4.3 · Principal civil divisions on St. John's Island

Holland's naming was much different than that of other British surveyors. In the late 1760s and early 1770s, James Cook named features as he encountered them along the coasts of New Zealand and New South Wales in what Paul Carter has memorably called the "light offing of travel."[31] The result was a mixture of names of prominent people and those reflecting the contingencies of the voyage. For every Mount Egmont and Queen Charlotte Sound there was a Cloudy Bay and a Cape Turn Again. Although prominent physical features were usually named after eminent people in Britain, Cook was never able to step back from the entire landscape and fix a hierarchy of names onto a rank order of physical features. When the *Endeavour* stopped for victualling at Batavia, Cook made some alterations to his maps and journals, but wholesale renaming of New Zealand and New South Wales would have been impractical. Holland's experience, on the other hand, was considerably different. He was not venturing along unknown coasts, assigning names to capes and bays as he encountered them, but standing in front of an entire systematic survey of St. John's Island. He could thus apply a rank order of names to physical features and civil divisions in much the same way that he had divided the island into a hierarchy of different-sized spaces.

Unlike English settlers in the thirteen colonies who frequently named places after their home towns and counties in England, Holland preferred to commemorate people rather than places. The names that he selected reflected his sense of the trans-Atlantic imperial world. He pulled names down from a constellation of imperial figures and spread them over the newly won territory of St. John's.[32] The largest civil divisions on the island were counties and these were assigned to royalty. They were named according to precedence, reading from left to right across the map: Prince, Queen, King. Each county had a town, again named after members of the royal family: Prince Town, Charlotte Town (after Queen Charlotte), George Town (after King George III). Parishes were named after cardinal directions (North Parish, East Parish); patron saints of the British union (St. Andrew for Scotland, St. David for Wales, St. George for England, St. Patrick for Ireland, and St. John named after St. John's Island); and prominent people: Queen Charlotte, Grenville (Prime Minister, 1763–65), Bedford (Lord President of the Council, 1763–65), Halifax (Secretary of State for the Southern Department, 1763–65), Hillsborough (First Lord of the Board of Trade, 1763–65), Egmont (First Lord of the Admiralty, 1763–66), and Richmond (Holland's patron).

Holland reinforced this spatial hierarchy by locating important civil divisions close to prominent physical features. The two men most closely involved with Holland's advancement were Hillsborough and Richmond. In gratitude, the surveyor named the two largest bays on the island after them, making clear his greatest debt by calling the largest bay "The Great Bay of Hillsborough." The parishes of Hillsborough and Richmond were located beside the respective bays. Other bays on the island were named after Egmont, Halifax, Bedford, and Grenville, and parishes named after these officials were usually close by.

Bays were especially important physical features for they provided safe harbours for shipping and thus were vital for the commercial development of the island. As Holland observed in his description: "Port Joy [Hillsborough Bay], Cardigan, and Richmond Bay, are without dispute the only places, where ships of burthen can safely enter, and consequently most proper to erect the principal towns and settlements upon."[33] Bays thus became sites for settlements. Each of the three bays navigable for ocean-going ships had a town-site: Charlotte Town on Hillsborough Bay, George Town on Cardigan Bay, and Prince Town on Richmond Bay. Holland identified Charlotte Town as the capital for the following reasons: "The capital called Charlotte

Town, is proposed to be built upon a point of the harbour betwixt, Yorke and Hillsborough rivers, as being one of the best and nearly a centrical part of the island, has the advantages of an immediate and easy communication with the interior parts of the island, by means of the three fine rivers of Hillsborough, Yorke and Elliot."[34]

Yet bays were more than simply sites for towns: They were also organizing spaces for clusters of related names (figure 4.4). The "Great Bay of Hillsborough," named after the First Lord of the Board of Trade and the most conspicuous coastal feature on the island, was also the most appropriate site for honouring the entire Board. The First Lord was again commemorated by Hillsborough River, which flowed into the bay and was the longest river on the island. Of other Board members, George Rice was honoured by Rice Cove and Rice Point, Edward Bacon by Bacon Cove and Bacon Point, Edward Eliot by Edward Creek and Eliot River, John Yorke

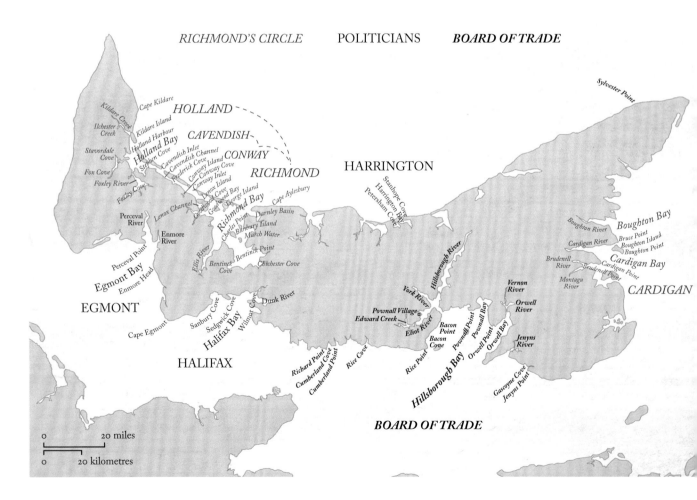

130 · SURVEYORS OF EMPIRE

by York River, Lord Orwell by Orwell Point, Orwell Bay, and Orwell River, Soame Jenyns by Jenyns Point and Jenyns River, and Bamber Gascoyne by Gascoyne Cove. John Pownall, Secretary of the Board, with whom Holland would have so much correspondence, was commemorated by Pownall Point and Pownall Bay. Edward Vernon, uncle of Lord Orwell, gave his name to Vernon River.

Holland created a second cluster of names at Richmond Bay, the second-largest bay on the island. Named after Charles Lennox, 3rd Duke of Richmond, Earl of March, and Earl of Darnley, whose country seat was at Goodwood, a few miles from Chichester in West Sussex, England, the bay offered great toponymic possibilities. Within the bay, Holland named Charles Point, Lenox Island, Lenox Channel, March Water, Darnley Basin, Goodwood Bay, Goodwood Cove, Chichester Cove, and George Island (after the Duke's father). The Duke's sister, Lady Sarah Lennox, had married Sir Thomas Charles Bunbury, which resulted in Charles Point and Bunbury Island.[35] Among the Duke's closest friends were the Bentincks, a leading Dutch family, whom Holland probably knew, which gave rise to Bentinck Point and Bentinck Cove. Ellis River was named after Welbore Ellis, Secretary of War, friend of Richmond, and supporter of the Grenville government.[36]

Holland named Cardigan Bay, the third largest bay on the island, after George Brudenell, 1st Duke of Montagu and 4th Earl of Cardigan, whose country seat was at Boughton in Northamptonshire. Brudenell was the son of Elizabeth Bruce, daughter of Thomas Bruce of Elgin. In 1757, Charles Lennox married Mary Bruce, daughter of Charles Bruce, 4th Earl of Elgin, thus cementing the link between the Lennox and the Montagu families. Holland was happy to celebrate the tie by naming Cardigan Bay, Cardigan Point, Cardigan River, Brudenell River, Brudenell Point, Montagu River, Boughton Island, Boughton Point, Boughton Bay, Boughton River, and Bruce Point. Although Holland most likely named George Town on Cardigan Bay after George III, the surveyor may also have been making a play on Brudenell's first name. A courtier and a fellow of the Royal Society, Brudenell petitioned the Board of Trade for land on the island in 1764 and may have been in contact with Holland before he left England that year.[37]

Holland also emphasized family ties through proximate geographic features. On the northwest coast of the island, the surveyor created a series of names that were associated with the Duke of Richmond. Among the Duke's associates was Henry Seymour Conway, who served as Secretary of State for the Southern Department (1765–66). He was commemorated by Conway Island, Conway Inlet and Conway Cove.[38] Conway married Caroline, widow of Charles, Earl of Aylesbury in 1747, a family memorialized in Cape Aylesbury. The Conway place names and Cape Aylesbury lay either side of Richmond Bay. Farther west, Holland honoured Frederick Cavendish, a friend of Conway and Richmond, in Frederick Cove, Cavendish Inlet, and Cavendish Channel. Yet these geographic features were relatively minor compared to Holland Bay, the most westerly bay on the north coast. Holland obviously saw the bay as a natural pairing with Richmond Bay and named it not for himself but for Henry Fox (1705–1774), 1st Baron Holland of Foxley (1763), who had married Lady Caroline Lennox (1723–1774), a sister of the Duke of Richmond. The Richmond and the Holland families were thus united through marriage and commemorated by two neighbouring bays. Fox was one of the leading politicians in England in the early 1760s and had played a prominent role in convincing the House of Commons to approve the Treaty of Paris that ended the Seven Years War. For that service, he was elevated to the House of Lords and created Baron Holland. The surveyor general commemorated the elevation by the names Holland Bay, Holland Harbour, Foxley River, and Foxley Cove. Fox's brother, Stephen Fox-Strangways (1704–1776), Lord Ilchester and Stavordale, was also honored by Stephen Cove, Fox Cove, Stavordale Cove, and Ilchester Creek. The remaining part of the Richmond cluster of names commemorated another of the Duke's sisters, Lady Emilia Mary Lennox, who had married James Fitzgerald, 1st Duke of Leinster, later created Marquess of Kildare. The surveyor named Cape Kildare, Kildare Creek, and Kildare Island after this family link.

Three lesser bays honoured other important political figures. Egmont Bay was named after John Perceval, 2nd Earl of Egmont and Baron Lovel and Holland of Enmore, who served as First Lord of the Admiralty. In Egmont Bay, Holland included Cape Egmont, Perceval River, Perceval Point, Enmore Head, and Enmore River. Halifax Bay was slightly more complicated. Named after George Montague Dunk, 2nd Earl of Halifax, who served as Secretary for the Southern Department, the bay included Dunk River and Sanbury Cove (Halifax had been Viscount Sunbury until he succeeded his father as 2nd Earl of Halifax in 1739; figure 3.1). The bay also included Sedgwick Cove for Edward Sedgwick, solicitor and Clerk of Reports for the Board of Trade (1753–63) and under-secretary to Halifax; and Wilmot Cove for Montague Wilmot, uncle of the 1st Earl of Halifax and governor of Nova Scotia. Finally, Harrington Bay was named after William Stanhope (Stanhope Cove), Viscount Petersham (Petersham Cove), and 2nd Earl of Harrington, who served in the Grenadier Guards and must have been known to Holland

With the main geographic features on the map named after prominent politicians, Holland used the remaining space to commemorate the military services and personal friends and acquaintances (figure 4.5). On the north coast of the island, Holland created an Admiralty cluster cen-

5 · Military and other place names on St. John's Island

tred on Grenville Bay and Grenville Parish. Grenville was not only British Prime Minister between 1763 and 1765 but had served as First Lord of the Admiralty between 1762 and 1763. Hans Stanley, Lord of the Admiralty, Member of Parliament, and supporter of Grenville, was commemorated by Stanley River which flowed, appropriately, into Grenville Bay.[39] A few miles to the east, Holland laid out Harris Bay after James Harris, another Lord of the Admiralty and supporter of Grenville. Part of the bay included Hunter River and Orby Head, which were named after Thomas Orby Hunter, yet another Lord of the Admiralty. Whitley River flowed into the bay, the name honouring Thomas Whately, Secretary to the Treasury during the Grenville administration and one of the more influential formulators of policy for the colonies.[40] Holland also commemorated commanders of the North America station by placing their names along the east coast of the island: Rear Admiral Alexander Colvill (Colville Bay,

Plans and Descriptions · 133

Colville River), Commodore Richard Spry (Spry Point, Spry Cove), and Vice-Admiral Philip Durell (Durell Point). Captain John Deane, one of the senior captains on the Halifax station who had helped Holland in his survey of St. John's Island, was rewarded with Dean Point on the east coast and Dean Cove on Hillsborough River. As Deane captained the *Mermaid* and his lieutenant was William Johnston, Dean Cove sat between Mermaid Farm and Johnstone River. With this series of toponyms, Holland had run down the hierarchy of naval officials, from the First Lord of the Admiralty, through the commanders of the Halifax station, to the captain and lieutenant of the vessel patrolling the Gulf of St. Lawrence.

Holland anchored the northern extremities of the island by cardinal points – North Cape, East Cape – but allocated the southern extremities to his commanders at Quebec. The western end was dedicated to James Wolfe (Cape Wolfe, Wolfe Marshes, Wolfe Inlet), while the eastern end was given to James Murray (Murray Harbour, Murray Islands, Murray River). With the Admiralty dominating the central part of the north coast, Holland assigned army names to much of the west, south, and east coasts of the island. On the west coast, Robert Monckton and Thomas Gage, two of the brigadiers at Quebec, were honoured by Monckton Cove and Cape Gage; on the south coast, Frederic Haldimand (Haldimand River), Gordon Graham (Gordon Cove, Gordon Point, Graham Head), Henry Bouquet (Boquet Cove, Boquet Point), Guy Carleton (Guy Cove, Carleton Cove, Carleton Point), Jeffery Amherst (Amherst Point, Amherst Cove), and Augustin Prevost (Prevost Cove, Prevost Point); and on the east coast, William Maule, Earl of Panmure (Panmure Island, Panmure Point, Panmure Basin), Richard Maitland (Maitland Point), William Howe (Howe Bay, Howe Point), Archibald Montgomery, 11th Earl of Eglinton (Eglinton Cove, Eglinton Point), Andrew Rollo, 5th Lord Rollo (Rollo Bay), and William Hervey (Hervey Cove, Hervey Point). Holland also included his fellow surveyor, J.F.W. Des Barres (Debarres Point).

Finally, Holland commemorated friends and colleagues whom he had known in London by clustering their names at the centre of the south coast of the island. Richard Cumberland, Clerk of Reports at the Board of Trade, served as Holland's agent, handling financial transfers from the government to the surveyor's account, and also as colonial agent for Nova Scotia and Quebec.[41] Holland preserved his name at Richard Point, Cumberland Cove, and Cumberland Point. Richard Brocklesby, formerly surgeon general of the army and in private practice in London in 1763, presented one of Holland's scientific papers to the Royal Society. He was honoured by Brokelby's Cove, Brokelby's River, and Brokelby's Head. The last member of the group was William Tryon, who had served in the army during the Seven Years War (but not in North America) and was in London in 1763 lobbying for a governor's appointment in America. In April 1764, he was appointed Lieutenant-Governor of North Carolina.[42] At some point during his sojourn in London, Holland must have met Tryon for he was commemorated by Tryon Cove, Tryon River, and Cape Tryon. Indeed, the friendship must have had some significance because Holland located Cape Tryon on the north coast and Tryon Cove and Tryon River almost exactly opposite on the south coast; a line drawn between these points neatly bisected the island.

Holland's innate modesty and sense of position in the social and military hierarchy of late eighteenth-century Britain prevented him from placing his own name on the map of the island, but he did locate Observation Cove beside Fort Amherst and Surveyors Inlet and Surveyors Point near the East Point of the island. Unlike Cook, Holland included virtually no memento of the day-to-day travails of the survey, except for Salutation Cove, on the south coast, which was probably the place where young Haldimand was saved from starvation.

If Holland refrained from commemorating himself, his map of St. John's Island was saturated with names from the British imperial world of the late 1750s and early 1760s. His map commemorated veterans of the Seven Years War, naval and army commanders, members of the Admiralty and Board of Trade, senior government officials, patrons, and the royal family. The surveyor had created a trans-Atlantic genealogy of power. He had integrated the former French island into a British cultural and imperial sphere rather than into the parochial world of settlers, land speculators, and provincial politicians in Nova Scotia. The map read across the Atlantic to Britain, not around the coast to the provincial capital of Halifax. The surveyor general's mental world focused on the metropole, not the colonial periphery. His systematic practice of naming, far more thorough than Cook's in the Pacific, should be seen as the beginning of the comprehensive British imperial naming of colonial possessions that so flourished in the late eighteenth and nineteenth centuries.

NAMING CAPE BRETON ISLAND AND NOVA SCOTIA

Holland's next plan was of Cape Breton Island, a territory he was determined to toponymically claim for Great Britain by renaming it the "Island of Cape Britain."[43] The surveyor general continued the cadastral and naming practices implemented on St. John's Island. The southern half of Cape Breton was divided, as per the instructions of the Board of Trade, into counties, parishes, and townships (figure 4.6). Three counties were named after members of the royal family: Louisa for one of George III's sisters; Chester for George, Prince of Wales and Earl of Chester; and York for Edward, Duke

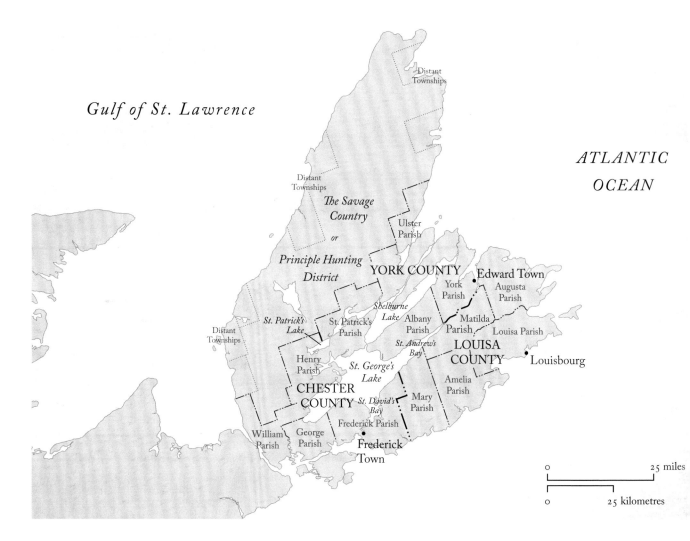

of York. Parishes in Louisa County were named for female royals: Mary, Amelia, Louisa, Matilda, and Augusta; those in Chester County took the names of the Prince of Wales: George, Frederick, William, Henry; and those in York County took the titles of the Duke of York: York, Albany, Ulster. In addition, York County had St. Patrick's parish, which was proximate to St. Patrick's Lake (St. Patrick's Channel), an arm of the Bras d'Or Lakes. As on St. John's Island, Holland laid out town sites with royal names: Frederick Town in Chester County, Edward Town in York County, and the existing town of Louisbourg served, appropriately, in Louisa County. Holland also named prominent physical features. In the southern half of the island, the Bras d'Or Lakes provided Holland with significant naming op-

4.6 · Principal civil divisions and place names on Cape Breton Island

portunities. The surveyor general honoured the British union by calling the principal lake St. George's Lake, and naming subsidiary lakes St. Andrew's Bay, St. David's Bay, and St. Patrick's Lake. Great Bras d'Or Lake was renamed Shelburne Lake, after the secretary of state who had helped instigate the general survey. Numerous coastal features around the island were named after many of the same people commemorated on St. John's Island; Holland also created similar family groupings.

While Holland was busy renaming St. John's Island and Cape Breton Island, Des Barres was pursuing a similar strategy along the coasts of peninsular Nova Scotia. Unlike Holland, who refrained from naming parts of North America after himself, Des Barres had no such scruples and affixed his name to the Desbarres River (River Hebert), which separated his land holdings at Minudie and Maccan at Chignecto, and to Joseph Harbor and Frederick Bay (Tatamagouche Bay), which lay alongside another extensive land holding. Yet Des Barres's principal homage was less to himself than to the port towns of Britain, government officials, and the royal family. His most systematic naming took place along the Atlantic coast of Nova Scotia; its many harbours, capes, and points offered numerous toponymic opportunities. On his manuscript chart of the coast from St. Margaret's Bay to Canso, he named harbours and points after the towns of Leith, Folkstone, Dungeness, Exeter, Bristol, Southampton, Winchelsea, Cockermouth, Berwick, and Torbay, and after provincial and British government officials including Colvill, Sandwich, Spry, Wilmot, and Mauger.[44] But the most complete cluster of names centred on Mahone Bay and St. Margaret's Bay, which lay to the west of Halifax, and formed the two largest bays on the coast (figure 4.7). Appropriately, Des Barres honoured the royal family, creating King's Bay (Mahone Bay), Charlotte Bay (St. Margaret's Bay), Mecklenburgh Bay (Queen Charlotte's family name), and Prince Harbour (Mahone Harbour), as well as Royal George Island (Big Tancook Island), Prince of Wales Island (Cross Island), Princess Royal Island (Backmans Island), William Henry Isle (Coveys Island), Frederick Island (Rafuse Island), Edward Isle (Mason Island), and Crown Point.[45]

Both Des Barres and Holland placed Hanoverian royal names on their maps, but their other toponyms differed. Whereas Holland's place names reflected his movement in metropolitan circles and the patronage of the Duke of Richmond and the Earl of Hillsborough, Des Barres's names reflected his much more provincial connections in Nova Scotia. He made gestures to Britain by placing names of royalty and port towns on his maps, but his real allegiance was to local patrons, politicians, and speculators, such as Colvill, Wilmot, and Mauger. It was not until Des Barres moved to England in 1773 and became much more closely dependent on the patronage of the Admiralty that metropolitan names began to appear on his maps.

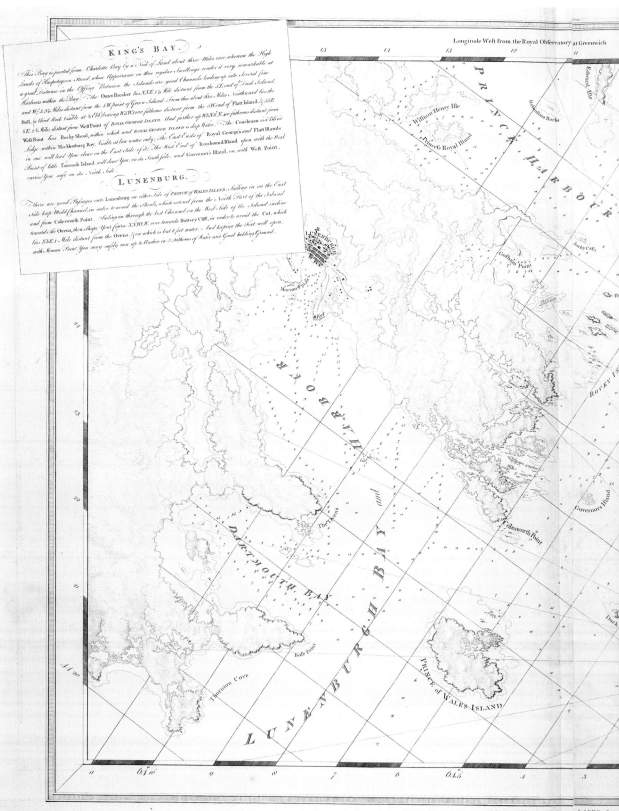

KING'S BAY.

This Bay is parted from Charlotte Bay by a Neck of Land about three Miles over, whereon the High Lands of Haspotagoen Stand, whose Appearance in three regular Swellings render it very remarkable at a great Distance in the Offing. Between the Islands are good Channels leading up into Several fine Harbours within the Bay. The Outer Breaker lies N.N.E. 1½ Mile distant from the S.E. end of Duck Island, and W. ¾ S. 3½ Miles distant from the S.W. point of Green Island. From this about three Miles Northward lies the Bull, (a blind Rock Visible at ¼ Ebb) bearing W.S.W. 200 fathoms distant from the S.W end of Flatt Island, & S.S.E. ¾ E. 2¾ Miles distant from West Point of ROYAL GEORGE ISLAND. And farther up W.b.N. & N. 100 fathoms distant from West Point lies Rocky Shoal, within which and ROYAL GEORGE ISLAND is deep Water. The Coachman is a blind Ledge within Mecklenburg Bay, Visible at low water only. The East Ends of Royal George's and Flatt Islands in one will lead You clear on the East Side of it. The West End of Ironbound Island open with the West Point of little Tancook Island will clear You, on its South Side, and Governor's Island, on, with West Point, carries You safe on its North Side.

LUNENBURG.

There are good Passages into Lunenburg on either Side of PRINCE of WALES ISLAND. Sailing in on the East Side keep Midd Channel, in order to avoid the Shoals, which extend from the North Part of the Island and from Colesworth Point. Sailing in through the best Channel, on the West Side of the Island, incline towards the Ovens, then shape Your Course N.N.b.W. over towards Battery Cliff, in order to avoid the Cat, which lies N.N.E. 1 Mile distant from the Ovens, & on which is but 8 feet water. And keeping the Fort well open, with Moreau Point You may safely run up to Anchor in 3 fathoms of Water and Good holding Ground.

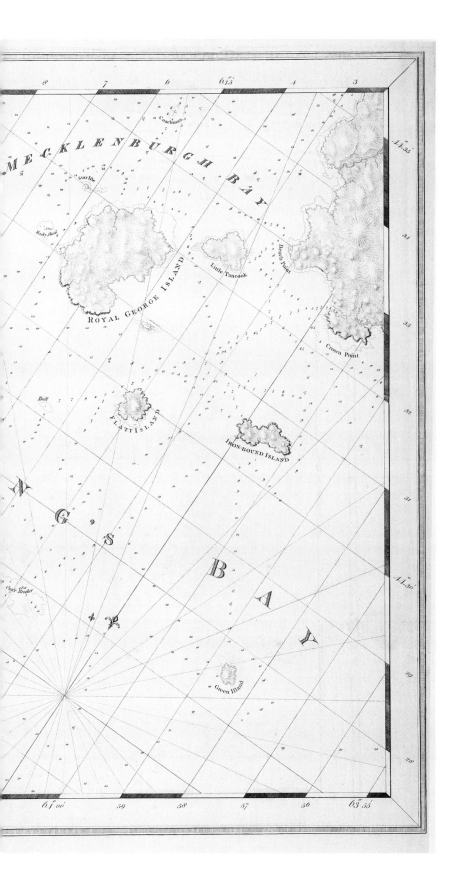

4.7 · *King's Bay [and] Lunenburg* from *The Atlantic Neptune*. Among place names on the chart are King's Bay, Prince Harbour, Mecklenburgh Bay, Royal George Island, Prince of Wales Island, Princess Royal Island, and Crown Point. © National Maritime Museum, Greenwich, London, HNS 35.

ERASURE AND PERSISTENCE

In several vigorously argued essays, geographer Brian Harley drew attention to the power of the map. According to Harley, "the power of the surveyor and the map maker was not generally directly exercised over individuals but over the knowledge of the world made available to people in general."[46] Cartographers, usually working for government and the military, could create their own particular view of the world and erase others. Harley was particularly sensitive to the absence of Native place names on European maps of colonial America. Surveyors and map makers, he argued, frequently failed to record indigenous names, producing "toponymic silence" and effectively erasing Native toponyms by excluding them from official maps.[47] A clear demonstration of such erasure was John Smith's map of New England, published in 1616, a forerunner, in some respects, of maps published in *The Atlantic Neptune*. Harley recounts how Smith presented his map of New England to Charles, Prince of Wales, and requested that the young prince replace the indigenous names with English ones. "Transplanting England in the paper landscape was an easy game to play," Harley writes. "The young prince named 'Cape James' (Cape Cod) for his father; 'Stuart's Bay' (Cape Cod Bay) after the reigning family; 'Cape Elizabeth' for his sister; 'Cape Anna' after his mother; and 'The River Charles' for himself." "Few of these royal inventions survived," Harley notes, "but Smith's map became a paradigm for further Anglicization."[48]

At first glance, the manuscript maps produced by Holland and Des Barres appear a clear case of Anglicization. The two surveyors superimposed a complex layer of British imperial names over many Mi'kmaq and French place names. Despite employing Native guides, Holland rarely recorded Mi'kmaq names, only locations with native associations, such as Indian Point, Indian Island, and Indian Rocks on St. John's Island, and Indian Bay on Cape Breton Island. Holland and Des Barres were more respectful of the French toponymic heritage. Some French names had been in use since Champlain's mapping of the region in the early 1600s and were generally accepted by European map makers; others continued to be used because they named French settlements along the St. Lawrence, around the Gaspé peninsula, and at the head of the Bay of Fundy. But on St. John's Island and Cape Breton Island, which had lost much of their former French colonial population, Holland simply noted the old French name below the new British name, as in "Cardigan Bay [called] by the French Trois Riviers" (figure 4.1) or "Harris Bay [called] by the French Grand Rastico."[49] On these two islands, the surveyor general assumed that British names would come into common usage.

Yet for all the names so lavishly bestowed on manuscript and engraved maps, only a few survive today. On Prince Edward Island, 92 of the 191 names given by Holland are still used; on Cape Breton Island and the coast of Nova Scotia, probably a bare handful.[50] The reasons for this poor rate of survival are relatively straight-forward. Essentially, Holland and Des Barres created an imperial toponymy on manuscript maps that only circulated among government officials in London. Even when the maps were engraved and published in *The Atlantic Neptune*, they were used mainly by naval officers on the North American station. The maps did not circulate among Acadian settlers along the shores of Nova Scotia, Cape Breton Island, and Prince Edward Island, who had little need for them. As a result, maps and the toponyms they carried were never embedded in the customary, vernacular world of the region's inhabitants. When new maps were made during the early nineteenth century, provincial surveyors recorded place names in local usage rather than decades-old imperial names from *The Atlantic Neptune*. Acadian areas of settlement on Prince Edward Island continued to be called by their French names (for example, Rustico instead of "Harris"), while Scottish settlers on Prince Edward Island, Cape Breton Island, and along the Eastern Shore of Nova Scotia introduced many new names from the Highlands and Islands of Scotland.

Institutional changes also led to the erasure of the surveyors' place names. On Prince Edward Island, parishes were not used as civil units and so their names never had any institutional foundation. On Cape Breton Island, Holland's division of parishes and counties was discarded in the late eighteenth century and replaced by a much simpler structure of counties. Although not one of Holland's original counties, modern Richmond County preserves the Duke of Richmond's name, a faint echo of the surveyor's designs for the island. In sum, Holland and Des Barres imposed names on the landscape but that was no guarantee that local settlers would use them. As remnant Acadian populations expanded and British settlers poured into the region in the late eighteenth and early nineteenth centuries, the imperial toponymy created by Holland and Des Barres was swamped by a plethora of new names.

British names that were adopted on St. John's Island and elsewhere soon took on their own cultural meanings. Few, if any, Native peoples and settlers knew the reasons why Holland and Des Barres named particular places as they did. For local inhabitants on St. John's Island, a place such as Stanhope Cove did not commemorate an English aristocrat but was a usable cove on the north shore. In other words, the meaning of names changed, from the British imperial context to the local and vernacular. Harley's emphasis on the top-down imposition of power through the map fails to accommodate the customary usage of local people, the vagaries of institutional change, and the cultural appropriation of names. The continued use of vernacular names – some Native, others French – after the publication of *The Atlantic Neptune* can be seen as a form of resistance to British attempts to rename the northeast.

Descriptions

In its instructions to General Murray, the Board of Trade stipulated that descriptions or reports on the new territories had to be submitted along with maps. The Admiralty also expected reports on the areas that were being surveyed. Holland's principal reports covered St. John's Island and Cape Breton Island, but he also wrote accounts of the lower St. Lawrence River and the District of Maine. Among his deputies, Haldimand produced a report of the Magdalen Islands. Des Barres wrote a lengthy description of Sable Island.[51] Such reports provided explanatory information about the maps and additional information about the geography, resources, economy, and population of existing settlements and proposed townships. In essence, the reports were inventories or gazetteers, providing as full an account as possible of the prospects for commerce and settlement in the newly acquired territories.

The two principal reports produced by the General Survey covered St. John's Island and Cape Breton Island and followed the format stipulated by the Board of Trade's instructions. In the St. John's report, Holland described the island's soil and produce; timber; birds, beasts, and fishes; location of proposed towns; and the nature and effects of the climate. In an accompanying table, he listed the sixty-five townships and the three county town lots, noting their acreage, boundaries, quality of lands and woods, cleared land and houses, and provided additional remarks, usually on topography. The report was directly referenced to the large scale manuscript map of St. John's Island. Indeed, Holland also copied the table of townships and county town lots onto the map.[52] The "Description of the Island of Cape Britain relative to the Plan surveyed" was an even fuller report than that of St. John's. After a general introduction, Holland gave a detailed account of Cape Breton Island, beginning at the Gut of Canso and proceeding round the coast and through the Bras d'Or Lake. Each place was described in terms of its accessibility by water, topography, soils, timber, minerals, climate, and potential for commerce, industry, and settlement. He also included quantitative and qualitative information about the former French fishery, and a table – exactly like the one for St. John's Island – listing the seventy townships and two county town lots. (Louisbourg was not covered in the table for it had been described at length in the account of the French fishery). As the title of the report suggests, it was to be read in conjunction with the large-scale manuscript map of the island. Although Holland's large-scale map of Cape Breton Island has not survived, the surveyor general most likely followed the arrangement of the St. John's map by placing the table of townships and geographic attributes onto the Cape Breton map. He provided an abstract of the table on a small-scale map of the island that does exist.[53]

The two reports revealed Holland's keen sense of observation, considerable knowledge, and methodical recording. His observations were descriptive and matter-of-fact, reflecting his surveyor's training. He noted depths of water, quality of harbours, and tidal range along the coasts; soils, terrain, minerals, timber, and climate inland. He also commented on flora and fauna. The surveyor had an eye for economic development, noting suitable places for the fishery, saw mills, and coal mining. He appears to have interrogated his Acadian guides well, for both reports contained detailed knowledge of French settlement and fishing. All the information that he accumulated was categorized and logically arranged in the reports. Of the two islands, Cape Breton impressed more. The French had exploited the large Cape Breton cod fishery for decades, and Holland saw plentiful opportunities for an English fishery of even greater scale. He also recognized the potential of the coal mines that outcropped along the east coast. "All rich & plentiful," he noted, "& which without any other resource or advantage, would distinguish this island."[54] In the nineteenth century, this became true.

Holland also produced a series of meteorological observations and records at Observation Cove on St. John's Island for the period from 8 October 1764 to 5 April 1765.[55] He recorded the day, time of day, temperature, wind direction, and weather conditions. Predating William Wales's observations at Fort Churchill by three years, these records may well have been among the earliest meteorological observations carried out by the British in North America and testified to Holland's scientific interests.[56] A copy of the records was submitted to the Board of Trade in 1765. Holland sent no further records but most likely continued to make observations for his own personal interest.

What Holland had produced in the maps and reports was almost a three-dimensional grid of the two islands. The maps, produced at different scales, covered two dimensions; the reports, particularly the tables, listed a series of geographic, economic, and demographic attributes that could be stacked up on each of the spatial divisions on the maps. The meteorological record also gave an indication of the daily weather and seasonal climate on St. John's Island. Although Holland's knowledge tailed off in the forested interiors of the two islands, he had produced systematic accounts of the coasts, turning undifferentiated spaces into fairly well known places. Such knowledge was presented through a combination of maps and reports. In the modern archive, the General Survey's plans and descriptions have been separated into different categories of record, but the surveyor general envisaged them as an inter-textual account of the newly acquired islands. In a sense, these two forms of text – cartographic and written – were a geographical information system that could be easily accessed by the Lords and civil servants of the Board of Trade.

SILENT SPACES

The combination of plan and description raises another question about Brian Harley's analysis of early cartography. One of the central claims of Harley's work was that European maps, through their failure to show Native settlements and place names, effectively erased Native peoples from their lands and presented blank spaces available for European settlement.[57] According to Harley, "many eighteenth-century map makers preferred blank spaces to a relict Indian geography."[58] "This sort of cartographic silence becomes an affirmative ideological act," he wrote. "It serves to prepare the way for European settlement. Potential settlers see, on the map, few obstacles that are insurmountable. Least of all does the map reflect the presence of indigenous peoples and their imprint on the land … in short, such maps are ethnocentric images, and part of the apparatus of cultural colonialism. It is not only that they offer a promise of free and apparently virgin land – and empty space for Europeans to partition and fill – but that the image offered is of a landscape in which the Indian is silent … Through these silences, the map becomes a license for the appropriation of the territory depicted."[59]

Holland's two major surveys of St. John's Island and Cape Breton Island were undoubtedly for the purpose of measuring and dividing land in preparation for European settlement. Although he was well aware of the remnant Acadian presence on St. John's Island and noted their settlements on the large-scale map of the island, he did not record the Mi'kmaq, perhaps because they were not resident during the survey. Holland was more aware of the Native people on Cape Breton Island. British soldiers had skirmished with Mi'kmaq during the Louisbourg campaign in 1758, and Holland encountered them during his survey. He hired some as guides, and also wrote to the Board of Trade recommending that a priest be sent from Quebec to administer to their spiritual needs.[60] More significantly, he set aside the northern, mountainous half of the Island and labelled it "The Savage Country or Principle Hunting District" on the small-scale manuscript map. In the later, engraved version of the map, reproduced in *The At-*

lantic Neptune, the region was called "Hunting Country." As the surveyor general explained to the Board of Trade, the region was "stiled the savage or hunting country as it is fit for nothing else, being without harbours, or rivers, & very mountainous: This however if put to the use, the title implies, may be brought to great value, by affording a trade in furs, with the Indians in return for English manufactures." Moreover, "the body of men this would maintain, might also be an auxiliary force notwithstanding they are Savages, in case of an invasion."[61] Holland appears to have been following Board of Trade policy, codified in the Proclamation of 1763, by setting aside land for Native peoples and fostering the fur trade. In sum, Holland and the Board of Trade were well aware of the Mi'kmaq, even though their presence was not always shown cartographically.

Furthermore, Holland did not always know what existed back from the coast. The British surveying effort was focused on delineating coastline, not marking out interior lakes and mountain ranges. When Acadians or Mi'kmaq were unavailable to guide surveying parties inland, areas were left unsurveyed and appeared blank on manuscript maps. On Cape Breton Island, Holland not only missed the upper reaches of the Margaree River and Lake Ainslie, but did not have a good sense of the agricultural potential of the interior. The surveyor's townships were a geometrical grid that bore little relation to the topography and soils of the island.[62] He also added a range of hills that wriggle like a hairy caterpillar across the eastern part of the island, but which have no existence in reality. Holland's knowledge of geography and population was partial at best and the silent spaces or blanks on the map reflect that ignorance. As geographers Felix Driver and Luciana Martins have noted about British naval surveyor John Septimus Roe in South America in the early 1800s: "the imperial eye appears not as transcendent, all-knowing, global, but instead as situated, partial, local."[63] The same could have been said of Samuel Holland.

The plans and descriptions produced by the General Survey of the Northern District and the Admiralty Survey of Nova Scotia were radical simplifications and abstractions of the 15,000 miles of coastline that Holland and Des Barres had surveyed over ten years. Enormous geographies and complex topographies had been miniaturized into small- and large-scale maps, while detailed information on terrain and resources had been abstracted into tables and lists. All this was necessary in order to transfer information across the Atlantic and make it comprehensible to politicians and civil servants in the centre of calculation in Whitehall.[64] The maps and reports submitted by Holland and Des Barres joined those from Cook, Gauld, De Brahm, Pittman, and others in the files of the Board of Trade, Admiralty, and Ordnance Office in London, a significant contribution to the rapidly growing imperial archive and a foundation of information on which to develop government policy. The plans and reports could be combined with other maps and information to create an overall graphic and statistical picture of the newly acquired territories. With this information to hand, the imperial government could evaluate the economic potential of "unsettled" areas of the continent and develop policy for their settlement.

The maps and reports sent by Holland and Des Barres not only created a particular form of knowledge but also helped shape the material landscape, a point often lost in the recent scholarly emphasis on maps as texts.[65] In particular, the General Survey's maps and descriptions made considerable impress on the landscape, particularly through subdivision of land and naming of places. Some of this textual assemblage grabbed the land for good, particularly on St. John's Island: Holland's cadastral divisions and many of his names still mark the island's landscape today. But on Cape Breton Island, the assemblage failed to take hold. The outbreak of the American Revolution halted the implementation of Holland's plan, leaving his paper landscape and report in the files of the Board of Trade. Nevertheless, Holland's systematic surveying and naming on St. John's Island and Cape Breton Island had produced a generic type of survey that was to be applied to other parts of the empire after the end of the War of American Independence.

Surveyors as Proprietors

5

After the French capitulation at Montreal in September 1760 and the collapse of the Indian threat to frontier settlements, American settlers and land speculators moved into Native and former French territories in search of land. This push was felt all along the frontier of the thirteen continental colonies and in peninsular Nova Scotia. Colonial populations were growing rapidly through natural increase and immigration from the British Isles and the German states, and land in the older settlements along the seaboard was in short supply and increasingly expensive. The western frontier acted as a safety valve for this build-up of population.[1] Yet expansion inevitably threatened Native peoples. To avoid a frontier war, the British government, through the Proclamation of 1763, attempted to restrict westward expansion by creating a boundary along the height of land of the Appalachian Mountains and encouraging the westward push to turn north to Nova Scotia and south to Florida. The General Surveys of the Northern and Southern Districts were intended, in large part, to survey and divide land in readiness for incoming settlers.

In the Northern District, New Englanders pushed north and east in search of land. Many moved up the Connecticut, Kennebec, and Penobscot river valleys; some along the eastern Maine coast; and others across the Gulf of Maine to Nova Scotia to occupy former Acadian settlements. After the forcible deportation of the Acadians in the late 1750s, provincial officials in Halifax were keen to settle the backcountry with loyal subjects and made land available along the Annapolis and Avon river valleys, the Bay of Fundy marshlands, and along the Atlantic coast.[2] The newly acquired territories of Cape Breton Island, St. John's Island, and the remains of French Acadia (modern New Brunswick) offered further potential for settlement. In the Southern District, East and West Florida provided enormous tracts for settlers spilling out of Georgia and the Carolinas.

As the newly acquired territories fell under the authority of the Crown, many British officials and military officers were keen to grab what they could. In late January 1763, a month before the signing of the Treaty of Paris, Amherst wrote to Egremont declaring that there was "an allmost universal desire amongst the officers and soldiers here to have grants of land to settle on this continent."[3] This "desire" was kept in mind as British officials drew up policy for governing and settling the newly acquired territories. The Proclamation, published in October 1763, included a provision for rewarding officers and soldiers with land grants. Army officers who had served in North America and private soldiers who had disbanded and were still resident on the continent were entitled to generous awards. Field officers were given 5,000 acres; captains, 3,000 acres; subalterns or staff officers, 2,000 acres; non-commissioned officers, 200 acres; and privates, 50 acres. Likewise, naval officers, who had served at Louisbourg and Quebec, were entitled to equivalent amounts.[4] The Board of Trade instructed provincial governors to make out the grants without fee or reward, but after ten years lands were again subject to the usual quit rents and conditions of cultivation and improvement. Such grants provided a considerable means of advancement for military men. For officers resident in Britain, land grants in North America could be sold for monetary gain; for officers and men who stayed on the continent, grants provided a means of making a living, either through direct cultivation or by bringing in tenants to work the land.

Both Holland and Des Barres were well aware of the importance of land ownership. Holland's career had been advanced by his patron, the Duke of Richmond, who owned large estates in England, while Des Barres had rubbed shoulders with enough titled military men to know the value of landed property. As "foreign Protestants," however, the two surveyors stood outside the social and military hierarchy of England and had no estates to fall back on in the British Isles. For both surveyors, the lands of North America represented an immense opportunity to establish themselves as independent gentlemen. Holland and Des Barres were also well aware of the speculative frenzy surrounding them. Holland's senior officers, Murray and Haldimand, had both acquired land in Quebec, and the Duke of Richmond, perhaps through Holland's prompting, had submitted a memorial to the Board of Trade for a grant to all of Cape Breton Island in early 1764.[5] Holland, too, hoped to gain land in Nova Scotia, submitting a memorial in conjunction with Benjamin Hallowell, comptroller of customs in Boston, and one Joseph Peach, to the Board of Trade for a township along the Saint John River.[6] Des Barres was also after land in Nova Scotia. In June 1764, the Board of Trade recommended to the Crown that twenty-six memorialists, including Des Barres and Holland, should be awarded grants of 10,000 and 20,000 acres in Nova Scotia.[7] Such was the interest in land in the early 1760s that Des Barres would later claim that "the surveys done by Capt Hollands people … seem chiefly entended, by the Lords of Trade and Plantations, for purposes of speculations of lands."[8] Both surveyors knew that their survey expeditions offered opportunities for personal enrichment, but the way they used the surveys to advance their respective interests differed markedly.

SAMUEL HOLLAND AS PROPRIETOR

The Board of Trade directed Holland to survey St. John's Island first, which reflected considerable interest in the island back in London.[9] British knowledge of St. John's Island had begun to accumulate after the siege of Louisbourg in July 1758. In August and September that year, Amherst sent two search and destroy missions into the Gulf of St. Lawrence. One, led by Wolfe, wrecked the French fishery at Gaspé; the other, led by Lord Rollo, captured the French fort at Port-la-Joye on St. John's Island. After taking the fort, British forces rounded up more than three thousand Acadian settlers for deportation. The British officers, almost all from landed backgrounds in England and Scotland, must have quickly weighed up the agricultural possibilities of the island, especially as the cleared land, rolling woodlands, and marshy inlets contrasted so markedly to the barren Atlantic shores of Cape Breton Island and Nova Scotia. New England merchants also had their eyes on the cod fishery. New Englanders had been involved in the Nova Scotia and Newfoundland banks fisheries for more than a century and must have had a good idea of the potential of the Gulf of St. Lawrence.[10] Although the French were allowed fishing rights to the west coast of Newfoundland under the terms of the Treaty of Paris, the west side of the Gulf was firmly in British hands. St. John's Island thus offered both good agricultural land and a convenient base for exploiting the cod fishery.

Soon after the Treaty of Paris was signed in February 1763, proposals for settling the island began circulating in London. The most prominent and most extraordinary came from John Perceval, 2nd Earl of Egmont and First Lord of the Admiralty.[11] Already involved in land speculation in Nova Scotia and East Florida, Egmont submitted his first proposal to the king in December 1763. He requested a grant of the entire island – then thought to contain two million acres – and proposed to create a feudal society consisting of himself as Lord Paramount, forty Capital Lords of Hundreds, four hundred Lords of Manors, and eight hundred Freeholders. Each Hundred or Barony was to be eight miles square and to contain a castle or blockhouse to serve as a local capital and place of defense. Vassals were expected to provide military service to their lord. By January 1764, the proposal had acquired the backing of thirty gentlemen, mostly military officers, who had claims on the government. Egmont also recruited Holland, who was then fitting out his expedition for North America. Egmont expected Holland to carry out the survey of St. John's Island according to his plan. The Board of Trade, however, quickly turned down Egmont's proposal, noting that it contained little provision for the advancement of commerce and that its model of settlement and tenure was contrary to prevailing practice in North America. Undaunted, Egmont submitted a second memorial, this time to the king, but received no official reply. A third memorial, which also included Holland, as well as Perceval family members, senior naval and army officers, members of parliament, and government officials, was sent to the Board of Trade in March 1764, and was again rejected. By this time, the Board had drawn up its policy for the division of St. John's Island and was not to be deflected. After the whole matter had been referred to the Privy Council, a final decision came down on 9 May, stating "That no grants be made of lands in the said Island of *Saint John*, upon any other principles than those comprized in the said report of the Lords Commissioners for Trade and Plantations."[12]

By that time, Holland and his party of surveyors had left London onboard the *Canceaux*. Yet Egmont was still hopeful that his plan would be adopted and kept in communication with the surveyor. In October 1765, he wrote to Holland, declaring:

> I think it proper to let you know that a petition will be again presented to His Majesty in a few days for a grant of the Island of Saint John, upon the very same plan as that proposed before, which I have new reason to expect will meet with better success than the former. The same persons very nearly will be concerned, those only excluded who were drawn away by proposals and grants elsewhere by the Board of Trade, in order if possible to defeat my scheme. For yourself, you may be assured of your Hundred, as formerly intended, if I have anything to do in the direction of the affair, – which probably I shall have in the same mode and manner. Whether the grant may be made before the arrival of the survey or not I cannot certainly say, but we wait patiently for

it, and hope it will be done accurately as to Hundreds, Manors, Freehold Villages, Towns, and Capitals, that a moment's time may not be lost afterwards in proceeding to draw the lots, and then in proceeding to erect the Blockhouses of the Hundreds on a determined spot, which is the very first work to be put in execution, and agreed to be completed by all the chief adventurers within one twelvemonth after the grant shall be obtained.[13]

Even in the middle of the survey, Holland remained entwined with Egmont's proposal. Nevertheless, the surveyor general could hardly ignore his instructions from the Board of Trade, and carried out the survey according to the Board's directives. The survey was sent to London in October 1765 and considered by the Lords the following May.[14] At some point, Egmont, who was still in charge at the Admiralty, must have seen a copy of the survey and expressed his displeasure to Holland. In August, the surveyor general wrote to Pownall at the Board of Trade, explaining:

> I find my Lord Egmont is greatly displeased with me for not sending him a plan of St. Johns Island, divided according to his Lordships project as his Lordship desired me by a letter wherein he likewise promised me that I should have one of the Hundreds of it, but I answered his Lordship that by my commission I was ordered and directed to follow the orders and instructions from the Honourable the Lords Commissioners and could not give any plans, nor divide them without their Lordships commands, and I took the liberty to point out some inconveniences that would arise in dividing the Island by his Lordships project. – I hope his Lordship will not carry his resentment so far as to hurt me in the opinion of the Right Honble. Board, of which I think my self honoured in being an officer belonging thereto, and tho I fear I shall lose the Barony I should be much more concerned, that the connection my business has with the Navy should not find that readiness in the forwarding of it.[15]

Knowing full well that he had alienated a powerful patron, Holland hoped to be rewarded by the Lords of Trade and urged them to make a quick decision about the settlement of the island. In November 1766, Holland wrote to the Board explaining that his land agent, a Mr. Urquhart, who had been hired to look after the surveyor's expected land holdings on the island, had informed him that settlers were leaving because the distribution of lands lay unresolved.[16] Holland's letter was considered by the Lords in March 1767.[17] As he was the Board's principal source of information about the island, his

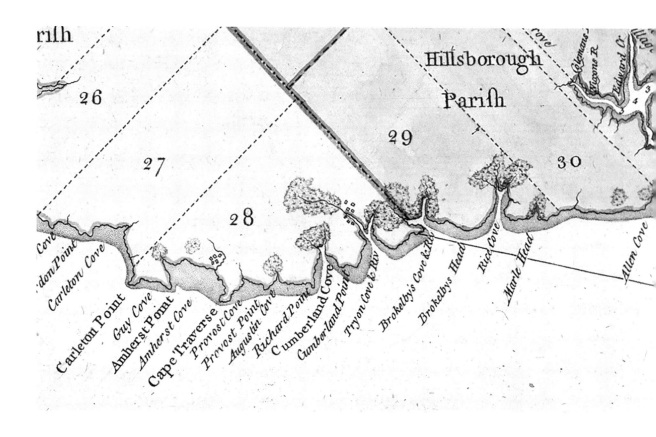

testimony no doubt carried some weight and over the summer the Lords drew up a plan for the allocation of land.

In July 1767 the Board released its plan for the allocation of land on St. John's Island.[18] The Lords decided on a lottery, which was advertised in the *London Gazette*, and attracted many of the supporters of the Egmont proposal. Holland applied for Lot 35 at Tracadie Bay on the north side of the island, mainly because of its location for the Gulf fishery, but in the draw got Lot 28 at Cape Traverse on the south side (figure 5.1).[19] As the surveyor explained to Haldimand in January 1768, "my lott is none of the best as having no fishery, but the lands are good & has about 600 acres cleared at Tryon River & Cape Traverse."[20] In his description of the island, Holland had described the 20,000 acre lot has having "very good" agricultural land, although the woods "along the coast are mostly spruce and indifferent."[21] As lots varied in their suitability for agriculture and the fishery, the Board set quit rents at different rates. Holland's lot had little commercial potential, so the quit rent was set at two shillings per one hundred acres after five years, and four shillings per one hundred acres after ten years. Even so, Holland

5.1 · Holland's lot 28 on St. John's Island. Detail from *A Plan of the Island of St. John … Survey'd by Capt. Holland, 1775*. Library and Archives Canada, NMC 23350[1].

5.2 · Hollandville, New Hampshire. Detail from Holland's *Topographical Map of the Province of New Hampshire*. Library of Congress, Geography and Map Division, G3740 1784.H6 Vault.

152 · SURVEYORS OF EMPIRE

found the terms "rather hard" and found himself in financial difficulties. He had just purchased a farm next to one owned by General Murray at Sainte-Foy outside Quebec for £450 so that he could provide his family with a home during his surveying expeditions. But the outlay left him short of funds to pay for the settlement of his new lot on St. John's Island. With "my mony matters in a bad situation," the surveyor confided to Haldimand, he felt obliged "to get leave to sell out of the Army ... by disposing of my Company [in the 60th Regiment], I could leave something for them [family] in case an accident was to happen to me."[22] By early 1773, he had laid out more than £300 sterling on the settlement of his St. John's property. Not wishing to incur further large expenses, he decided to sell 10,000 acres in order to provide capital to develop the remaining 10,000 acres.[23] Holland's speculation on St. John's Island had not turned a ready profit.

The surveyor general was more sanguine about his prospects in the continental colonies. In summer 1769, he travelled to New York to settle the provincial boundary dispute with New Jersey, and quickly realized the agricultural potential of the Mid-Atlantic colonies. In the fall, he wrote to Haldimand regretting that he had "layed out so much money in Canada in building & purchasing land next to General Murray's lands." "I could wish" the Surveyor continued, "I had layed it out in this Province [New York] or in New Jersey."[24] But the surveyor general was to have another opportunity. In 1770, he relocated from Quebec to Portsmouth, New Hampshire, and soon after carried out the first official survey of the province. Through surveying, Holland not only acquired first-hand knowledge of the new lands

Surveyors as Proprietors · 153

being settled in the northern and western parts of the province but also gained the gratitude of Governor Wentworth. In March 1773, the governor rewarded Holland with a sliver of land comprising 3,105 acres between the townships of Rumney and Compton (figure 5.2).[25] Named Hollandville, the tract occupied fertile intervale land in the Baker Valley amid the White Mountains. After he had received the grant, Holland wrote to Haldimand explaining that he had been "offered a dollor pr. acre but I am advised not to sell it but to improve a 100 acres of it & let the rest lay to increase in value … by laying out 400 dollors on it, it's [sic] value in 5 years will increase to two dollor an acre."[26]

Holland's claims against losses incurred during the American Revolution reveal in detail the process of improving land.[27] In April 1776, the surveyor general had leased out three parcels. The first required Edward Everett, innkeeper of the General Wolfe Tavern in Hollandville, to provide labour for clearing and fencing thirty-five acres of land, sow ten acres with wheat, and raise livestock in return for renting 150 acres of land for five years.[28] The second lease to Peter Mellony had similar terms but comprised only a hundred acres for five years. The third lease to Edward Kendall required clearing and fencing twenty-five acres in return for renting a hundred acres for four years. In April 1776, Holland reckoned that 550 acres had been cleared and improved and three houses and barns erected worth an estimated £1,989.15s.[29] A year earlier, Wentworth had written to Holland informing him that "your estate at Rumney is in a prosperous way, and your tenant industrious, prudent & honest."[30] Through the labour of tenants, Holland was greatly increasing the value of his property.

Yet Hollandville was only one part of the surveyor general's land holdings in northern New Hampshire and neighbouring New York.[31] His second and third tracts, acquired through purchase, lay in neighbouring Rumney Township; one comprised two full rights to land worth £135, the other thirteen-and-a-half undivided rights worth £306.[32] A fourth tract consisted of one third of Corinth Township in New York (modern Vermont), containing 7,308 acres, worth £1,099.10s. His sons John and Henry also had shares in the township, comprising 1,050 acres, making a total of 8,358 acres. His fifth tract was 1,600 acres – two-thirds of Topsham Township, which adjoined Corinth Township – valued at £1,650. Finally, he had two rights to land in the townships of Fairlee and Barnet, also in New York, of 350 acres each, making a total of 700 acres. At the outbreak of the Revolution, Holland was a substantial man of property, owning more than 22,000 acres in St. John's Island, New Hampshire, and New York. He also owned the farm at Sainte-Foy and a town lot in Quebec (figure 5.3).

Holland had carefully amassed these holdings through grants and purchases. In the early 1770s, he sold his Company in the 60th Regiment for

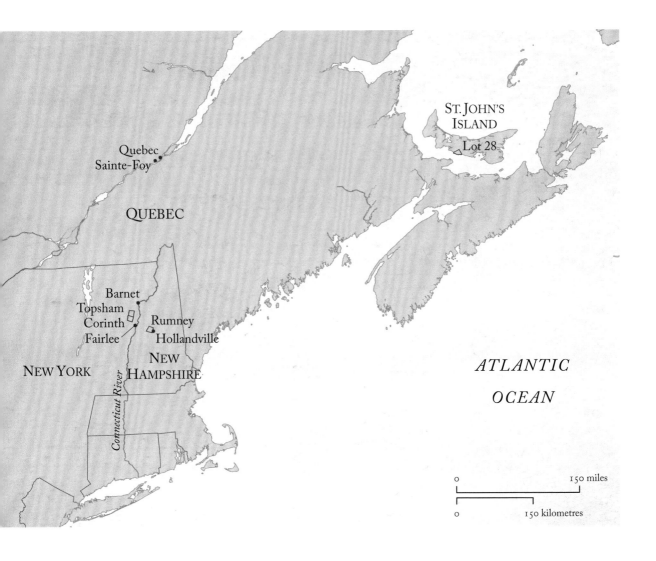

3 · Holland's land
oldings in 1775

£1,500, and invested the money in land and mortgages, providing capital to settlers in the "new townships."[33] At the time, it seemed a wise move. Holland reckoned that the 319 acres of prime intervale in Hollandville cost him eight dollars an acre to be cleared and four dollars an acre for the first ploughing. "The land proved so fertile" he wrote later, "that the first 2 years produce repaid me above half the sum expended thereon." Moreover, the money lent to neighbouring settlers "enabled them to improve and stock their lands, and make the roads to my estate & from thence to Haverhill on the Lower Cohass." Hollandville was being connected to the outside world, which would further enhance its value. The surveyor general also lent £50 sterling to John Hurd, clerk and register of Grafton County, to build a court house and jail. For Holland, these lands were "what I once thought to make a numerous family happy with, after my exit," but unfortunately for the surveyor general the Revolution intervened.

J.F.W. DES BARRES AS PROPRIETOR

Like Holland, Des Barres used his position as a government surveyor to advance his interests. In 1760, he was posted to Nova Scotia to assist Chief Engineer Bastide in Halifax, and the following year was ordered on "reconoitering [sic] expeditions into several districts of the colony." Des Barres prepared a brief "Essay examining the expediency, benefits, & advantages that might arise from a speedy adoption of measures for opening roads of communication practicable for carting" for the provincial government, and offered his assistance in the opening of a road from Halifax to Windsor on the Fundy side of the province.[34] At the time, Windsor and surrounding area were the focus of considerable speculative activity. Windsor was part of fertile marshland along the Avon River that had been dyked and cultivated by Acadians but now lay empty after the deportation of the Acadians during the Seven Years War. In the early 1760s, it was at the centre of several townships that were being surveyed to accommodate incoming settlers from New England.[35]

On 8 June 1763, Des Barres was granted 500 acres in Falmouth Township, which lay across the Avon River from Windsor. Des Barres was one of four officers from the 60th Regiment granted land in the township, and

5.4 · Castle Frederick, Falmouth Township, Hants Co., Nova Scotia. House on right, observatory on left behind trees. Library and Archives Canada, Acc. No. 1959-004-1.

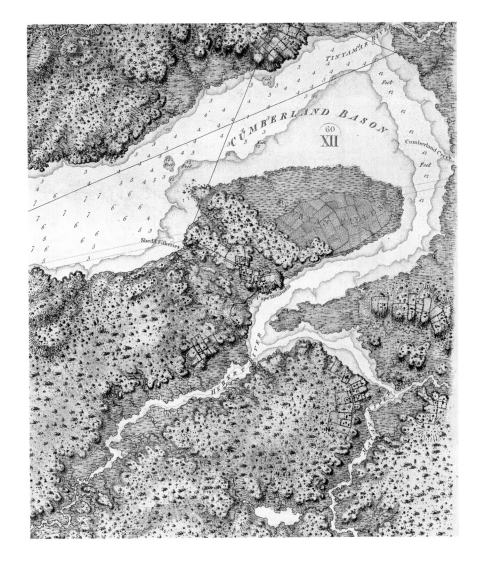

Elysian Fields, Minudie Marsh, Chignecto. The map also shows farms on Des Barres's holdings along the Maccan and Nappan rivers. Detail from *Chignecto Bay* from *The Atlantic Neptune*. © National Maritime Museum, Greenwich, London, NS 17.

soon acquired further land in the area. Estimates of the size of the Castle Frederick estate vary, ranging from 4,135 to 6,000 acres.[36] Des Barres divided the land into seven farms, setting aside one farm, near the head of the tide on the Avon River, as a home farm, where he built a modest, one-story Georgian dwelling named "Castle Frederick" (figure 5.4). During the years of the Nova Scotia survey, Des Barres used "Castle Frederick" as his base; he had a drafting office for working on the survey maps during the winter months and constructed an observatory for taking astronomical observations.[37] By 1770, the estate also included a grist mill, built on the West Branch of the Avon River in Upper Falmouth. Des Barres reckoned that he had invested "near four thousand pounds" in developing the estate.[38]

Surveyors as Proprietors · 157

Under the terms of the Proclamation, Des Barres, as an army lieutenant, was entitled to 2,000 acres, but he soon amassed much more. In 1765, he put together his largest land holdings, all in the northern periphery of the province. On 5 May, he and seven other applicants were granted a tract of land, comprising 8,000 acres, known as the "Elysian Fields" on Minudie Marsh at the head of Cumberland Basin (figure 5.5). Twelve days later, Des Barres purchased most of the shares of his co-grantees and became proprietor of seven-eighths of the tract (i.e., 7,000 acres).[39] Des Barres reckoned that he invested "nearly £3,000 alone for dyking and draining the marshlands."[40] He also purchased a second estate of 8,000 acres situated along the nearby Maccan and Nappan rivers.[41] On 25 August, the surveyor acquired a grant of 20,000 acres at Tatamagouche overlooking Northumberland Strait.[42] Most likely this grant was the one authorized by the Board of Trade the previous year. In 1775, he completed his major land holdings by purchasing two tracts, totalling 40,000 acres, between the Memramcook and Peticodiac rivers.[43] Apart from these large holdings, Des Barres purchased four rights of land totalling 2,000 acres in Cumberland Township and one right of 500 acres near Bay Verte, and claimed two rights totalling 2,000 acres in the Township of Maugerville on the Saint John River.[44] The surveyor's rights in Maugerville cost £1,200.[45] By 1775, Des Barres had amassed approximately 81,000 acres and was one of the largest landlords in Nova Scotia (figure 5.6).

As these holdings were subject to quit rents and settlement requirements, Des Barres had to attract tenants and establish them on the land. On his Falmouth estate, he brought in "some indigent (but industrious) English, North of Scotland and German families, together with disabled & discharged soldiers, who were supplied with provisions, utensils, tools and requisite implements, stock of cattle, &c."[46] By 1770, Des Barres had ninety-three tenants and their families on his estate – by far the largest tenantry in Falmouth – including twenty-four Irish, twenty-one Scots, seventeen Germans, fourteen English, ten Acadians, and seven Americans.[47] The Acadians were particularly important. Although many had been deported from Nova Scotia in the late 1750s, they had started dribbling back to the province during the 1760s, burdened with "helpless families and standing in need of means of existence." Des Barres spoke fluent French and was ready to provide protection for the Acadians "when they were almost hunted down by all others."[48] As many knew how to maintain marshland, Des Barres put them to work on his Falmouth and other estates, "dyking and draining marshlands, in the improvements of which they evinced remarkable intelligence and industry."[49] He also hired Acadians, who were "alert, dexterous and [of] obedient demeanor," to crew survey boats. In this way, wages paid by the Navy Board to Acadian crewmen flowed to Des Barres as rent.

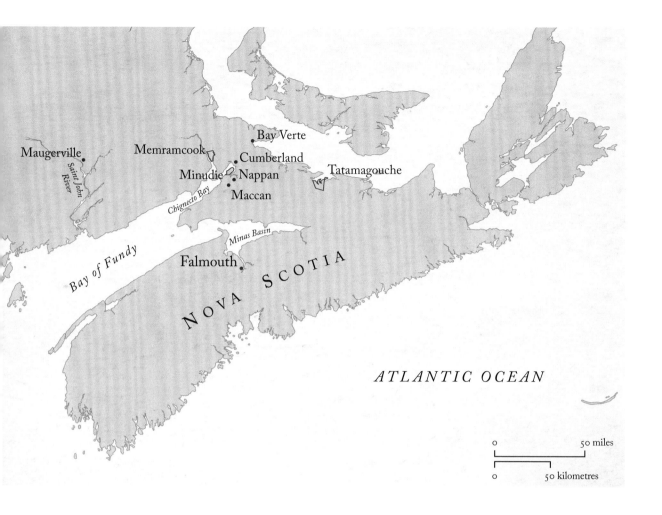

Des Barres's land holdings in 1775

After the purchase of the Elysian Fields, Des Barres was even more in need of Acadian assistance. At the centre of the tract was some 3,000 acres of marsh which had to be protected from tidal inundation. This required "deep canals and cross-drains abutting to proportionate aboiteaux for the discharge of freshets."[50] In Spring 1766, Des Barres settled a group of Acadians on neighbouring upland, where they erected temporary habitations, enclosed parcels of cleared land, and then began "the work of the dykes." The proprietor visited the settlement in summer 1773 and was "particularly delighted in beholding one of the most abundant crops, perhaps ever of marshlands, which it was computed would not yield less than 4,000 bushels of wheat beside a comparable proportion of other grains." In 1795, Des Barres, then governor of Prince Edward Island, commissioned Captain John Macdonald, a proprietor on the island, to survey his properties.[51] Of all the estates, Minudie made the most impression. "I found myself upon the

Surveyors as Proprietors · 159

edge of a great green plain," Macdonald wrote to Des Barres, "extending as far in all directions as my eye could distinguish. This was your great Marsh of Minudie, which you call the Elysian fields. I must say my astonishment was raised, nor has it subsided to this day. I had never observed the like before, and did not conceive there was such any where."[52]

On the Maccan and Nappan estates, Des Barres settled fourteen Acadian families in the late 1760s, and fifteen from Yorkshire in 1772, providing them each 200 acres of upland and ten acres of marsh.[53] The estate soon became a "very flourishing settlement." In going up the two rivers in 1795, Macdonald "was struck with the beautiful appearance of the few farms under the management of the Yorkshire men – The houses and barns look well as well as the considerable extent of verdant fields which they have made. The marsh extends along the sides of the rivers – each has the portion thereof belonging to his front dyked."[54] Des Barres also established settlers on his Tatamagouche grant. In 1770, he brought in eighteen families from Lunenburg on the South Shore of Nova Scotia, and judging from their last names they were mostly Acadians.[55] He allocated sixty acres, cattle, and provisions to each family in return for a rent of five shillings in the seventh year, ten shillings in the eighth year, one pound (twenty-four shillings) in the ninth year, and three pounds in the tenth year. In addition, Des Barres expected one half of all reared stock. Farther north, at the Memramcook-Peticodiac tract, twenty-five Acadian families settled in 1768, and were "no expense to the proprietor."[56] By 1784, when Des Barres' agent Mary Cannon visited the property, the tenants had made extensive improvements, dyking the marshlands along the Memramcook and Peticodiac rivers. A decade later, Macdonald was immensely impressed and fully understood its agricultural importance. "In the residue of Nova Scotia, New Brunswick, Cape Breton and St. John Island," he declared to Des Barres, "you may meet with small spaces of fine soil; but there is no where in them any thing of a tract of country comparable in the quality of soil to Memramcook. Nor have I seen any thing like it between Philadelphia and this."[57] For Macdonald, the Memramcook and Peticodiac estate was "truly a diamond hid in the rubbish of a mountain," and the Scot was certain that Des Barres "had enough in Nova Scotia to make you as happy as a Duke if judiciously attended to."[58]

Yet Des Barres, like many another landlord in colonial North America, was faced with the problem of developing his estates in an environment where land was plentiful, labour in short supply, and agricultural markets weak.[59] In such circumstances, land was virtually worthless in its uncleared state and rents had to be set low in order to attract settlers. Des Barres was savvy enough to realize that the Acadians would prefer to take their chances with a French-speaking landlord who would allow them to settle in groups well away from English settlers. Even so, rents had to be kept low. In his assessment of Des Barres's Maccan and Nappan estate in 1795, Macdonald reported that the agent had been forced to lower rents in order to prevent settlers leaving for Governor Franklin's neighbouring estate. Macdonald noted that the new rent of £60 "however trifling and short of your original demand, is still something and infinitely preferable to losing the lands by the total want of settlers."[60] Under the terms of the Crown grant, Des Barres needed settlers to improve the land, otherwise he faced losing the land through the Court of Escheats. Macdonald also noted that 1,500 acres had been set in reserve "for being lett upon better terms at some future period, when the province is better peopled & new lands in greater demand," but that was little help to Des Barres. In the late 1760s, the Nova Scotian economy, like that of much of the Atlantic world, was in recession and there was little demand for land; indeed, the situation was so bad that the province may have seen more emigration than immigration during those years.[61] Des Barres found himself caught in a credit squeeze. During the 1760s and early 1770s, he had spent several thousand pounds acquiring and improving his estates, but lacked the rental income to cover his mortgage payments. By the 1780s, creditors in Halifax and London were chasing him for debts. Further exacerbating the situation was Des Barres's carelessness about financial matters. According to Holland, Des Barres was "indifferent to his finances" during the Nova Scotia survey, so much so that Admiral Montague, the naval commander on the Halifax station, refused to authorize the surveyor's accounts.[62]

One way out of his financial difficulties was to publish the charts from the Nova Scotia and other surveys. Certainly, Des Barres must have had this in mind when he met with Holland at Liverpool in summer 1770 and drew up the agreement for sharing charts. Holland, too, was well aware of Des Barres's situation. Writing to Haldimand three years later, the surveyor general noted that he was sending his surveys to Des Barres "to insert in his General Chart, to inhance the value of his publications which I hope may turne out to his advantage as he is much in want of cash."[63] For Des Barres, engraving the survey charts and publishing them in a nautical atlas offered a financial life-line.

In their different ways, Holland and Des Barres exploited the surveys of northeastern North America to advance their own personal interests. Holland used the surveys of St. John's Island and New Hampshire to identify good tracts of land, which he later claimed or purchased. Through careful oversight, he accumulated substantial land holdings scattered across the provinces of Quebec, St. John's Island, New Hampshire, and New York. Although he never held as much land as Des Barres, he was never mortgaged to the hilt; indeed, he appears to have had spare capital to invest in the mortgage market. Des Barres made the bulk of his acquisitions before the Survey of Nova Scotia began, but he soon became so over-extended that he had to fall back on the survey as a way to make money. Holland generously helped him out by providing maps from the General Survey, which could be incorporated into *The Atlantic Neptune*. In these contrasting ways, the cartographic and written texts that Holland and Des Barres produced on government service were woven through the material circumstances of their individual lives.

The Atlantic Neptune

6

By the late eighteenth century, London had become "the universal centre of cartographic progress."[1] The empire's territorial expansion during the Seven Years War and maritime expansion through Cook's explorations of the Pacific had produced a mass of new geographic information and generated great demand from government, military, merchants, ship captains, and the educated public for printed maps. Thanks to the greater use of astronomical observation and chronometers as well as the growth of marine surveying, these maps were increasingly accurate and usually superior to those produced on the European continent. Although the British government financed much of the surveying that lay behind cartographic production, the publishing of maps and charts was left in private hands. During the 1750s and early 1760s, Thomas Jefferys, Geographer to the King, was at the centre of the London map trade.[2] He profited considerably from the demand for maps during the Seven Years War and made a specialty of producing maps of North America. By the early 1760s, he turned his attention to producing maps of English counties, financing several county surveys. But his involvement in large-scale surveying soon led to insolvency. His publishing business was taken over by William Faden, while many of the copper plates were purchased by print seller Robert Sayer.[3] Faden and Sayer now became the principal map makers in London.

Des Barres arrived in this hive of map making late in 1773, still unsure how to launch his great publishing project. Events across the Atlantic, however, soon lent urgency to his venture. In May 1774, Boston was blockaded; in April 1775, conflict broke out at Lexington and Concord. The crisis in the American colonies now fully occupied the government in London. Given the importance of the navy in enforcing the blockade and sustaining communications and transportation between Britain and North America, the need for up-to-date charts and maps of the western Atlantic was imperative. Des Barres could not have arrived at a better time to propose his project. By the time Holland reached London in December 1775, the publication of *The Atlantic Neptune* was well underway. The sheets of the nautical atlas would continue to be pulled from the printing press until 1802.[4]

PATRONAGE

Des Barres was determined to avoid entanglement with London's commercial map publishers. Knowing that he had the finest collection of manuscript charts of northeastern North America and that he had his own pressing financial problems to sort out, Des Barres resolved to publish the charts himself and take the credit and what he hoped would be considerable profit. Even so, engraving and printing such a mammoth atlas would be extremely expensive, and Des Barres lacked the necessary capital. While still in America, he sounded out close friend James Luttrell, a former midshipman on the Survey of Nova Scotia and now a lieutenant attached to the Royal Court at Windsor.[5] Raising funds through a subscription seemed the best option. "I think you propose a subscription from the Navy," Luttrell wrote in February 1773, "which with others might answer best for the publication of the large scale[.] You must be sensible how necessary a little *tinsel* is in all such things."[6] The "tinsel" Luttrell had in mind was the patronage of the Duke of Edinburgh (King George's younger brother, Prince William of Wales). Luttrell also suggested a "dedication … drawn under several heads. That of the encouragement approbation &c. of Admirals & Commander in Chief whilst under their inspection – captains of men of war &c. The countenance of the Admiralty & its first Lords – the consequences of a work that secures the glorious conquests of last war (may not be deemed unworthy &c.)." Luttrell thought that it might "be all arranged by the time you come to England & subscriptions might be began &c."

Yet Des Barres decided not to go with a subscription but with a direct appeal to the Lords of the Admiralty. As soon as he arrived in London in late November or early December 1773, he paid them a visit. At the time, John Montagu, 4th Earl of Sandwich, was the First Lord (figure 6.1).[7]

1 · *John Montagu, 4th Earl of Sandwich*, by Thomas Gainsborough, circa 1780. A particularly sympathetic portrait of Sandwich revealing something of his easy charm and affability. Captain Cook Memorial Museum, Whitby.

Sandwich had held the position since 1771, and also briefly in 1763 and for a longer period between 1748 and 1751. Apart from being a leading socialite, Sandwich was a formidable political operator and an extremely able administrator. Des Barres made his case well. Thomas Spry, master of the *Diligent*, later congratulated the surveyor "on your work being so much esteem'd, we are inform'd Lord Sandwich with the Board and the first nobles in town, was much pleas'd."[8]

To further his case, Des Barres enrolled Samuel Hood, former commodore on the Halifax station, with whom he had struck up a firm friendship during the course of the Survey of Nova Scotia. Just before relinquishing command and leaving for England in September 1770, Hood wrote to Des Barres expressing his support: "I entreat that you will do me the justice to believe that I will on every occasion that may offer, either before I go hence, or after I gett to England, exert myself in promoting, your interest & happiness, with as much zeal & pleasure as your own heart can suggest."[9] Hood was one of the navy's most promising younger officers, characterized by impressive ability, energy, and organization. During his command in North America, he had sorted out the dockyard in Halifax and played a central role in supporting the army in its occupation of Boston.[10] Des Barres kept Hood informed of developments with the Admiralty. In December 1773, Hood again offered his support and connections. The commodore "very sincerely rejoice[d] at the reception you mett with from the Admiralty. If you will call on my brother in Harley Street he will be glad to see you, and as he is intimate with Mr. Stephens [Secretary of the Admiralty], as well as Mr. Jackson [Under-Secretary]; he may be able to do you some service … I need not I think now repeat, how ready I am to render any service in my power."[11] Hood's elder brother, Alexander, had also made a name for himself in the navy, serving with distinction during the Seven Years War.

Early in 1774, Des Barres prepared a memorial and specimen chart to present to the Lords of the Admiralty. With an imprint date of 20 March 1774, the chart of "White Haven" (modern Whitehead Harbour, Nova Scotia) was the first map in *The Atlantic Neptune* to be engraved and printed (figure 6.2). The engraved chart served as a demonstration of Des Barres's skills as a surveyor and cartographer and showed off his engraver's considerable ability. The chart represented a portion of Nova Scotia's convoluted Atlantic coast just west of Canso, and included soundings, sailing directions, coastal profile, and legend. The chart stood comparison with the fine charts produced by Cook and Lane of Newfoundland and Labrador that had been engraved and published by Thomas Jefferys, "by permission of the Right Honourable the Lords Commissioners of the Admiralty," in 1770.[12]

The Lords of the Admiralty considered Des Barres's memorial on 25 May 1774, along with a memorial from Murdoch Mackenzie Sr., the leading

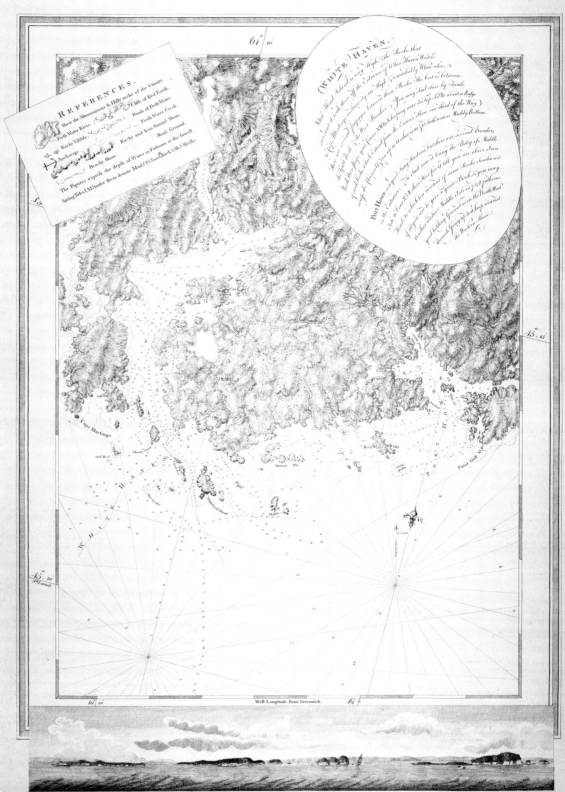

civilian hydrographer, who had been busy surveying the coast of Ireland. Both surveyors had requested that their respective surveys be published at "public expense, for the service of navigation in general." Mackenzie had asked for nearly £2,000, Des Barres almost £4,000. The Lords felt themselves unable to authorize "such expensive works without the sanction of Parliament," and recommended that the expenses be included in the Naval Estimates for the next year.[13] The following March, the estimates for engraving Des Barres's charts were laid before the House of Commons and approved. For a subvention of £3,711.15s., Des Barres promised to deliver two general charts on five plates, ten coasting charts on twenty-eight plates, thirty-four charts of pilotage "into fifty-two harbours, fit for the King's ships (twenty-five of which are fit for ships of the line) and one hundred and twenty-five for merchant and fishing vessels" on fifty-eight plates, and fifty-seven views of the shore on ten plates.[14] The same month, the Lords of the Admiralty approved Des Barres's schedule of publication.[15]

As the title page of Volume I of *The Atlantic Neptune* made clear, the first batch of charts was published "under the directions of the Right Honble the Lords Commissioners of the Admiralty" and Des Barres was keen to commemorate his patrons.[16] Although he had already submitted manuscript charts to the Admiralty in the late 1760s and early 1770s that contained numerous place names recognizing his superior officers (see chapter 4), he now had the opportunity to revise those names to better reflect current navy politics. When he began the Survey of Nova Scotia in 1764, Des Barres was beholden to his superior officer, Admiral Colvill, and had affixed his name prominently on the manuscript chart of the Eastern Shore.[17] Colvill had died in 1770, however, allowing Des Barres to reorganize the names on his Nova Scotia chart. Whereas Holland had drawn names of prominent people in Britain across the Atlantic to populate his maps of St. John's Island and Cape Breton Island, Des Barres now pulled the landforms of Nova Scotia across the Atlantic to London to commemorate his powerful patrons.

Most obviously, John Montagu, Earl of Sandwich, had to take pride of place. What had been Country Harbour and Franklin River on the manuscript map of Nova Scotia now became Sandwich Bay on the engraved chart of "Sandwich Bay" published on 8 January 1776. Around the bay, Des Barres clustered Port Montagu, Port Hinchingbroke, Monteacuto Lake, Huntington Cove, and Point St. Neots (figure 6.3). These names honoured Sandwich's family name and country house at Hinchingbrooke, near St. Neots in Huntingdonshire. In April 1775, Sandwich appointed his protégé, Sir Hugh Palliser, as Naval Lord on the Board of the Admiralty, and Des Barres recognized him by renaming Winchelsey Harbour on the manu-

2 · *White Haven* from *The Atlantic Neptune*. Des Barres's first engraved chart. © National Maritime Museum, Greenwich, London, HNS 56.

6.3 · *Sandwich Bay* from *The Atlantic Neptune*. Among place names on the chart related to Sandwich are Sandwich Bay, Port Montagu, Port Hinchingbroke, Monteacuto Lake, Huntington Cove, and Point St. Neots. © National Maritime Museum, Greenwich, London, HNS 54.

script as Port Pallisser [sic] on the engraved chart of "Spry Harbour, Port Pallisser, Port North, Beaver Harbour &c" published on 11 March 1776. Secretary Stephens and Under-Secretary Jackson were similarly honoured. What had been an unnamed bay (near modern Barren Island) on the manuscript became Port Stephens on the engraved chart of "White Islands, Port Stephen's, Liscomb Harbour &c." published on 15 January 1776. Appropriately, Cape Philip was attached to an island adjacent to Port Stephens. On the South Shore of Nova Scotia, Des Barres located Port Jackson (modern Port Medway), which was entered by Admiralty Head. This chart was published on 30 September 1777. Des Barres may well have been following the precedent set by Cook, who had placed Port Jackson (modern Sydney Harbour) and Port Stephens on his chart of New South Wales, which had been published in 1773.

Des Barres was particularly attentive to Hood. On a chart published sometime before March 1776, Des Barres gave the name Port Hood Island to an island off the northwest coast of Cape Breton.[18] In the vicinity of Port Hood, Des Barres affixed Samuel River to the entrance to Little Judique Harbour. Hood's wife was commemorated by Point Susannah, their only son by Henry Island, and the family's connections to Portsmouth by Portsmouth Point. Des Barres also honoured other prominent naval officers. The most complete roll call lay along the Eastern Shore of Nova Scotia with Keppell Harbour, Knowles Bay, Saunders Harbour, Deane Harbour, Spry Harbour, Port Pallisser, and Port Parker. As for senior government officials, Des Barres had inscribed Wilmot Bay and Harbour (after the governor of Nova Scotia between 1764 and 1766) on the manuscript chart, but replaced them with Port North (modern Sheet Harbour), for Lord North, the prime minister, on the engraved chart published on 11 March 1776. Frederick North, 2nd Earl of Guilford and MP for Banbury, inspired Des Barres to cluster Frederick Cove, Guilford Island, and Banbury (island) around Port North.

Yet for all Des Barres's attention to the First Lord of the Admiralty, he also curried favour with Admiral Lord Howe (figure 6.4). Although Sandwich was an obvious patron for the whole project, Des Barres took to heart advice given him by Luttrell, a later opponent of Sandwich, who warned "the present first Lord [Sandwich] is not likely to remain so till you begin to publish & having been three times in the Admiralty not likely to come there again in the course of his life therefore you would gain little from a connection with his fortunes."[19] In fact, Sandwich stayed at the Admiralty until 1782 and was instrumental in steering the navy through the War of American Independence and the even more important conflict with France which broke out in 1778, but Des Barres had no way of foreseeing this. At some point during 1774, the surveyor made an overture to Howe, showing

him his engraved charts. "They appeared so well executed," Howe's secretary later wrote, "that Lord Howe hardly distinguished them at first, from the drawings he has been allowed the pleasure of seeing at Mr. Desbarres lodgings last year [i.e., 1774]."[20] Des Barres must surely have shown Howe the chart of "White Haven" because it included "Port Howe … a good snug harbour."[21]

Des Barres's attention to the admiral was prescient. Howe was among the navy's most distinguished officers. He had served during the Seven Years War, commanding the lead ship at the Battle of Quiberon Bay. After the death of his esteemed elder brother, George, at Ticonderoga in 1759, Howe became Viscount Howe in the Peerage of Ireland. In 1762, he was elected MP for Dartmouth. Between 1763 and 1765, he served as the Naval Lord on the Board of the Admiralty, and between 1765 and 1770 as Treasurer of the Navy. At the end of this appointment, he was promoted to Rear Admiral, and then again, in 1775, to Vice Admiral.[22] Even though he had not served in North America, he immersed himself in the developing crisis between Britain and her colonies and in the hope of averting war conducted secret negotiations with Benjamin Franklin late in 1774 and early the following year. Although these talks came to nothing, Howe attracted the attention of senior members of the government, and in autumn 1775 he was appointed commissioner to open peace negotiations with the Americans. Early the next year, he was also appointed commander-in-chief of the naval squadron in North America. Meanwhile, his brother, General Sir William Howe, had been appointed as the second peace commissioner and commander-in-chief of the army in North America. The future of the American war lay in the hands of the brothers Howe.

Even before his appointment as commander-in-chief in North America, Lord Howe had taken a keen interest in marine surveying. In 1769, he urged Egmont, then first Lord of the Admiralty, to establish a hydrographic office, and suggested Alexander Dalrymple, an East India Company surveyor, to be the first naval hydrographer.[23] The suggestion was turned down, but Howe continued to show an interest in surveying, which was demonstrated when estimates for the General Surveys of North America were laid before Parliament in 1777. Of the £2,993.5s. budgeted for the surveys, £2,000 was granted "to Joseph Frederick Wallet Des Barres Esquire on account of his expences in engraving and publishing several plates from surveys taken of different parts of the coast of North America; being part of a general work undertaken by the direction of Lord Viscount Howe."[24] With Howe's patronage, the publication of the atlas was assured and further sums were granted by Parliament for engraving and printing: £1,835 in 1778; £1,504.5s. in 1779; and £1,541 in 1780 for "the completion of a general work undertaken by the direction of Lord Viscount Howe."[25]

Admiral of the Fleet Richard Howe, 1st Earl Howe, by John Singleton Copley, 1794. A late portrait depicting Howe as indomitable sea-dog. © National Maritime Museum, Greenwich, London, BHC 2790.

Des Barres acknowledged Howe's support on the title page of the volume of "Charts of the Coast and Harbours in the Gulph & River of St. Lawrence" published in 1778. According to Des Barres, these charts were "composed and published at the request of the Right Honourable Vice Adml. Lord Viscount Howe, for the use of the Royal Navy of Great Britain."[26] A similar acknowledgment to Howe appeared on the title page of the volume of "Charts of the Coast and Harbors of New England," also published in 1778. Yet by the end of that year, Howe had relinquished command of the North American squadron after a string of military setbacks, and returned to Britain to defend his reputation. Ever mindful of his superiors' changing fortunes, Des Barres continued to curry favour with Sandwich, still the First Lord. In May 1778, Des Barres wrote to Sandwich explaining that he "did not wish to bring out the work I am engaged upon under your Lordship's patronage, ere I have prepared proper impressions of it for his Majesty and for your Lordship, which I flatter myself I shall soon be able to accomplish."[27] Significantly, the title page of the volume of "Charts of several Harbours, and divers parts of the Coast of North America from New York southwestwards to the Gulph of Mexico," published in 1778, did not acknowledge Howe but "Command of Government."[28]

Publication

In an influential discussion of the transfer of knowledge from one part of the world to another, Bruno Latour has referred to manuscript maps as "immutable mobiles."[29] In contrast to a map drawn by a Native with a stick on a sandy beach, which will last only a few hours until the tide washes it away, a manuscript map is difficult to change and easily transported. Yet manuscript maps, such as those sent back to London by Holland and Des Barres, existed in only a few copies and were stored away in departmental archives. They were never intended for general distribution. Geographic knowledge was not truly mobile and immutable until manuscript maps were reproduced through engraving and print publication. In this regard, *The Atlantic Neptune* was a monumental success. Des Barres not only engraved and printed maps from the Survey of Nova Scotia, but also from Holland's General Survey of the Northern District, Cook's survey of Newfoundland and Labrador, De Brahm's General Survey of the Southern District, and Gauld's Survey of the Gulf of Mexico. In just a few years, cartographic coverage and geographic awareness of the eastern seaboard of North America were transformed.

The production of maps for *The Atlantic Neptune* went through several stages. At first, Des Barres had to gather manuscript maps and charts from

6.5 · Denmark Street (showing number 21) and Soho Square, London. Detail from Horwood's Plan of London, 1799.

the Admiralty, Board of Trade, and Board of Ordnance and have them copied. The estimates approved by Parliament included funds for copying. For each of the four years, £100 was included as "an allowance for defraying the expence of making copies, for the use of the public offices, of the maps and plans, returned by the surveyors & for preparing them for publication."[30] At least during 1775, Holland and his deputy surveyors must have been involved copying charts. Second, Des Barres had to have the maps and charts engraved and printed. We do not know who did this work, but it was almost certainly one or more of the craftsmen in London's "crowded corridor" of print and map dealers, which stretched from the political capital in Westminster along the Strand through Fleet Street to the docks in Wapping. Printers and engravers were particularly concentrated in the City north of St. Paul's Cathedral in the Peternoster Row area and around the Royal Exchange.[31] Des Barres took lodgings close to this east-west axis. After arriving in London in late 1773, he first took rooms at Mr. Gilbert's, Perfumer, in Poland Street, a short walk west of Soho Square, not far from the Strand. By early 1777, the surveyor was staying at 21 Denmark Street, a block east of Soho Square, where he remained until at least 1779 (figure 6.5).[32] By 1782, he had moved to Soho Square proper.[33] At the time, Soho was packed with Huguenot engravers and printers, just the sort of skilled

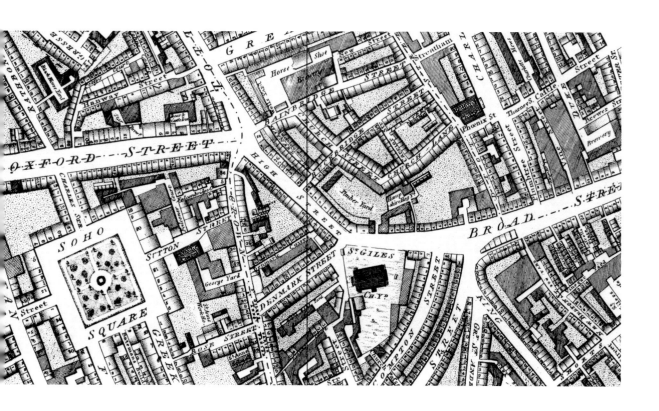

THE Atlantic Neptune,

PUBLISHED

For the use of the Royal Navy

OF

Great Britain,

By Joseph F. W. Des Barres Esq.r

Under the Directions of the

Right Hon.ble the Lords Commissioners of the

ADMIRALTY

Sunt ingeniorum monumenta quæ sæculis probantur. Liv.

VOL. I

LONDON.

MDCCLXXVII.

T: Tomkins Scripsit.   H: Ashby Sculpsit.

craftsmen who were needed for engraving such an ambitious project as *The Atlantic Neptune*.[34] Soho was also close to the Admiralty in Whitehall and senior naval officers who lived in London's developing West End.

Des Barres may have had the *Neptune* engraved by James Newton, a renowned map and globe maker who lived at 21 Denmark Street shortly before the surveyor arrived.[35] As the engravers and printers involved in the production of the *Neptune* worked for Des Barres, their names were not inscribed on the plates. In contrast, the general title pages of the four volumes were signed (figure 6.6). Des Barres commissioned the magnificent rococo designs from calligrapher T. Tomkins and engraver Harry Ashby. Although no records appear to have survived for Tomkins, Ashby was an engraver and printer of bill heads, bank notes, and specimen sheets of calligraphy. Between 1774 and 1784, he resided at Russell Court off Drury Lane, an easy walk from Denmark Street.[36] Des Barres also had designs done for individual volumes; the title for the volume containing "Charts of the Coast and Harbours in the Gulph & River of St. Lawrence" was engraved by Downes, possibly Charles John Downes, an engraver and printer in Fetter Lane, which runs between Fleet Street and Holborn.[37] At the end of the project, when Des Barres was lobbying the Admiralty for more compensation, he produced an illustrated text sheet entitled "Utility of the Atlantic Neptune," which was printed by Thomas Bensley, a printer at Bolt Court, a few yards east of Fetter Lane on Fleet Street.[38] All these engravers and printers lived within walking distance of Des Barres's lodgings in Soho.

In his memorial to the Lords of the Admiralty, Des Barres undertook to produce forty-six charts and fifty-seven views, but as the War of American Independence escalated and the demand for charts increased, many more charts were engraved and published. Imprint dates provide a guide to this publishing history (figure 6.7).[39] Not surprisingly, the greatest production of charts was in the early years of the war. Between April 1775 and November 1776, Des Barres published at least thirty-five charts and twenty-six coastal views. Of these charts, twenty-two were of Nova Scotia and thirteen of New England. The charts covered such important ports as Boston, Newport, and Halifax, as well as parts of the Nova Scotia and New England coast. The large-scale charts of Ipswich Bay and Cape Ann, Boston Bay, Massachusetts Bay, Buzzards Bay and Vineyard Sound, Rhode Island and Narragansett Bay, would have been immensely useful to naval vessels blockading the New England coast and sailing the treacherous waters around Nantucket Shoals. As the war expanded up and down the eastern seaboard, Des Barres widened his coverage. Between January 1777 and February 1778, he published nineteen charts, which included eight of Nova Scotia, five of the Gulf of St. Lawrence, four of New York southward, one of New England, and one general chart of the northeast from New

6 · Title page from *The Atlantic Neptune*. A superb demonstration of engraved rococo calligraphy. National Maritime Museum, Greenwich, London, HNS 79.

The Atlantic Neptune · 175

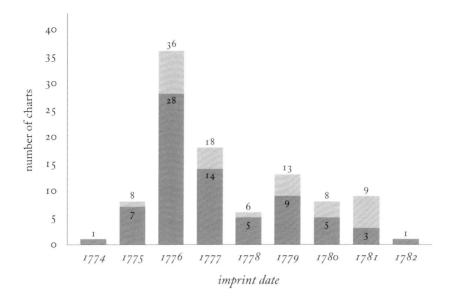

Jersey to Newfoundland. Several of these charts covered British campaigns at New York, Philadelphia, Charles Town, and up the Hudson River. Between January 1779 and August 1780, Des Barres published a further fourteen charts, including eleven of New York southward, two of New England, and one of the St. Lawrence. These included British attacks on forts Montgomery and Clinton on the Hudson River and on Charles Town, South Carolina. The charts also covered Delaware Bay and River, the northeast shore of the Gulf of Mexico, the lower Mississippi River, and Montego Bay, Jamaica. Between January 1781 and June 1782, Des Barres published his final four charts, which included two of Nova Scotia, one of the St. Lawrence, and one of New York southward. The last chart, perhaps appropriately, was "A plan of the posts of York and Gloucester in the Province of Virginia," published on 4 June 1782, which depicted the British surrender at Yorktown the previous October.

As the project expanded in the early years of the war, Des Barres was faced with the challenge of bringing some order to the various charts and views. In 1777, he published the first title page to include the name *The Atlantic Neptune*, suggesting that by then he had conceived of the individual charts and views as parts of a unified publication. The title was most likely a deliberate reference to France's great nautical atlas, *Le Neptune Francois*, first published in 1693 and reissued in a new edition in 1753.⁴⁰ All the charts and views were available as loose sheets or collected together into four volumes. These volumes began appearing in 1777 and covered different regional

6.7 · Imprint dates of *The Atlantic Neptune*

surveys: volumes I and II were devoted to Nova Scotia; volume III to New England; and volume IV was divided into two parts, one covering the Gulf and River St. Lawrence, the other New York southwestwards to the Gulf of Mexico. Des Barres drew on his survey for the two Nova Scotia volumes, Holland's surveys for the New England and St. Lawrence volumes, and De Brahm, Gauld, and other surveyors for the New York to the Gulf of Mexico volume. Given the long publishing history of the *Neptune* and the numerous states of many of the plates, individual volumes varied in the number of sheets they contained and their level of detail. Later editions of the *Neptune* were not necessarily more complete than earlier editions, simply because early plates and views were deleted or rearranged in later volumes.[41]

In producing these four volumes, Des Barres created an atlas of enormous geographic scale and reach (figures 6.8 and 6.9). At the smallest scale (1:2,000,000), Des Barres's most expansive sheet covered northeastern North America from New Jersey to Newfoundland and Labrador. Within a grid of relatively accurate latitude and longitude, Des Barres produced a map of remarkable precision and detail. It is the first map of the region that looks like a modern map. It can also be seen as the cartographic precedent for George Vancouver's map of the Northwest Coast published two decades later. In addition, Des Barres produced small-scale charts of the Gulf and River St. Lawrence (1:500,000), Cape Breton and St. John's Islands (1:450,000), Nova Scotia (1:500,000), the coast of New England (1:500,000), the coast from New York to North Carolina (1:620,000), the coast of Georgia (1:400,000), and the coast of Alabama, Mississippi, and Louisiana (1:500,000). At a larger scale, he produced sheets that covered the St. Lawrence River (1:130,000), the coasts of Nova Scotia (1:125,000), Cape Breton Island (1:250,000), the Magdalen Islands (1:150,000), Sable Island (1:120,000), Delaware Bay (1:150,000), the lower Mississippi River (1:200,000), and the northeastern Gulf Coast (1:130,000). An obvious absence among these small-scale charts was an individual sheet covering St. John's Island. In April 1775, Sayer and Bennett published Holland's map of the island (1:260,000), which was included in the *American Atlas*, a large compendium of maps and charts published the following year (figure 4.2). Des Barres no doubt thought it both impolitic and redundant to reproduce the map again.[42] At a larger scale (1:30,000 to 1:50,000), Des Barres published numerous charts of individual harbours, particularly those along the Atlantic coast of Nova Scotia. At the largest scales, the surveyor published town plans of Newport (1:6,000) and Halifax (1:5,000), and a map of the environs of Fort Cumberland at the head of the Bay of Fundy (1:3,000).

For small-scale charts, Des Barres used Mercator's projection, the standard projection at the time and one still used for making charts. Mercator's projection had the advantage of allowing a course to be plotted in any

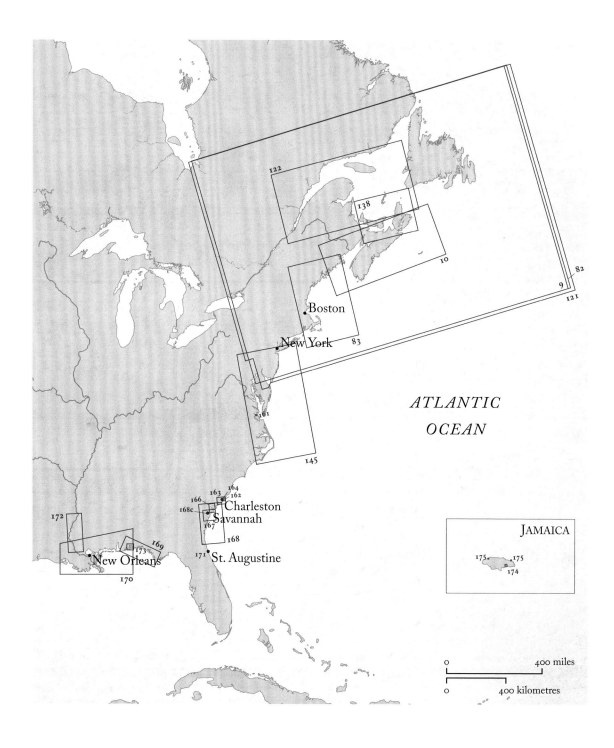

6.8 · Small scale coverage of *The Atlantic Neptune*. For chart reference numbers, see Appendix 3. Map based on chart index manuscript, National Maritime Museum, Greenwich, London.

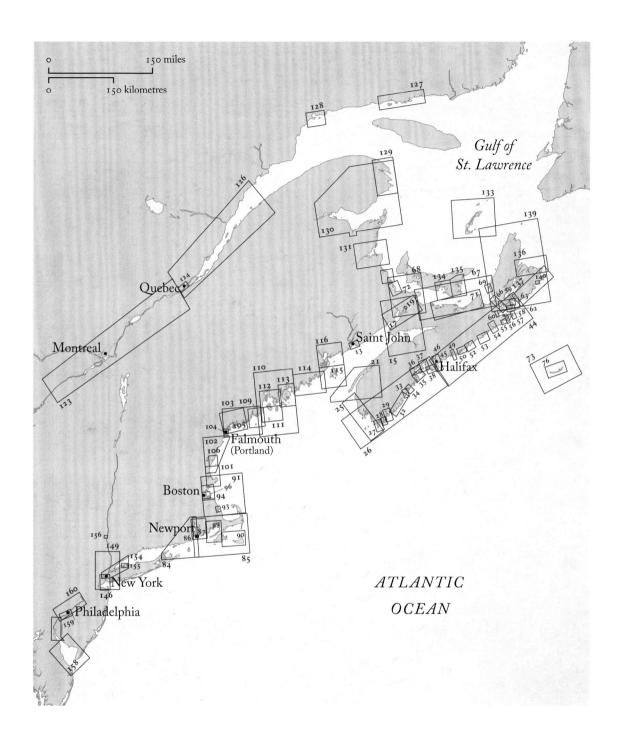

6.9 · Large scale coverage of *The Atlantic Neptune*. For chart reference numbers, see Appendix 3. Map based on chart index manuscript, National Maritime Museum, Greenwich, London.

direction on a constant bearing. To create large-scale charts, the surveyor used plane charts (charts that did not allow for the curvature of the earth) or simply enlarged small-scale charts.[43] A manuscript "Nautical Chart of the South Coast of Nova Scotia from Sambro Light House to Rugged Islands" created by Des Barres reveals the process.[44] To produce the large-scale map of Port Jackson, Des Barres or one of his assistants laid out a grid on the original chart and then reproduced it at a larger scale (figure 6.10). Such a process facilitated fairly rapid production of large-scale charts. All the charts included a border graduated with latitude and longitude, allowing a navigator equipped with a sextant and chronometer to fix a ship's location on the paper chart.

Many of the engraved charts in the *Neptune* went through numerous states. The first, most basic charts delineated coastline and located sound-

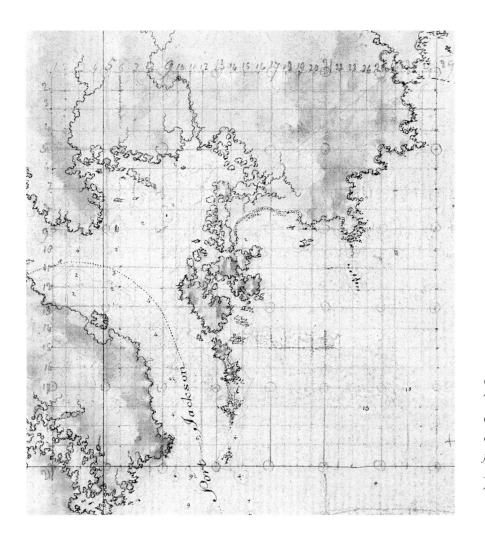

6.10 · Port Jackson, Nova Scotia. The grid allowed rapid copying and enlargement. Detail from *Nautical chart of the south coast of Nova Scotia from Sambro light house to Rugged Islands*. UK Hydrographic Office, Taunton z52/4.

ings (where available), creating a rather austere cartography. Later printings included more information, commonly topographical shading and more soundings, creating a much richer and fuller image. In some cases, topographic shading is so dense as to be oppressive, particularly on Des Barres's small-scale chart of Nova Scotia. At their best, the charts were masterpieces of design and content. This was well shown on sheets containing islands, such as the Magdalen Islands, Sable Island, Buzzards Bay, and Nantucket. Apart from Sable Island, these islands were surveyed by Holland's teams and their manuscript maps may well have dictated the layout of the engravings. The chart of "The Magdalen Islands in the Gulph of St. Lawrence," originally surveyed by Haldimand in 1765, demonstrated Des Barres's cartographic skill (figure 6.11). The chart combined rococo calligraphy in the title with a classical font for names on the map. As on modern maps, place names were in different point sizes to show the hierarchy of information on the chart. Shading of terrain was sensitively done, and beaches and soundings were clearly delineated. The chart was a model of elegance, clarity, and rigour. In terms of lettering, the chart of "Buzzards Bay and Vineyard Sound" was another striking demonstration of Des Barres's design. The curve of the place names paralleled the coastline creating a superb marriage of form and content (figure 6.12).

Judged by modern standards, many of the charts in *The Atlantic Neptune* were remarkably accurate. In a study of the Nova Scotia charts based on Des Barres's surveys, cartographer Walter K. Morrison found them to compare "very favourably" with the 1:50,000 Canadian National topographic maps.[45] The maps based on Holland's surveys in the Gulf of St. Lawrence reveal, however, the "greatest discrepancies in co-ordinate values," although even with these charts the discrepancies are relatively minor. Undoubtedly, both surveyors made errors. Des Barres placed Sable Island too far north and west, no doubt because he could not get an accurate fix on latitude and longitude. His engraved chart of *Canso Harbour*, published in April 1775, showed longitude increasing in value from left to right rather than the other way around, an elementary mistake corrected in later printings. More difficult to fathom are omissions. On the chart of Mount Desert Island, which was surveyed by one of Holland's teams, Northeast Harbor, Seal Harbor, and Otter Creek are missing (figure 6.13).[46] Historian Samuel Eliot Morison, who summered at Northeast Harbor in the early twentieth century, thought that fog must have concealed the harbour entrance on the day that Holland's surveyors sailed by.[47] Certainly, fog is common enough on the Maine coast during the summer months. The log book of the *Canceaux*, which was moored in neighbouring Cranberry Island Harbor for four days in June 1773 while the survey of Mount Desert Island was underway, records one day of "thick fogg."[48] However, it seems unlikely that Holland's team

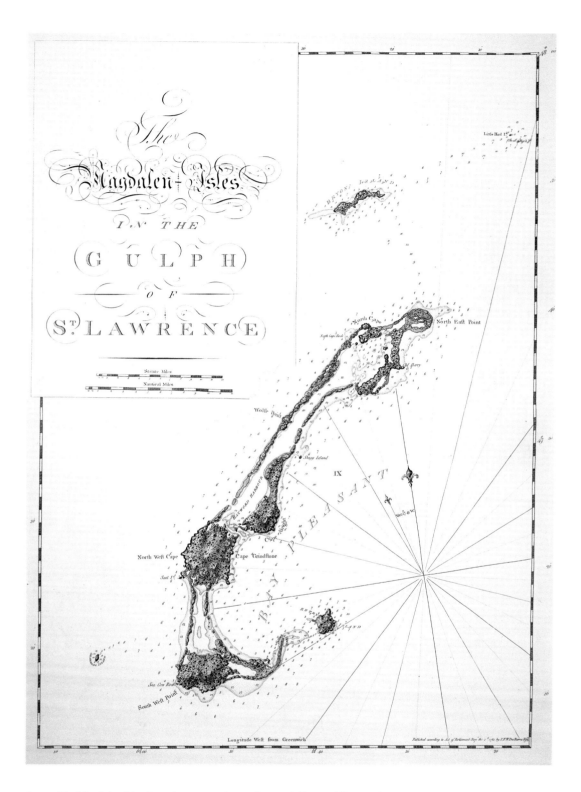

6.11 · *The Magdalen Islands in the Gulph of St. Lawrence* from *The Atlantic Neptune*. © National Maritime Museum, Greenwich, London, HNS 133.

6.12 · *Buzzards Bay and Vineyard Sound* from *The Atlantic Neptune*. © National Maritime Museum, Greenwich, London, HNS 88.

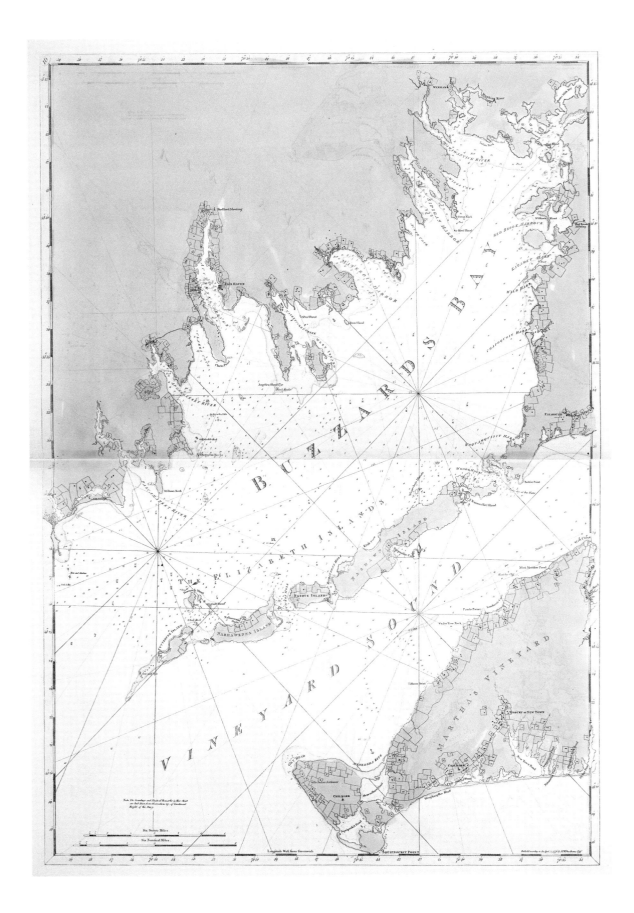

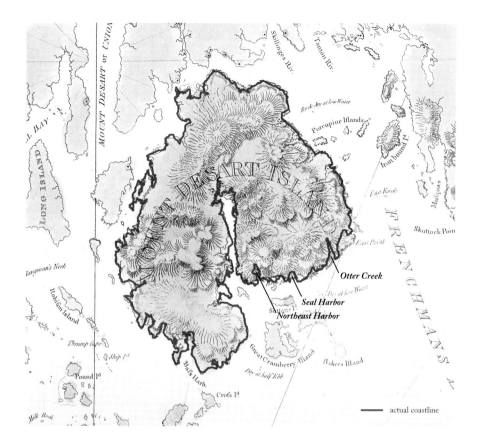

6.13 · Missing harbours on Mount Desert Island, Maine. Detail from *Penobscot Bay* from *The Atlantic Neptune*. © National Maritime Museum, Greenwich, London, HNS 110E.

would have triangulated the coast in fog. The suggestion by the Tafts and Rindlaub that the omission of Northeast Harbor was deliberate disinformation in order to conceal the location of the British fleet is simply fanciful.[49] The omission also existed on the manuscript chart drawn circa 1773, two years before the outbreak of the Revolution.[50] More likely, Holland's draftsman made an error transcribing information from surveyor's notebook to manuscript chart.

Apart from delineating coastlines and harbours, a particularly important aspect of *The Atlantic Neptune* charts was description of the seabed. Numerous charts of northeastern North America produced in the seventeenth and eighteenth centuries contained soundings, and mariners and fishermen had a rough idea of the location of the offshore banks and shallows, but no one had undertaken a systematic survey of the seabed until Des Barres and Knight. The results of their surveys appeared on both small- and large-scale charts. "A Chart of Nova Scotia," published on 1 November 1780, clearly showed tracks made by the *Diligent* as she sailed on a bearing from the coast out to the offing and then back again, a routine repeated many times along the coast of Nova Scotia (figure 6.14). In the process, Knight was able to delimit offshore banks and grounds and observe the character of the seabed in considerable detail. For the first time, a whole underwater geography became apparent. Significantly, the soundings and observations stop at the mouth of the Saint John River, terminus point of the Survey of Nova Scotia. The lack of soundings for the coast of Maine, northern Gulf of St. Lawrence, and the St. Lawrence River were major shortcomings of the General Survey and reflected Mowat's general lack of interest in sounding.

In addition to charts, *The Atlantic Neptune* contained approximately 145 coastal views.[51] These appeared on separate sheets and as insets or profile views at the top or bottom of charts. Des Barres's first published chart, "White Haven," included a view at the bottom of the sheet (figure 6.2). Des Barres produced sea-level views throughout the publishing project, but the greatest concentration was in 1777 when no fewer than seventeen appeared, no doubt reflecting the great demand for charts and views in the early stages of the war. Apart from views of Quebec, Charles Town, and Havana, the views covered coasts, rivers, harbours, and ports of Nova Scotia, New England, and New York. This geographic coverage suggests that members of the survey parties drew the views, most probably Wright and Peachey, who were able topographic artists.[52] Lieutenant William Pierie, active in the Royal Artillery from 1759 to 1777, was credited with "Three Views of Boston" and "A View of Boston taken on the road to Dorchester."[53] As for the Nova Scotia views, perhaps Des Barres himself drew them, as he had been trained in topographical drawing at the Royal Military Academy in Woolwich.[54] Alternatively, Knight or one of the midshipmen on the *Diligent* may have turned their hands to drawing.[55] Whoever the artist, the images fall into two types: views and coastal profiles.[56] Views commonly depicted port towns (Quebec, Louisbourg, Halifax, Annapolis Royal, Portsmouth, Boston, New York, Havana) or striking parts of coastline (Port Hood, Grand Passage in the Bay of Fundy, Sable Island). Such views gave navigators a visual sense of particular landfalls, but in other cases the views had little to do with navigation. A picturesque vignette of the Falls of Hinchinbroke River, in honour of Sandwich, and another vignette showing a settler's clearing and cabin on the chart of Egmont Harbor had more to do with flattering the Lords of the Admiralty than with providing nautical information about a dangerous seaboard. Even so, these various views of coasts and port towns were among the earliest visual images of northeastern North America, and contributed considerably to the British visual sense of the nation's Atlantic empire.[57]

In contrast, coastal profiles in the *Neptune* were much more utilitarian, designed to provide a navigator with a sea-view of a harbour or coastline. Des Barres created several profiles for leading ports such as Halifax, Boston,

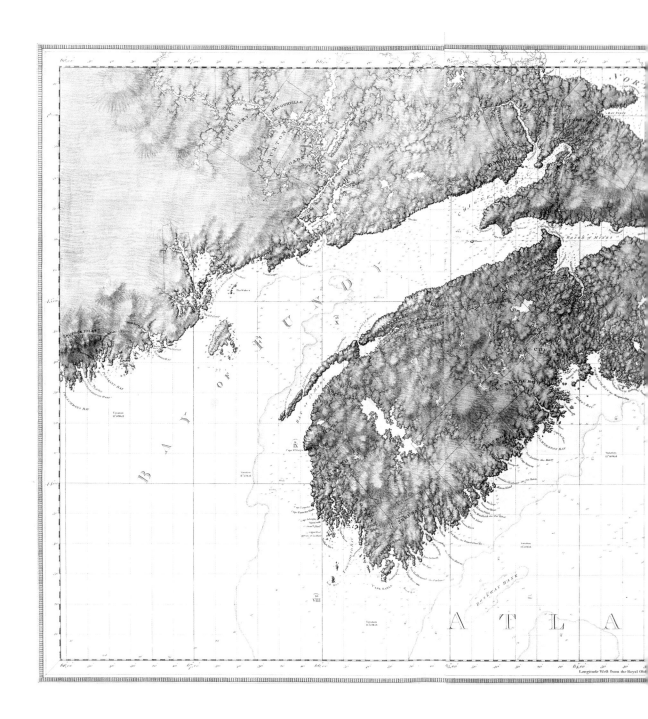

6.14 · Delineation of undersea banks on *A Chart of Nova Scotia* from *The Atlantic Neptune*. Soundings show the tracks of the survey vessel *Diligent*. © National Maritime Museum, Greenwich, London, HNS 10.

and New York (figure 6.15). Over the course of the printing, he also embellished some of the views and coastal profiles by adding ships and dramatic skies. This was well shown on different printings of "Views in the vicinity of Halifax." Apart from views and coastal profiles, Des Barres included sailing directions on large-scale maps of individual harbours and ports in Nova Scotia (figures 6.2–3). These included information on deep-water channels, anchorages, sea bottoms, depth of water, tides, and navigation markers. Because of Mowat's lassitude, similar information could not be included for harbours and ports in the Gulf of St. Lawrence and New England.

Although title sheets of the four volumes of *The Atlantic Neptune* state that Joseph F.W. Des Barres was the publisher (figure 6.6), the surveyor had entered into an agreement with Holland about sharing charts and acknowledging each other's work. As a result, Des Barres, on the title of the volume of "Charts of the Coast and Harbours in the Gulph & River of St. Lawrence," clearly acknowledged that they were "from Surveys taken by Major Holland Surv[eyo]r. Gen[era]l. of the North[er]n. Distr[ic]t. of North America." A similar acknowledgment was made on the volume of "Charts of the Coast and Harbors of New England." Within these volumes, Des Barres also acknowledged the work of surveyors such as Wright and Grant on individual charts. Similarly, in the volume of "Charts of several Harbours, and divers parts of the Coast of North America from New York southwestwards to the Gulph of Mexico," Des Barres recognized the contributions of Holland and Gauld in surveying and Knight and other naval officers for their work in sounding. Even so, authors of many individual charts were not acknowledged. As charts became detached from individual volumes or printed as stand-alone publications, the question of who did what became more difficult to ascertain, giving rise to the common impression today that Des Barres was responsible for all the surveys.[58]

As charts came off the printing press, Des Barres could have been justifiably proud of his achievement. Not only in their number but also in their design and content, the charts compared extremely favourably to other charts published in London during the same years. Murdoch Mackenzie Sr., who had spent more than three decades surveying the west coast of Britain and the coast of Ireland, brought out his two-volume *A Maritim[e] Survey of Ireland and the West of Great Britain* in 1776.[59] Whatever the accuracy of the charts in the collection, they looked quaint in comparison to those in the *Neptune*. Typography was amateurish and the hills used as navigation marks were reduced to symbols. Utilitarian in the extreme, Mackenzie's charts lacked the superb design displayed on so many of Des Barres's charts. James Cook had also produced a large corpus of charts by the mid-1770s. His "Chart of the West Coast of Newfoundland," published in 1768, was one of the largest from the Newfoundland survey of the 1760s

6.15 · Coastal profiles on a *Chart New York Harbour* from *The Atlantic Neptune*. © National Maritime Museum, Greenwich, London, HNS 146.

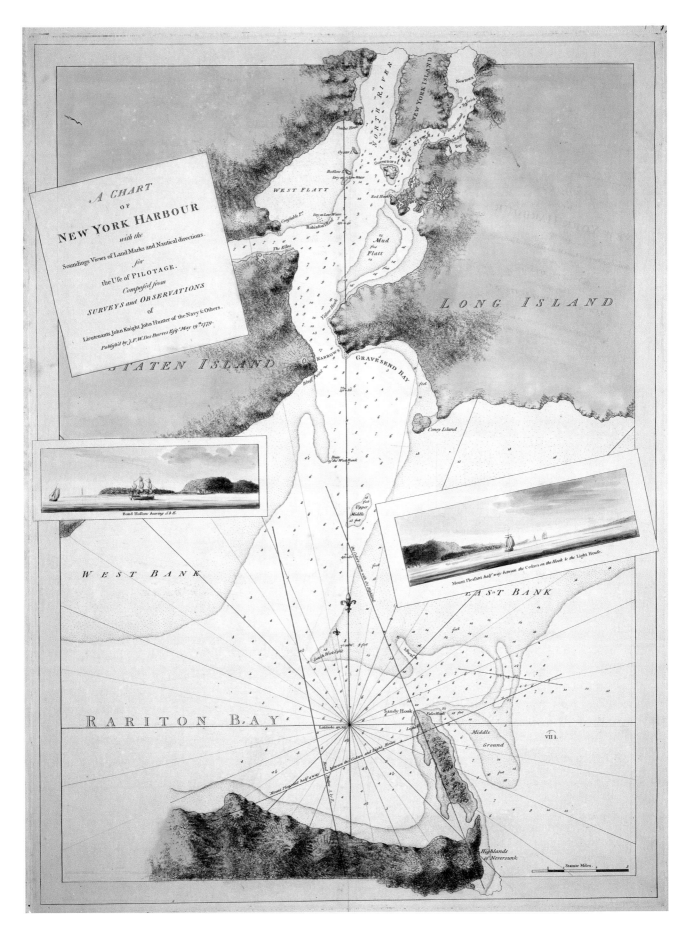

and contained a wealth of hydrographic information, including soundings, anchorages, and coastal profiles. Nevertheless, several loxodromes with heavy rhumb lines disfigured the chart, a feature also characteristic of Cook's manuscript charts of Newfoundland.[60] As for the first and second voyages to the Pacific, the extraordinary intrinsic importance of the charts outweighed their aesthetic qualities.[61] It is difficult, however, to find a manuscript or engraved chart from Cook's voyages that can compare to charts such as "Buzzards Bay" and "Isle of Sable" in the *Neptune*. As with Mackenzie's charts, Cook's charts of the Pacific lacked typographical skill and elegant design.

Distribution

Charts from *The Atlantic Neptune* were distributed to the navy as soon as they were printed. During spring and summer 1775, Des Barres was hard at work supervising the engraving and printing. By the autumn, he was starting to supply charts to the Admiralty. With Washington and the rebel forces besieging General Howe and the British army in Boston, the Royal Navy played a critical role in providing logistical support and maintaining contact with the outside world. Nevertheless, sailing conditions along the New England coast were often dangerous, particularly in winter. In December 1775, Vice-Admiral Graves wrote to Stephens observing that "their Lordships well know the situation of this coast in the winter, the prevailing winds SE and NW, hard gales each way, and with the former thick weather, rain, snow, and ice, without a friendly port to push for except Boston, the entrance of which is narrow and dangerous."[62] Good charts of Boston Harbor and Massachusetts Bay were thus essential. Anticipating the need, Des Barres published his chart of Boston on 5 August. A month later, on 7 September, Stephens wrote to Graves noting that "a box containing fifty impressions of Mr Des Barres chart of the Harbour of Boston, with as many copies of the nautical directions" was being sent down to Spithead for the first ship to sail. "I am commanded by my Lords Commissioners of the Admiralty," Stephens continued, "to recommend to you to distribute the said charts and nautical directions to such of the ships and vessels under your command as you shall judge proper."[63] Given the importance of Halifax and the Nova Scotia coast, Des Barres submitted to the Admiralty on 21 October a general chart of the province, which showed harbours that could be printed as individual charts in ten or twelve days time. Four days later, Stephens instructed him "to print them off accordingly as soon as possible."[64] The following month, Des Barres published "A chart of Nova

Scotia" and over the next five months produced at least fourteen charts of harbours in Nova Scotia, including Halifax Harbour.

Early in 1776, Admiral Howe went out to North America to take over from Graves. Before leaving, Howe met with Des Barres and "expressed an earnest desire that the ships of his squadron should be furnished with the best charts of the coasts that could be procured, and therefore wished me to accelerate by every ounce in my power not only particular parts of any one survey which his Lordship thought might be most useful but engaged me to engrave charts of the coast of America from surveys in possession of the Board of Trade." Des Barres applied himself with "unremitting diligence" and furnished Howe before he sailed with six books of charts of the coasts and harbours of Nova Scotia and New England. A further 12 sets were sent out in June 1776, and another 36 sets in December. Des Barres informed Howe that he had "sent by the Elephant store ship which is about sailing for New York thirty six setts of charts of the coast and harbors of Nova Scotia and New England, they are contained in four cases add[ressed] to Major Holland who will receive of your Lordship directions for the disposal of them[.]"[65] Holland had gone out with Howe as part of the force that took New York in September. Des Barres also kept in close contact with Stephens at the Admiralty. As the surveyor explained to Howe in February 1777, "I have waited often on Mr Stephens for the determination of my Lords of the Admiralty with regard to the number of books [i.e., volumes of *The Atlantic Neptune*] their Lordships may judge proper to order to be purchased for the use of his Majesty's ships."[66] Later that year, Des Barres dispatched to Howe forty-four general charts ("The Coast of Nova Scotia, New England, New York, Jersey, the Gulph and River of St. Lawrence" published in March 1777) "which are likely to be of utility to the fleet."[67] These charts were sent in "a small case, by this night's Portsmouth coach, with a sextant for my assistant Lt. Knight."

In addition, Des Barres supplied charts to government officials. In December 1776, as strategy was being worked out in Whitehall for the following year's campaign, William Knox, one of two under-secretaries of state for the colonies and a confidant of colonial secretary Lord George Germain, wrote to Des Barres requesting "a book of his maps which contain the Province of New York & the country to the eastward of it."[68] Given that Des Barres had not published any chart of New York, Knox's request may well have prompted the surveyor to bring out "A sketch of the operations of His Majesty's fleet and army under the command of Vice Admiral the Rt. H.ble Lord Viscount Howe and Genl. Sr. Wm. Howe, K.B. in 1776," which was published on 17 January 1777, and showed New York and the lower Hudson. But this map was hardly adequate. Despite its numerous inaccuracies, the best map of the province remained Holland's map originally published by Jefferys in 1768 and most recently revised in 1775. Des Barres could not produce a map of the "Province of New York & the country to the eastward of it" simply because it had not been systematically surveyed. If Gage's suggestion to Holland that he survey "the interior parts of this country" had been acted upon in 1767, then the British might well have been better prepared for the campaign, but this was not the case.[69] Des Barres's general chart of "The coast of Nova Scotia, New England, New York, Jersey, the Gulph and River of St. Lawrence," published on 29 March 1777, was the best he could produce, but it hardly provided the geographic and topographic information that government ministers needed to make informed decisions and by then it was too late. Earlier in the month, Germain had issued orders to General Burgoyne to take the army in Canada and march south down the Champlain-Hudson corridor to seize Albany and then attack the rear of New England. The lack of information about the terrain, as well as Burgoyne's inexperience with wilderness warfare in North America, certainly contributed to the ensuing British debacle at Saratoga. Holland, who knew that area of New York reasonably well, was dismayed. In January 1778, he wrote to Haldimand hoping and praying "to God that affairs may take a different turne, by the opening of an early campaign, upon a rational plan consistent with geography as well as tactiques."[70]

After the American victory at Saratoga, France threw in its lot with the rebels and declared war on Britain in February 1778. This dramatically changed the strategic situation. While fighting to preserve its far-flung colonial possessions, Britain now faced the possibility of a French invasion. Keeping most of the British fleet in home waters to deter the French, Sandwich dispatched a squadron to North America to intercept D'Estaing's warships.[71] The Admiralty now had even greater need for charts of the western Atlantic. On 1 May, the day Sandwich left London for Portsmouth to oversee the departure of the squadron, Des Barres wrote to the First Lord: "Yet as the public report says, it may be expedient to his Majesty's Service that a squadron of ships be ordered to sail for America; I have thought fit to send, for their use, some of the books I kept prepared in case they should be wanted, which I hope your Lordship will approve."[72]

As Des Barres was allowed to charge for each set of charts delivered to the Admiralty, his surviving accounts provide some insight into the distribution of *The Atlantic Neptune*. In May 1779, he submitted invoices to Stephens for thirty-six sets of charts of Nova Scotia and thirty-six sets of New England.[73] In July, he delivered twelve sets of charts of the coasts and harbours of Nova Scotia bound in volumes; three sets appear to have been for the Admiralty office, including one for Vice-Admiral Sir Hugh Palliser, the First Navy Lord. Des Barres also included two additional sets "in sheats,

Table 6.1 · Charts Supplied for His Majesty's Service sent to the Commander in Chief of His Majesty's ships in America, 24 July 1779.

44 General charts. The northern district of America from the Banks of Newfoundland to the Delawar[e]		4/- each	8.16.---
Two Atlantic Neptunes improved --- £7.7.--			14.14.---
Two sets of charts of the coast of New England, the harbours in the Gulph & River St. Lawrence, the harbours of New York, Charlestown, Port Royal, the coast of Florida, &c. ££ch			12.12.---
Sent to Commodore Hotham, viz.			
An Atlantic Neptune -----------------------------£7.7.---			
The charts of the Coast of New England, &c. 6.6.			19.19.---
Delivered for Vice Admiral Biron		Dto	19.19.---
Dto	Rear Admiral Parker	Dto	19.19.---
Dto	Rear Admiral Gambier	Dto	19.19.---
Dto	Rear Admiral Arbuthnot	Dto	19.19.---
Dto	Commodore Sir James Wallace	Dto	19.19.---
12 Atlantic Neptune at the Admiralty – bound in 2 vol.		7.7.---	88.4.---
			£206.4.---

viz: one for His Majesty, and one for my Lord Sandwich."[74] Later the same month, he submitted a detailed invoice for charts supplied to Commodore Hotham, commander-in-chief in North America, and his subordinate officers (table 6.1).[75] After Hotham was replaced by Rear Admiral Digby in 1781, another set of *The Atlantic Neptune* was supplied to the new commander-in-chief.[76] The final delivery to the Admiralty appears to have occurred in October 1784 when Des Barres delivered twenty bound sets of the *Neptune*, including additional charts and views, at a total cost of £460.8.6.[77]

Des Barres also distributed charts to the great and the good in British society. In December 1775, he presented charts of Boston to William Henry Lyttleton, MP for Bewdley and former governor of South Carolina and Jamaica, and his nephew Thomas Lyttleton, the "bad Lord Lyttleton," who then served as Chief Justice of Ireland.[78] In February 1777, Des Barres received a request from Lady Campbell, wife of Lord William Campbell, former governor of Nova Scotia and last royal governor of South Carolina, for "A chart of the harbour of Rhode Island and Naraganset Bay."[79] She observed that Des Barres "distributes [charts] to his friends" and that Lord Campbell "claims a friendship with him"; she also noted that she had seen a

Table 6.2 · Distribution of A sketch of the operations before Charlestown the capital of South Carolina, *1780*.

POLITICAL FIGURES	MILITARY FIGURES
Lord Dartmouth	Philip Affleck (Navy)
George Dempster, MP	Lord Amherst (Army)
William Dolben, MP	J. Barrett (Army)
William Eden, MP	General Fitzroy (Army)
Welbore Ellis, MP	Lord Howe (Navy)
Lord George Germain, Prime Minister	Sir William Howe (Army)
Soame Jenyns, Board of Trade	Commodore Sir William James (Navy)
Duke of Northumberland	Augustus Lord Keppel (Navy)
Lord Percy	Lieutenant General Keppel (Army)
Governor Thomas Pownall	Major General Raynsford (Army)
Duke of Richmond	
Lord Shelburne	OTHER
Benjamin Thompson, Under Secretary	___ Lewis, War Office
William Henry Lyttelton, Lord Westcote	___ Roberts, Pell Office
	William Savage

copy of the chart that Des Barres had "lately sent the Duke of Argyll." With government funding for the publication of *The Atlantic Neptune* ending in 1780 Des Barres began searching for further remunerative work and became far more aggressive in distributing charts to people of influence. In April 1780, the surveyor sent charts to William Eden, MP for New Woodstock and a commissioner of the Board of Trade, who thanked him for his "very valuable and curious work."[80] The following month, the capture of Charles Town provided welcome news to the war-weary British and presented an excellent publishing opportunity for Des Barres. On 17 June, he brought out "A sketch of the operations before Charlestown the capital of South Carolina," and sent copies to numerous public figures, including Dartmouth, Germain, Shelburne, Howe, Keppel, Eden, Amherst, the Duke of Richmond, and the Duke of Northumberland (table 6.2).[81] Such contacts would eventually pay off with Des Barres's appointment to a colonial governorship after the end of the war.

As Des Barres printed charts, engravers and publishers in London eagerly acquired them and began copying. Although copyright laws existed, charts could be copied if they included improvements or obvious changes. In July 1776, Des Barres published "A chart of the harbour of Rhode Island and Naraganset Bay"; a year later, Faden, the leading map publisher in London, brought out "A Topographical Chart of the Bay of Narraganset."[82]

The Faden sheet was strikingly similar to Des Barres's, and, indeed, admitted its common ancestry by acknowledging that it was based on a survey by Charles Blaskowitz, one of Holland's surveyors. But the two charts differed in one crucial respect. Whereas Des Barres's chart was dedicated to Lord Howe, the Faden chart was dedicated to Hugh, Earl Percy. In the twelve months that had passed between the publishing of the two charts, much had changed. Percy had been in command at Newport but resigned over differences with General Howe and returned to London in 1777. In dedicating the chart to Percy, Faden was following the changing winds of war. He might also have had access to manuscript charts of Narragansett Bay from Percy.[83] Faden's shop at Charing Cross was directly across the street from Percy's Northumberland House. The new dedication also absolved Faden of copyright infringement.

By the end of the century, commercial chart publishing houses were regularly copying *The Atlantic Neptune*. In 1794, Laurie and Whittle of Fleet Street brought out "A new and correct chart of the coast of New England and New York with the adjacent parts of Nova Scotia and New Brunswick, from Cape Sable to the entrance of Hudsons or North River by Captain Holland."[84] The chart reproduced part of Des Barres's small-scale chart of "The coast of Nova Scotia, New England, New York, Jersey, the Gulph and River of St. Lawrence," published in 1777, with additional soundings and new coastal profiles, most likely done by Knight in 1778.[85] The many rivers and hills on the Des Barres chart were not reproduced, presumably because Laurie and Whittle considered such information unnecessary on a mariner's chart. Although Holland's name was attached to the new chart, it is doubtful that the surveyor had any role in its publication. In the early 1790s, Holland was ensconced in Quebec as surveyor general of the province and had retired from the army as a major. If this new chart had been produced by Holland, it would, surely, have reflected his changed status. As with his much earlier map of New York, Holland's name was being used as a "catchpenny." Scaled and embedded in a grid of latitude and longitude measured from the Equator and Greenwich, charts from *The Atlantic Neptune* could be readily combined to produce a variety of charts at different scales of the western Atlantic and North America. Laurie and Whittle's publication marked the beginning of commercial chart makers using *The Atlantic Neptune* to construct charts of northeastern North America, which would continue into the early nineteenth century.[86] In this way, the scientific surveys of Holland and Des Barres made their way into the public realm.[87]

Publication of *The Atlantic Neptune* marked a watershed in the cartography of North America and the British Empire. For the first time, a large body of charts covering an extensive area of North America had been published that were based on systematic scientific surveys tied to the Greenwich meridian.[88] With the problem of establishing longitude resolved, the shape of eastern North America represented on Des Barres's charts looked strikingly different to that depicted on earlier maps; his charts marked the beginning of the modern mapping of the continent.[89] Des Barres also had produced an extensive body of large-scale maps. Such maps had been made before but almost always in manuscript form. Many of these maps still reside in government department archives in France, Spain, and Britain, known only to archivists and scholars. But by printing so many large-scale maps of port towns and harbours, Des Barres began to change the mapping history of North America. After the publication of *The Atlantic Neptune*, large-scale engraved maps became much more common.

Publication of the *Neptune* also demonstrated the power and influence of the Admiralty in British political and military affairs. The Admiralty, rather than the Board of Trade, took the initiative in getting the surveys published, and immediately made use of the charts in the War of American Independence. Nevertheless, the Admiralty's *ad hoc* arrangements for publishing charts were increasingly anachronistic. Leaving publication of charts to individuals, such as Des Barres, was hardly the most efficient way of creating and organizing knowledge. Des Barres's near neighbour in Soho Square, Joseph Banks, had become the "centre of calculation" for natural knowledge about the empire, but Des Barres was no Banks and the trend was increasingly towards institutional organization.[90] As the problems of managing the flow of hydrographic information increased in the last decades of the century, the Admiralty established its own Hydrographic Office in 1795.

More broadly, publication of the *Neptune* revealed the collaborative nature of publishing, or what historian Robert Darnton has called the "communication circuit."[91] Although Des Barres trumpeted his role as publisher, the contents of the atlas relied heavily not only on his charts but also those of Holland, Cook, De Brahm, Gauld, Knight, and other naval officers. The contents were also determined by the demands of the Admiralty, which required charts of different locations and at different scales. The actual production and form of the atlas plates owed much to Des Barres's engravers and printers: Their skills were essential to the look of the final sheets. At every stage of the publishing process, the atlas was pushed along by a combination of different people.

The "communication circuit" was also extremely elastic. The atlas may have been published in London but its conception lay in Liverpool, Nova Scotia, and its contents were produced through surveys of 15,000 miles of North American coast by, appropriately, Holland's "travelling academy" and the cruises of the *Diligent*. The contents of the atlas had been created at multiple sites on either side of the Atlantic and through almost continuous circulation of commands, information, and knowledge, some of it extremely localized, other parts trans-oceanic. Production also rested on assemblages of men, equipment, and techniques drawn together from even more places for the specific tasks of surveying northeastern North America and printing the atlas in London. The *Neptune* was less a creature of the enlightened metropole or the provincial periphery than of an ever-changing series of nodes and networks encompassing Britain and North America. As Des Barres no doubt realized, the atlas was, above all, an *Atlantic Neptune*.

EPILOGUE

Beyond the Surveys

If the genesis and publication of *The Atlantic Neptune* stretched back and forth across the Atlantic in a complex piece of imperial circuitry, so the later careers of the surveyors wove through the empire in an intricate web of interconnections.[1] Knight, Mowat, Des Barres, Holland, and deputy surveyors Wright and Sproule all made lasting contributions to the empire. With their specialist knowledge of northeastern North America, the surveyors were of immense value to the British army and navy during the War of American Independence. After the end of the war, several surveyors played important roles demarcating new territories in British North America and laying out lands for Loyalist refugees. Over the longer term, the surveyors contributed to the growth of marine surveying in Britain, the development of the Admiralty Hydrographic Office, and the cartographic representation of the British Empire.

The first of the survey personnel who became enmeshed in the American conflict was Lieutenant John Knight. As the Revolution broke out in spring 1775, Knight was surveying in the Bay of Fundy, taking soundings that Mowat had not done and that Des Barres, back in London, desperately wanted in order to complete his charts. As the *Diligent* sounded along shore, Knight was probably unaware of the revolutionary fervour that was spreading from Boston along the downeast coast and that would soon en-

gulf him. That spring, the British army, besieged in Boston, contracted with a merchant from Machias, Maine, for firewood and lumber.² After receiving threats from local patriots, the merchant requested and received naval protection. On 2 June 1775, two cargo vessels, *Unity* and *Polly*, defended by naval sloop *Margaretta* commanded by a young midshipman, entered Machias River. At first lading of the two vessels was conducted peacefully, but trouble broke out after patriots attempted to capture the British naval crew during Sunday service. Although the crew managed to get back to the *Margaretta* and cast off, patriots boarded the vessel in the river and in the resulting melée killed the midshipman. The Machias patriots had pulled off their own maritime version of Lexington and Concord, but there was more to come. On 15 July, Knight sailed the *Diligent* and the surveying shallop *Tatamagouche* into Machias Bay, probably to wood and water, and put ashore at Buck's Harbor where he was promptly seized and the two vessels taken without a fight.³

The capture of three naval vessels, however small they may have been, was obviously a great triumph for Machias, a place James Warren called "an obscure small town in the easternmost parts of this colony [Massachusetts]."⁴ The implications of the capture were not lost on Admiral Graves in Boston. In a letter to the Admiralty in August, he observed "the rebels having possession of the *Diligent* not only deprives the squadron of her assistance, but will I fear for some time hinder the publication of such of Mr Des Barres draughts as wait only for the soundings."⁵ Even more import-

7.1 · *French squadron entering Newport, 8 August 1778* (detail), by Pierre Ozanne. Knight's survey of these waters in 1778 proved invaluable to Howe and the British fleet in their pursuit of the French squadron that summer. Library of Congress, Prints and Photographs Division, LC-USZ62-900.

7.2 · *John Knight Esq. Rear Admiral of the White Squadron* © National Maritime Museum, Greenwich, London, F61323.

200 · SURVEYORS OF EMPIRE

antly, Knight possessed manuscript charts of a good part of the coast from New England to Nova Scotia, potentially of immense value to the Americans. Nevertheless, the patriots, no doubt intoxicated by their victory, overlooked the surveys. In January 1776, after Knight had been transferred to Cambridge, Massachusetts, James Lyon, Presbyterian minister in Machias, explained to revolutionary leader James Otis that "we generously, perhaps too generously, gave the officers taken in the schooner Diligent, all their pr[ivate pr]operty, & among other things all the plans of this continent, in their possession, which oversight we greatly regret & for which we can make no apology but our distress & confusion at the time, which would not admit of our attending to this matter as its vast importance required. Lieut Knights goods are all sent away."[6] At the end of 1776, Knight was exchanged back to the British.[7]

Once back on the British side, Knight resumed surveying. In 1778, he was employed charting and sounding New York Harbor and Long Island Sound eastward to Nantucket Shoals, a coast of great strategic interest that year because of the French naval threat to New York and Newport, Rhode Island.[8] During the course of the summer, Knight served on Howe's flagship, guiding the British fleet from New York through Long Island Sound to Rhode Island to thwart D'Estaing's attack on Newport (figure 7.1) and, later, in a vain attempt, to bring the French fleet to battle before it reached safety in Boston Harbor.[9] Knight returned to England that fall and soon after joined Samuel Hood's flagship *Barfleur*. Naval historian N.A.M. Rodger has argued that senior officers had followings of junior officers, and there seems little doubt that Knight was a follower of Hood, having served under him at Halifax in the late 1760s.[10] Knight served with Hood almost continuously throughout the rest of the war, and in the aftermath resumed marine surveying. Admiral Sir John Jervis, Earl St. Vincent, considered Knight "to be the most accurate [surveyor] in H.M. service"; in 1800, St. Vincent assigned him the exceedingly difficult and dangerous task of surveying the treacherous waters at the entrance to the heavily defended French naval base at Brest.[11] The charts were of immense value to the blockading British fleet during the Revolutionary and Napoleonic wars. Knight retired from the navy as an Admiral and KCB (figure 7.2), and spent his remaining days at Wood End, which lies just over Portsdown Hill from Portsmouth Harbour. He died at his home in 1831 and was buried in the local parish church at Soberton.

Knight's capture at Machias was to have a significant impact on the other naval officer involved in the surveys, Henry Mowat. As the officer most familiar with the coast and harbours of New England, Mowat was the obvious person to bring retribution down on the heads of the rebels. As Admiral Graves explained to Mowat in October 1775: "whereas from your having been employed on the survey of the coast to the eastward of this harbor [Boston], you cannot but be qualified to carry on this service from your knowledge of all the harbors, bays, creeks, shoals." "My design," Graves continued, "is to chastize Marblehead, Salem, Newbury Port, Cape Anne Harbour, Portsmouth, Ipswich, Saco, Falmouth in Casco Bay, and particularly Mechias where the *Margaretta* was taken, the officer commanding her killed, and the people made prisoners, and where the *Diligent* schooner was seized and the officers and crew carried prisoners up the country, and where preparations I am informed are now making to invade the Province of Nova Scotia."[12] In addition to the *Canceaux*, Mowat was given command of three smaller armed vessels and ordered to "go to all or to as many of the above named places as you can, and make the most vigorous efforts to burn the towns, and destroy the shipping in the harbors." The "flying squadron" left Boston on 9 October, determined to begin chastisement at Cape Anne Harbour (modern Gloucester). Wind and weather, however, prevented the flotilla from making landfall until Townshend Harbor (modern Boothbay Harbor). From there Mowat worked westward to Casco Bay and entered Falmouth (modern Portland) on the 16th. Mowat was already familiar with the port. In April 1775, he had put in to provide protection to a leading loyalist merchant who was loading a vessel. The British presence attracted the attention of local patriots, who managed to capture Mowat and the ship's doctor while they were taking a turn on shore. After the master of the *Canceaux* threatened to burn the town, Mowat and the doctor were released.[13]

By October, after the insults at Machias, Mowat must have been spoiling for a fight. On the same day that the *Canceaux* anchored in Falmouth Harbor, Mowat delivered a letter to the townspeople warning that the "rod of correction" was about to be applied and that "you [should] remove without delay the human species out of the … town, for which purpose I give you the time of two hours, at the period of which, a red pendant will be hoisted at the maintop gallant."[14] In the event, Mowat delayed for two days while parlaying with town leaders took place, but after fruitless discussions the bombardment began. The ships' cannons and a landing party of marines and seamen soon had the Anglican church, residences, stores, warehouses, wharves, and eleven vessels in flames (figure 7.3). The town was still burning three days later.[15] Mowat would have inflicted similar destruction on other ports in the area if his artillery stores had been adequate, but having run low of powder and shot he returned to Boston. The destruction of Falmouth stunned both sides. The resolve of the New England patriots was undoubtedly strengthened, while the British government's attempt to find a peaceful resolution to the conflict was severely weakened.

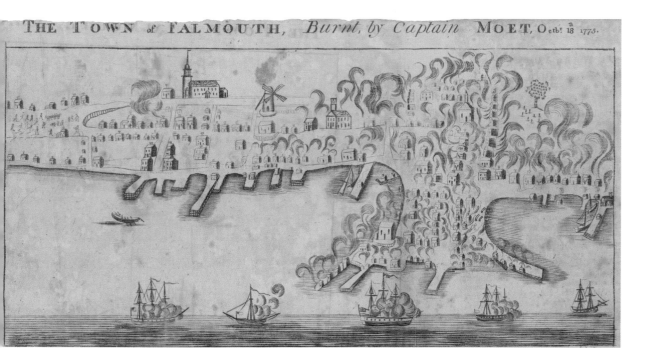

3 · *The Town of Falmouth burnt by Captain Moet, Octr. 18, 1775* Courtesy of the John Carter Brown Library at Brown University.

Mowat's knowledge of the coasts of New England and Nova Scotia kept him, as he later wrote, "chained down to stations, in which he was supposed to be necessary & the most calculated for being useful."[16] Appointed commander of the armed merchantman *Albany* in 1776, he was left to guard the harbour at Canso, protecting the fishery, for the next three years. In January 1778, he complained to Sandwich about his "servitude on the American coast [which] might have been applied to a better use."[17] The following year, Mowat was given orders to convoy an army detachment from Halifax to the mouth of the Penobscot River in order to establish a fort on a peninsula overlooking Majabigwaduce Harbor (modern Castine). Such a move was intended to secure the eastern bank of the Penobscot and provide a haven for Loyalists. Although the landing was unopposed, an American fleet from Boston soon put to sea to drive the British out. Two days later, forty American vessels with a thousand militiamen began laying siege to the peninsula.[18] In a remarkable display of tactical *nous*, Mowat anchored the *Albany* and two armed sloops broadside across the narrow entrance to Majabigwaduce Harbor, thereby protecting the transports in the inner harbour and the southern flank of the fort. Mowat and the army garrison held off far superior numbers for three weeks until a British squadron from New York appeared and routed the besieging forces (figure 7.4). All the American

Epilogue: Beyond the Surveys · 203

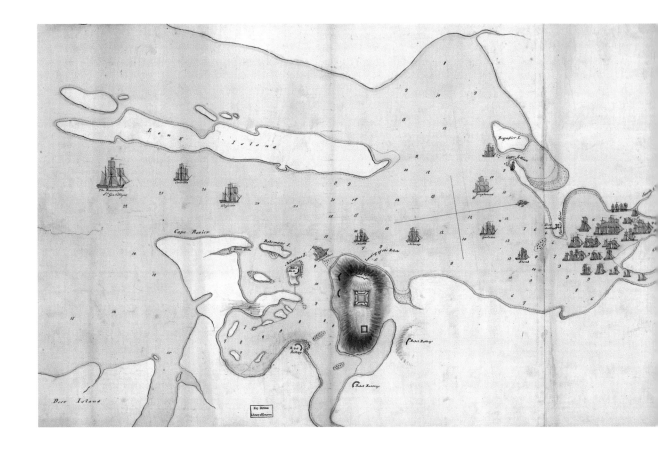

vessels were captured, burnt, or scuttled. After contributing to "the severest blow received by the American naval force during the war" (in fact, the worst naval defeat ever suffered by the United States), Mowat expected to be handsomely rewarded, but the Falmouth episode still hung over him like a dark cloud. When the British government decided to give up the eastern shore of the Penobscot in 1783, he was left to transport Loyalists to their new homes in Nova Scotia.[19] Mowat continued to serve on the North American station for a further fifteen years. He died, after a fit of apoplexy, while in command of the *Assistance* off the coast of Virginia in 1798, and was buried in the graveyard of the Episcopal church in Hampton. His gravestone recorded that he was "universally lamented" – a cry scarcely heard among the people of Falmouth, Maine.[20]

Among the crew of the *Canceaux*, midshipman Thomas Hurd, future Hydrographer of the Navy, cut his surveying teeth along the Maine coast. In August 1771, Hurd was transferred from the *Salisbury* to the *Canceaux* in order to provide an extra midshipman to command small boat crews for surveying the complicated Maine coast.[21] The following year, survey parties

7.4 · *Penobscot River and Bay, with the operations of the English fleet, under Sir George Collyer, against the division of Massachusetts troops acting against Fort Castine, August 1779* (detail). This map shows the British fleet giving chase to the American ships. Mowat moored the *Albany* and two other vessels across the entrance to Castine harbour in order to prevent an American landing. Library of Congress, Geography and Map Division, G3732.P395S3 1779.P4 Faden 101.

began working along the downeast coast, from the Penobscot to the Saint John, and Hurd was appointed to command the small boat crew conveying Thomas Wright's party and to carry out inshore soundings.[22] During the course of the survey, Hurd contributed soundings to Wright's map of Passamaquoddy Bay and also produced two maps of the Bay of Fundy, from Campobello Island to Beaver Harbour, and Maces Bay to Saint John. A year or two later, he drew a small-scale map of the coast from Penobscot to Saint John, and in 1776 one of the coast westward from Penobscot Bay.[23] Hurd was discharged from the *Canceaux* in July 1774.[24] The following year, he passed the lieutenant's exam and served on various ships during the American conflict.[25] At the end of the war, he was appointed surveyor general to the new colony of Cape Breton, but ran afoul of Des Barres, the new governor, and was dismissed. In 1789, the Admiralty sent him to carry out a marine survey of Bermuda which took nine years to complete and a further three years to map. By then, the Admiralty Hydrographic Office had been established under Dalrymple, and Hurd became increasingly involved in the office. In 1808, Hurd succeeded Dalrymple as hydrographer, the first naval officer to hold the position.

At the outbreak of the Revolution, surveyor general Samuel Holland knew as much about northeastern North America as anyone on the British side and soon found himself back in the king's service. In March 1776, he was appointed aide-de-camp to General von Heister, commanding officer of the Hessians, and promoted to major.[26] He accompanied the British expeditionary force to New York and served under the commands of Sir William Howe, Sir Henry Clinton, and General Tryon. While in New York, he raised a corps of guides and pioneers, a specialist unit for surveying and making maps.[27] Holland's guides surveyed the town of New York, possibly Manhattan Island, and forts Clinton and Montgomery.[28] After the capture of the forts in October 1777, there was little local surveying left to be done, and Holland must have been delighted to receive an invitation from Haldimand, the new governor of Quebec, to resume his service as surveyor general of the province. "I am resolved to joyn you there as early as possible," Holland replied in January 1778, "I shall bring my whole family with me with an intention to remain in that province."[29] Yet it was not until September that he managed to find a vessel leaving for Quebec, and even then the ship was "twice forced back by severe storms out of the Gulf of St. Lawrence," and had to put back into Halifax. Holland remained there with his family for the winter, calling it "the dearest place on the continent."[30] He reached Quebec the following year and resumed his position as surveyor general.

Holland spent the rest of his life in Quebec, settled on his estate at Sainte-Foy (marked by modern Holland Avenue), serving on the Legisla-

tive Council, and working in the provincial surveyor's office in town. As the American war wound down, he became increasingly involved in preparing land for Loyalist refugees. In summer 1783, he ranged west from Montreal along the upper St. Lawrence as far as the former French Fort Frontenac at Cataraqui (modern Kingston) (figure 7.5).[31] He was accompanied by Captain Joseph Brant of the Six Nations, who had allied with the British during the war and was now looking for land to occupy. Holland and his deputy surveyor John Collins started laying out townships at Cataraqui and along the north shore of Lake Ontario and the upper St. Lawrence in fall 1783 and through the following year (figure 7.6). These townships followed the dimensions that had been laid out by the Board of Trade in its instructions to Murray in 1763. In honour of the royal family, they were given defiantly Hanoverian names such as Charlottesburg, Williamsburg, Edwardsburg, Augusta, Kingston, Fredericksburg, and Adolphustown.[32] As on St. John's Island, Holland was creating landscapes of loyalty. After the exertions of surveying the townships, "where insects are in such multitudes," Holland returned to Quebec in summer 1784, leaving field surveying to his deputies.

7.5 · *A view of the ruins of the fort Cataraqui*, by James Peachey, June 1783. Mohawk Joseph Brant may be the figure seated to the right of the fire. The survey boat and camp are shown left background. Library and Archives Canada, Acc. No. 1989-217-1.

7.6 · *Encampment of the Loyalists at Johnston, a new settlement, on the banks of the River St. Laurence in Canada*, by James Peachey, 6 June 1784. Library and Archives Canada, Acc. No. 1989-218-1.

Between 1784 and 1791, when Quebec was divided into the two provinces of Upper and Lower Canada, Holland's deputies were busy laying out townships along the upper St. Lawrence, the lower Ottawa, the Bay of Quinte, the north shore of Lake Ontario, and the west bank of the Niagara River. In effect, Holland was directing settlement of all British territory west of Montreal, as well as being responsible for all surveying east to Gaspé.

Surveying and laying out townships appears to have been thoroughly done. An American traveller to Upper Canada commented on the surveyor general as "a gentleman of liberal education, good information, and indefatigable in the duties of his office, by which means he has collected notes, from the different field-books of his deputies, of the soil, timber, and streams, of all that country, and in such parts as I went over, I found his notes very correct, and by no means exaggerated."[33] Holland was maintaining the standards set in the surveys of St. John's Island and Cape Breton Island. With the political division of Quebec in 1791, the surveyor general shed responsibility for Upper Canada; two years later, he supervised the survey of the new provincial boundary.[34] His deputy surveyors also laid out townships along the Lower Canada–U.S. border in the 1790s. In the early 1800s, Faden in London published two maps that summarized all this survey work and stand as monuments to Holland and the survey office in Quebec. The first, "A Map of the Province of Upper Canada," compiled by William Chewett, formerly of Holland's office, and David William Smyth, surveyor general of Upper Canada, showed a continuous band of townships from the Upper/Lower Canada border to the Detroit River, the cadastral foundation of modern Ontario.[35] The second, "A New Map of the Province

of Lower Canada," compiled by Holland, showed seigneuries, townships, and grants of land. Like the map of Upper Canada, it served as the cartographic foundation of modern Quebec.[36] The surveys of Upper and Lower Canada represented the second great achievement of Holland's career.

The surveys of Upper and Lower Canada also place in context Holland's much earlier survey of St. John's Island. This survey, conducted in 1764–65 and published by Sayer and Bennett in 1775, has seemed anomalous in the history of the Canadian Maritime provinces. There is no other survey in the region quite like it. In the scholarly literature, geographer Andrew Hill Clark interpreted the survey as part of the settlement history of Prince Edward Island; historians have seen it has the beginning of the proprietary system on the island.[37] British failure to implement the companion survey of Cape Breton Island reinforced the uniqueness of the St. John's survey. Nova Scotia provincial archivist D.C. Harvey compiled all the documents related to the survey of Cape Breton in the early 1930s but failed to locate any of Holland's plans of the island.[38] Although the large-scale plan is still missing, small-scale maps of Cape Breton have since surfaced in the Hydrographic Office and in the Gage Collection at the William L. Clements Library.[39] From these, it is clear that Holland developed a model of scientific surveying and imperial naming on St. John's Island which he then replicated in Cape Breton. Two decades later, Holland reproduced the same survey model in Upper Canada, which may have influenced later British surveying in New South Wales.[40] The Survey of St. John's Island thus takes on much greater significance; it marks the beginning of British imperial surveying in widely different parts of the world.

By the early 1790s, Holland was in his sixties and a weary man. While on his way to take up his position as first lieutenant-governor of Upper Canada, John Graves Simcoe, son of Captain John Simcoe of the *Pembroke*, passed through Quebec and met the surveyor general. "Poor Holland" he wrote to Evan Nepean in April 1792, "that good and faithful servant of the Crown, is worn out in body, tho' in full possession of his intellect."[41] But like many an old soldier, Holland was always happy to show friends and visitors the great battlefield on the Plains of Abraham. In a letter to Simcoe, he declared that he had "frequently fought the battle over with gentlemen who have had the curiosity to view the ground of action."[42] Holland died in Quebec in 1801, still in harness as surveyor general, and knowing that he had surveyed and mapped more of North America than anyone else (figure 7.7).

Of Holland's deputy surveyors, George Sproule and Thomas Wright made substantial contributions to surveying in British North America. Both surveyors were still with the General Survey in early summer 1774 when the survey parted company with the *Canceaux* in Boston Harbor.[43] Soon after, Sproule was appointed surveyor general of New Hampshire, but could

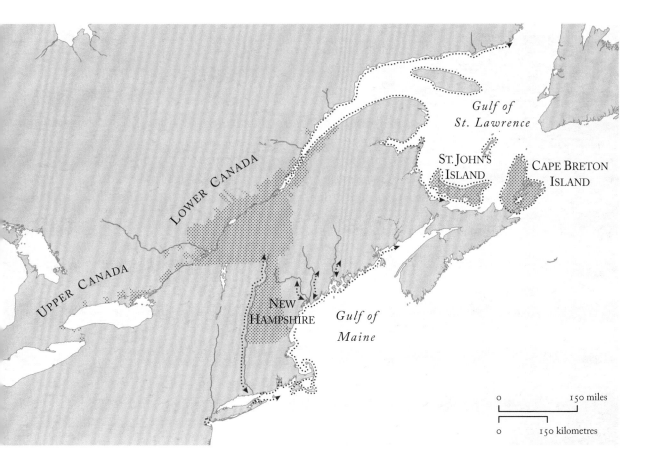

· Areas surveyed by Samuel
olland and his deputies, 1756–1801

hardly have taken up the appointment before the Revolution broke out.[44] He rejoined the army in Boston as an assistant field engineer and served throughout the war. Having lost property and a government position in New Hampshire, he was compensated by the British government after the war by being appointed surveyor general of the new Loyalist colony of New Brunswick in September 1784, a position he held until his death in 1817. As surveyor general, Sproule was responsible for settling Loyalists in the infant colony, laying out counties and townships along the shore of the Bay of Fundy and up the Saint John River valley.[45] Leading New Brunswick Loyalist Edward Winslow described Sproule as "that correct, faithful & devoted officer," attributes that he displayed throughout the General Survey.[46] Sproule's colleague Thomas Wright was appointed surveyor general of St. John's Island in 1773, and probably took up the appointment the following year.[47] Because proprietors, rather than the Crown, held most of the land, there was little for Wright to do, allowing him time to serve on both the Council and Supreme Court. His most important contribution to surveying

Epilogue: Beyond the Surveys · 209

came in 1796 when Great Britain and the United States appointed a joint commission to settle the international boundary through Passamaquoddy Bay. At the Peace of Versailles in 1783, British and American negotiators agreed on the St. Croix River as the international boundary, but no one was quite sure which of the two rivers flowing into Passamaquoddy was the "true St. Croix."[48] Wright served as astronomer on the British side and worked with Sproule and others in locating the correct river. Having surveyed the region twenty-four years earlier for the General Survey, the irony of having to do so again for an international boundary commission could hardly have been lost on Wright.

Of all the surveyors, Des Barres had the most remarkable later career (figure 7.8). Well-placed in London to lobby for a government position after the end of the war, he secured the lieutenant-governorship of the new Loyalist colony of Cape Breton Island in 1784.[49] He arrived in Cape Breton in January 1785, the island that Holland had so carefully surveyed twenty years earlier. In the intervening period, little had been done to settle and develop the colony, and Des Barres had grand plans. Among his first acts was to shift the capital from the ruins of Louisbourg to a peninsula in Spanish River, which he named Sydney, after the Home Secretary (figure 7.9). Holland had advocated such a transfer in his report on the island. But such reorganization was not enough. Described by a visiting army officer as "a great surveyor … but a most eccentric genius," Des Barres soon ran into difficulties administering the nascent colony and was recalled to London in 1787.[50] He remained in England for the next seventeen years, defending his reputation and lobbying the government for further compensation for the Survey of Nova Scotia and publication of *The Atlantic Neptune*. The endless stream of petitions had some effect, for he was awarded some of his claims and eventually, at the age of 76, the lieutenant governorship of Prince Edward Island (formerly St. John's Island). He arrived in July 1805. Although more circumspect in dealing with proprietors on the island than with Loyalists in Cape Breton, Des Barres again ran into administrative difficulties and was recalled in 1812. He spent his remaining years in Nova Scotia, ending his days in Halifax, where he died, in his 96th year, in 1824. He was buried in St. George's Anglican church, in the heart of the garrison town.

Among Des Barres's legacies were the copper plates from *The Atlantic Neptune*.[51] Despite their enormous weight (about 4,600 lbs.), Des Barres took some 251 full-size plates and numerous smaller ones with him to Cape Breton in 1785. With no printing press in Sydney, the plates were effectively useless and Des Barres sold them to one of his creditors, who promptly shipped them back to London. After his recall to England, Des Barres managed to repossess the plates and stored them in a warehouse near St.

7.8 · *Joseph Frederick Wallet Des Barres*, by James Peachey, 1785. Drawn in London and perhaps commemorating his appointment to the governorship of Cape Breton, this portrait depicts Des Barres as a scholarly Enlightenment gentleman. Library and Archives Canada, Acc. No. 1989-216-1 Gift of Mrs. Doreen Desbarres Bate.

7.9 · *Founding of Sydney, Cape Breton Island, 1785*, by Lieutenant William Booth. Des Barres's new capital set amid the forests of Cape Breton. John Clarence Webster Canadiana Collection W 1710, New Brunswick Museum, Saint John.

Paul's Cathedral. Desperate for funds, Des Barres instructed his son James Luttrell Des Barres (named after the surveyor's great friend) to hawk them to the Admiralty. But by then, Hurd had become Hydrographer and was in no mood to be generous to his former antagonist. After years of haggling, the Admiralty purchased the plates for their scrap value in 1820. Although many plates were re-planished and re-engraved during the nineteenth and early twentieth centuries, sixty-four still remained in the Hydrographic Office in 1946. As a gift to Britain's wartime allies, thirty were sent to Canada, where twenty-nine eventually reached the National Archives, and thirty-four were presented to historical societies and libraries in the United States. In 2002, the Massachusetts Historical Society, recipient of sixteen plates covering parts of the New England coast, published re-strikes from the original plates.

By the time these surveyors of empire passed away in the 1790s and early 1800s, the state of British military surveying had been completely transformed. Although Cook's voyages to the Pacific have tended to overshadow the development of marine and land surveying by Holland and Des Barres, the two surveyors were instrumental in carrying out the largest military survey ever conducted by the British in the eighteenth century and helped lay the foundations for later surveys across the empire. At the beginning of their work in the mid-1760s, the government adopted an *ad hoc* approach to surveying the coasts of North America, but very quickly the Admiralty realized the shortcomings of its marine surveying and began to implement institutional changes. Cook's appointment to command the Pacific voyages was one; the establishment of the Hydrographic Office in 1795 was another.[52] By the time of Vancouver's voyage to the northwest coast of North America in 1791, the Admiralty had figured out a highly efficient system of marine surveying, which stood it in good stead during the great imperial expansion of the nineteenth century.[53] Land surveying, too, was placed on a firmer scientific and institutional footing, well able to deal with the massive surveys that would be needed in Canada, Australia, New Zealand, India, and parts of Africa. Although British policy makers failed to make the best use of the General Survey of the Northern District in the 1760s and early 1770s, the long term impact of the General Survey and the Survey of Nova Scotia was considerable: The two surveys laid the foundations for much of the systematic scientific surveying of Britain's colonial possessions and thus a significant part of the modern world.

APPENDIX 1

Samuel Holland, "List of Plans sent to Government, from the General Survey of the Northern District of North America." 14 February 1776.

+ In possession of Holland in 1776
∗ In the office of the Board of Trade
× Not found in the office of the Board of Trade

3 plans of St. John's Island, by a scale of 2 miles to an inch
1 ...... St. John's Island .... 4,000 feet ....
1 ...... Magdalenes .... 4,000 feet ....
1 ...... ditto .... 2 miles ....
1 ...... Canada, the settled parts, with Lt. Haldimand's survey of the lower parts, 2,000 feet ∗
1 ...... Cape Britain .... 4,000 feet .... ∗
3 ...... ditto .... 2 miles ....
1 ...... Anticosti .... 4,000 feet .... ∗
1 ...... Cape Britain with additions .... 4,000 feet ....
1 ...... Canada from the River Ottawa to Baye Mille Vaches & Ile au Bic .... 4,000 feet
1 ...... Gaspey Bay .... 4,000 feet .... +
1 ...... Chaleur Bay .... 4,000 feet ....
1 ...... Canada North Shore .... 4,000 feet .... +
1 ...... West Coast of the Gulph of St. Lawrence from Chaleur Bay to Bay Verte, 8 sheets ....   4,000 feet .... +
1 ...... Connecticut River from Hinsdale to its source .... 4,000 feet .....
1 ...... Winnipissiokee & Smith Lakes .... 4,000 feet ..... ∗
1 ...... Casco Bay, Kennebec & Amossescoggan Rivers .... 4,000 feet .....
1 ...... West Coast of the Gulph of St. Lawrence .... 2 miles ....
1 ...... Piscataqua Harbor .... 2,000 feet ......
1 ...... Sea Coast from Cape Anne to Kennebec River, 3 sheets .... 4,000 feet ..... ∗
1 ...... Sea Coast from Cape Anne to Kennebec .... 2 miles ....
1 ...... Sea Coast from St. John's River to Grand Manan Island .... 4,000 feet .....
1 ...... Sea Coast from Penobscot to Pleasant River .... 4,000 feet ..... +
1 ...... Sea Coast from Kennebec to Pemaquid .... 4,000 feet .....

1 ...... Sea Coast from Cape Elizabeth to St. John's River .... 2 miles ....
1 ...... Sea Coast from St. John's River to Kennebec .... 4,000 feet .....
1 ...... Boston Harbor .... 4,000 feet ..... +
1 ...... Sea Coast from Cape Elizabeth to Newport in Rhode
Island with New Hampshire .... 2 miles .... *
1 ...... Plymouth Harbor .... 4,000 feet ..... +
1 ...... Sea Coast from Merrimack River to Newport
Rhode Island .... 4,000 feet ..... +

Source: NA Audit Office 3/140/60-61.

APPENDIX 2

Cartobibliography of extant manuscript plans and charts related to the General Survey of the Northern District of North America by Samuel Holland and his deputies, and the Survey of Nova Scotia by J.F.W. Des Barres.

BRITISH LIBRARY

Maps K.Top. 119.23 – A map of the river St. Lawrence, reduced from the actual surveys of Samuel Holland, Esq. (Holland & Lewis, 1773)

Maps K.Top. 119.50 – Plan of the Coast from the West Passage of Passamiquodi Bay to the River St. John in the Bay of Fundy (Wright, 1772). 4,000 feet to an inch.

Maps K.Top. 120.18 – Sea Coast from Cape Elizabeth, on the west side of Casco Bay, to St. John's river, in the Bay of Fundy (Sproule, Grant, Blaskowitz, Wright, circa 1772). 2 miles to an inch.

Maps K.Top. 120.19 – Plan of the sea coast from Cape Elizabeth, to the entrance of Sagadahock, or Kennebeck River, including Casco Bay with all its islands, harbors, &c. (Sproule, 1771). ¾ mile to an inch.

Maps K.Top. 120.20 – Plan of the coast from Kennebec River to Round Pond, on the west side of Muscongus Bay, including the islands, rivers, etc. (Sproule, 1772). 1⅓ mile to an inch.

Maps K.Top. 120.21 – Plan of the coast from Pleasant River to the Penobscot Bay (Blaskowitz & Grant, 1772). 1⅓ miles to an inch.

Source: British Library Integrated Catalogue

UK HYDROGRAPHIC OFFICE, TAUNTON

9/73 – Narrangansett to Hampton Harbour, including Cape Cod (Holland)
10/83 – Pemaquid Point to Ile au Haut, including Penobscot Bay and River (Holland)
260 Hist Pr – Island of Anticosti (Holland, 1767)
264 Ah1 – Magdalen Islands, Gulf of St. Lawrence (Holland 1767)
265 Ah1 – Cape Breton Island (Holland, 1767)
267 Ah1 – Cape Britain [Breton] Island (Holland, 1767)
276 (2) Ah2 – Island of St. John (Mowat/Holland, 1765)

284/1 5f – Bay of Fundy, Campobello Island to Beaver Harbour (Hurd, 1772)
284/2 s Ah2 – Bay of Fundy, Mace[s] Bay to St. John (Hurd, 1772)
289 5b – Coast westward of Penobscot Bay (Hurd, 1776)
395 88 – Boston Harbour & Casco Bay from Cape Elizabeth to Renebre River (Holland)
B253 Ah3 – Survey of part of the north shore of the River St. Lawrence, Bay of Seven Islands (Sproule, 1768)
C83 88 – Original survey of River St. Lawrence (Holland)
D601 37a – Gulf of St. Lawrence and St. Johns River (Wright)
E468 88 – Coast of North America from Penobscot to St. Johns (Hurd)
z52/1 Hist Pr – Part of the Sea Coast of Nova Scotia between the entrance of St. Margaret's Bay and Canso (Des Barres)
z52/2 Hist Pr – Part of the Sea Coast of Nova Scotia between the entrance of St. Margaret's Bay and Canso (Des Barres). Another version of z52/1 varying in a few details and with more names.
z52/3 Hist Pr – Chart of the sw coast of Nova Scotia from Isle Hope to Cape Sable (Des Barres)
z52/4 Hist Pr – Chart of the south coast of Nova Scotia from Sambro Lighthouse to Rugged Islands (Des Barres)
z52/5 Hist Pr – Chart of the northern coast of Nova Scotia from River Philip to Thoms Head (Des Barres)
z52/6 Hist Pr – Port Amherst and Townsend Harbour with sailing directions (Des Barres)
z52/7 Hist Pr – Ramshick Harbour to Moll Poole (Des Barres)
z52/8 Hist Pr – Port Hood with sailing directions (Des Barres)
z52/9 Hist Pr – Richmond Harbour and Inlet with sailing directions (Des Barres)
z52/10 Hist Pr – St. Peters Bay, Mass Cove and Bay of Rocks with sailing directions (Des Barres)
z52/11 Hist Pr – Crow Harbour with sailing directions and two views (Des Barres)
z52/12 Hist Pr – Unfinished chart of Sable Island (Des Barres)
z52/13 Hist Pr – Four views in Bay of Fundy: Cape Blowmedown and Cape Split; Spencers Island and Entrance of Mines Bason; Isle Haut; and Cape Chignecto (Des Barres)
z52/14 Hist Pr – Torbay with sailing directions and view
z100/1 – Plan of the Province of Quebec from Little Fox River on South Side of River St. Lawrence to beyond Little Pabos (Holland)
z100/2 – Untitled chart from Montreal to Point Mille Vache (Holland)
z100/3 – Untitled chart from River Godfrey to Anticosti Island (extending on the north to Natashquan Point and on the south to Chaleur Bay) (Holland)
z100/4 – Untitled chart from Mosie Bay to Clearwater Point (Holland)
z100/5 – Untitled chart from Manicougan Bay to Pointe des Monts (Holland)
z100/6 – Untitled chart from Clearwater Point to Natashquan Point (Holland)
z100/7 – Untitled chart of Saguenay River (Holland)
z100/8 – Untitled chart of St. Johns [Saint Jean] River (Holland)

z100/9 – Untitled chart of Lake St. Johns (Holland)

Source: A.C.F. David, *Supplement to Geographical Index: North & South America, West Indies & Polar Regions for Items relating to Canadian and adjacent waters recorded up to 1850 and held in the Hydrographic Office, Taunton* (Taunton: UK Hydrographic Office, 1993)

NATIONAL ARCHIVES (UK)

- CO 700 New Brunswick 1 – St. John River (Holland, 1758). About $2^{2/5}$ miles to an inch.
- CO 700 New Brunswick 5 – Miramichi Bay (Wright & Blaskowitz, 1770). 4,000 feet to an inch
- CO 700 New Brunswick 6 – Miramichi Bay (Wright & Blaskowitz, 1770). 2 miles to an inch.
- CO 700 New Brunswick 11 – Passamaquoddy Bay (Wright & Hurd, 1772, 1785). ¾ mile to an inch.
- CO 700 Nova Scotia 42 – Gut of Canso (Wright, 1766). ½ mile to an inch.
- MR 1785 – Prince Edward Island (Holland, 1765). 2 miles to an inch.
- MR 1881 – Prince Edward Island (c.1765). Copy of MR 1785. 2 miles to an inch.
- CO 700 Canada 25 – Chaleur Bay (Pringle, 1766–67). 2 miles to an inch.
- CO 700 Canada 27 – Magdalen Islands (Haldimand, 1766). About ¾ mile to an inch.
- CO 700 Canada 28 – St. Lawrence River (Sproule, 1768). About ¾ mile to an inch.
- CO 700 Canada 29 – Quebec, Seven Islands (Sproule, 1769)
- CO 700 Canada 30 – Quebec, Mingan Islands (Sproule, 1769). 4,000 feet to an inch.
- CO 700 Canada 31 – Manitougan Bay (Sproule, 1770)
- CO 700 Connecticut 3 – Connecticut River (Grant, 1770–71). ¾ mile to an inch.
- CO 700 Maine 15 – Maine: Ogunquit River to Cape Elizabeth (Grant, 1771). About ¾ mile to an inch.
- CO 700 Massachusetts Bay 13 – Massachusetts, New Hampshire: Cape Ann to Hampton (Grant & Wheeler, 1771). About ¾ mile to an inch.
- MPG 346 – New Hampshire, Maine, New Brunswick (Holland, 1770).
- CO 700 New Hampshire 6 – Piscataqua Harbour (circa 1770). About 2,000 feet to an inch.
- CO 700 New Hampshire 22 – Lake Winnipesaukee (Wheeler, 1770). 4,000 feet to an inch.
- CO 323/29 – Delaware River (Holland & Rittenhouse, 1774). 50 chains to an inch.

Source: John Brian Harley and Minda C. Phillips, *Manuscript Maps Relating to North America and the West Indies*. Part 1: *The Revolutionary Era in the Public Record Office, London* (East Ardsley: EP Microform, 1974)

WILLIAM L. CLEMENTS LIBRARY, UNIVERSITY OF MICHIGAN

Gage Papers – A plan of Louisbourg (Sproule 1767). 200 feet to an inch.

Gage Papers – A plan of the Harbour of Louisbourg (Holland & Sproule 1767). Approximately 1,980 feet to an inch.

Gage Papers – A plan of the Island of Cape Britain (Sproule 1767). 4 miles to an inch.

Gage Papers – A plan of the sea coast from Gage Point to Cumberland Cape (Holland & Goldfrap, 1767). ¾ mile to an inch.

Gage Papers – A plan of part of the Province of Quebec from the Lake de Deux Montagne to River Batiscant (Holland & Goldfrap, 1767). 2 miles to an inch.

Gage Papers – A plan of part of the Province of Quebec from the River St. Anne to the Island of Coudre (Holland & Goldfrap, 1767). 2 miles to an inch.

Gage Papers – A plan of the Magdelain. Brion. Bird and Entry Islands (Haldimand & Goldfrap, 1766). 2¼ miles to an inch.

Gage Papers – A plan of the settled part of Canada reduced from the large survey made in the years 1760 & 1761 by order of General Murray governour of the Province (Holland & Sproule, 1767). 6 miles to an inch.

Gage Papers – New Hampshire with part of Quebec and Maine c. 1773 [probably by Holland]

Clinton Papers. Clinton Map 29 – The coast from St. Johns, New Brunswick to Goldsborough Bay, Maine [probably related to Holland's survey]. Approximately 2 miles to an inch.

Clinton Papers. Clinton Map 30 – The coast of Maine from modern Portland to Mt. Desert Island c.1770 [probably related to Holland's survey]. Approximately 2 miles to an inch.

Clinton Papers. Clinton Map 31 – Falmouth [probably related to Holland's survey]. Approximately 2 miles to an inch.

Clinton Papers. Clinton Map 32 – Penobscot Bay [probably related to Holland's survey]. 4,000 feet to an inch.

Clinton Papers. Clinton Map 58 – Narragansett Bay and the surrounding shores (Wheeler). 3,960 feet to an inch.

Source: Christian Brun (compiler), *Guide to the Manuscript Maps in the William L. Clements Library* (Ann Arbor: University of Michigan, 1959)

APPENDIX 3

Catalogue of the Henry Newton Stevens Collection of *The Atlantic Neptune* at the National Maritime Museum, Greenwich, London.

The Henry Newton Stevens Collection is the most comprehensive collection of charts and views from *The Atlantic Neptune*. Numbers of charts are keyed to Figures 6.8 and 6.9. For a more detailed catalogue, see http://www.nmm.ac.uk/collections/explore/index.cfm/category/90437

VOLUME I

1. General Title to Volume I
2. Title: "The Sea Coasts of Nova Scotia"
3. Utility of the Atlantic Neptune
4. Preface
5. References
6. General Remarks
7. Tables
8. Contents of the Charts and Views of Nova Scotia
9. Coast of Nova Scotia, New England, New York, Jersey, &c.
10. A Chart of Nova Scotia
11. Two Coloured Views on One Sheet
    i. North Point of Grand Manan Island in the Bay of Fundy;
    ii. View of Campobello, Passamaquady Bay
12. Two Views on One Sheet
    i. Mechios River near the Mills; ii. A Sketch of Mechios Mills, July 31, 1777
13. River St. John
14. Five Views
    i. The Wolves; ii. Grand Manan Island; iii. Shore West of River St. John; iv. Entrance of River St. John; v. Entrance of Pasamaquady Bay
15. Isthmus of Nova Scotia
16. Four Views on One Sheet
    i. Cape Blowmedown; ii. Cape Split; iii. Entrance to Mines Bason; iv. Ile Haut and Cape Chegnecto

17. Chignecto Bay
18. Six Views on One Sheet
    i. Isle Haut; ii. Cape Dore; iii. Cape Baptist; iv. Entrance into Bason of Mines; v. Isle Haut bearing W.N.W.; vi. Cape Blowmedown
19. Environs of Fort Cumberland
20. A View of Partridge Island
21. Chart of S.E. Coast of Bay of Fundy
22. Five Views of Annapolis Bason
    i. Entrance into Annapolis Bason; ii. North Entrance of Grand Passage; iii. Eden and Gascoyne Rivers; iv. Annapolis Royal; v. North Entrance of Petit Passage
23. View of Annapolis Royal
24. Three Views
    i. South Entrance of Grand Passage; ii. Cape St. Marys; iii. St. Marys Bay
25. Chart of West Coast of Nova Scotia
26. Chart of S.W. Coast of Nova Scotia
27. Chart of Barrington Bay
28. Chart of Port Amherst and Port Haldimand
29. Chart of Port Campbell
30. Four Views on Two Sheets
    i. South West Coast of Nova Scotia; ii. South West Coast of Nova Scotia; iii. Seal Isle; iv. Boston Bay
31. Three Views on Two Sheets
    i. South West Coast of Nova Scotia; ii. Cape Sable; iii. Coast of New Hampshire
32. Chart of Port Mills, Port Mansfield, Gambier Harbour
33. Chart of Liverpool Bay
34. Chart of Port Jackson
35. Chart of Kings Bay and Lunenburg
36. Chart of Mecklenburgh Bay
37. Chart of Charlotte Bay
38. Chart of Leith Harbour
39. Five Views on One Sheet
    i. Cape Prospect; ii. High Lands of Haspotagoen; iii. The Ovens; iv. Cape Sable and Barrington Bay; v. Cape Sable N.E. by N.

VOLUME II

40. General Title to Volume II
41. Title: "The Sea Coasts of Nova Scotia"
42. Contents of the Charts and Views of Nova Scotia
43. References
44. The South East Coast of Nova Scotia
45. Halifax Harbour

46. Plan of Halifax
47. View of Halifax
47*. View of Town & Harbour of Halifax
48. Halifax Harbour, Six Views on One Sheet
    i. View of Halifax Harbour; ii. The shore eastward of Halifax Harbour; iii. Shore of Halifax Harbour; iv. Sambro Lighthouse bearing West; v. Sambro Lighthouse; vi. Chebucto Head
49. Chart of Egmont Harbour
50. Chart of Keppell Harbour, Knowles Harbour &c.
51. Six Views on One Sheet
    i. Cape Egmont; ii. Egmont Harbour; iii. Keppell Harbour; iv. Falls of Hinchinbroke River; v. Chisetcock Inlet; vi. Dartmouth Shore in Halifax Harbour
52. Chart of Spry Harbour, Port Pallisser, Port North, Beaver Harbour, &c.
53. Chart of White Islands Harbour, Port Stephens, Houlton Harbour, &c.
54. Chart of Sandwich Bay
55. Chart of Torbay
56. Chart of White Haven
57. Chart of Canso Harbour
57*. Printed Directions for Canso Harbour & St. Peters Bay
58. Chart of Crow Harbour &c.
59. Chart of St. Peters Bay
60. Chart of Milford Haven
61. Eight Views on Two Sheets
    i. Land from White Islands to St. Marys River; ii. Entrance of Milford Haven; iii. Entrance of Port Bickerton; iv. Entrance of Beaver Harbour; v. View of Beaver Harbour; vi. The Shore westward of Canso; vii. The Beaver Islands; viii. S.E. Point of Nova Scotia from Canso Island
62. Chart of Conway Harbour & Port Aylesbury
63. Chart of Lenox Passage, Richmond Isles, &c.
64. Six Views on One Sheet
    i. Louisbourg Harbour; ii. Shore to W. of Gabbarrus Bay; iii. W. Shore of Richmond Isle; iv. S.W. Shore of Cape Breton Island; v. Ramea Isles; vi. Sta. Maria Island
65. View of Louisbourg
66. Chart of the Gut of Canso
67. Chart of North East Coast of Nova Scotia
68. Chart of the Coast from Bay Verte to Buctush
69. Chart of Port Hood
70, 70A, 70B. Two Views on One Sheet
    i. Port Hood; ii. Plaister Cliffs
    70A. i. A View of the Entrance of Port Hood
    70B. ii. A View of the Plaister Cliffs on the West Shore of Georges Bay
71. Chart of Frederick Bay, Pictou Harbour, &c.
72. Chart of Port Shediack and Cocagne

73. Chart of the Isle of Sable
74. Five Views of the Isle of Sable
    i. A View of the East End; ii. The Eastern End from the Southward; iii. A View from the South Side; iv. A View from the Ridge; v. A View of the North Shore
75. Views and Coastal Profiles of the Isle of Sable on Two Sheets
    i. Entrance of the Pond; ii. Wreckers Den near the Pond on the Isle of Sable; iii. North Shore; iv. North shore of Isle of Sable 2 miles distant
76. Chart of the Isle of Sable
77. View from the Camp at the East End of the Naked Sand Hills on the Isle of Sable
78. Remarks on the Isle of Sable

VOLUME III NEW ENGLAND

79. General Title Page to Volume III
80. Title: "Charts of the Coasts and Harbours of New England"
81. Contents New England
82. References
82*. Coast of Nova Scotia, New England, New York, Jersey, &c.
83. Coast of New England
84. Chart of the Coast from Point Judith, R.I. to Great Bay, Long Island
85. Chart of Coast from Rhode Island to Cape Malabar
86. Chart of the Harbour of Rhode Island and Narragansett Bay
86*. Directions for Sailing into Rhode Island Harbour and the Sound of Long Island
87. Plan of the Town of Newport
88. Chart of Buzzards Bay and Vineyards Sound
89. Six Views on One Sheet
    i. Cape Poge; ii. Sandy Point; iii. Gay Head and Nomans Land; iv. Gay Head bearing S.E.; v. Sankoty Head bearing S.b.W.; vi. Sankoty Head bearing S.W.
90. Chart of Nantucket Island & Martha's Vineyard
91. Chart of Massachusetts Bay
92. Two Views on One Sheet
    i. Cape Cod; ii. Point Shirley
93. Chart of Plymouth Bay
94. Chart of Boston Bay
95. Four Views of Boston Harbour
    i. Boston seen between Castle William's and Governor's Island; ii. High Lands of Agameticus; iii. Boston Bay; iv. Entrance of Boston Harbour
96. Chart of Boston Harbour
97. View of Boston taken on the Road to Dorchester
98. A View of Boston

99 & 100. Six Small Views of Boston on Two Sheets
    i. Boston from Willis Creek; ii. Long Island; iii. A View of the Country towards Dorchester; iv. A View of Boston from Dorchester Neck; v. A View of the Harbour from Fort Hill; vi. A Front View of the Lines

101. Chart of Coast Marble Head to Hampton Harbour
102. Coast from Newbury to Cape Elizabeth
103. Chart of the Coast from Cape Elizabeth to Musketo Island
104. Chart of Falmouth Harbour
105. Chart of the Harbours and Rivers between Portland Point and Stage Island
106. Chart of Piscataqua Harbour
107. A View of Portsmouth in New Hampshire
108. Two Views on One Sheet
    i. Castle William in Boston Harbour; ii. A View of Newcastle with the Fort
108*. Large View on One Sheet
    Castle William in Boston Harbour
109. Chart of Coast frm Pemaquid Point to Owls Head Bay
110. Chart of the Coast from Musketo Island to Skuttock Point
111. Chart of the Entrance into Penobscot Bay
112. Chart of Penobscot Bay (Upper Part)
113. Chart of the Coast Burntcoal Island to Skulock Point
114. Chart of the Coast Goldsborough to Moose Harbour
115. Chart of Grand Manan Island
116. Chart of Pasamaquody Bay

VOLUME IV (PART I): ST. LAWRENCE

117. General Title Page to Volume IV
118. Title: "Charts of the Coast and Harbours in the St. Lawrence"
119. Contents of the St. Lawrence Section
120. References
121. Chart of the Coast of Nova Scotia, New England, New York, Jersey and the Gulph and River of St. Lawrence
122. General Chart of the Gulph and River of St. Lawrence
123. Chart of River St. Lawrence from Chaudiere to Lake St. Francis
124. Plan of Quebec and Environs
125. View of Quebec from the Southeast
126. Chart of the River St. Lawrence from Cock Cove to Quebec
127. Chart of the N.W. Coast of the Gulph of St. Lawrence from the River St. John eastwards to St. Genevieve
128. Chart of the Bay of Seven Islands
129. Chart of the Bay of Gaspey
130. Chart of the Bay of Chaleurs
131. Chart of the Bay of Miramichi

132. Chart of the Harbours of Rishibucto & Buctush
133. Chart of the Magdalen Islands
134. & 135. The S.E. Coast of the Island of St. John
136. Chart of the N.E. Coast of Cape Breton Island
137. Chart of the S.E. Coast of Cape Breton Island
138. Chart of Cape Breton and St. John Islands
139. Chart of the Island of Cape Breton
140. Chart of the Harbour of Louisbourg
141. View of Louisbourg from the North East

VOLUME IV (PART II): NEW YORK

142. Title: "Charts of Several Harbours and divers parts of the Coast of North America from New York southwestward"
143. References
144. Contents of New York Section
145. Chart of the Coast of New York, New Jersey, Pennsylvania, Maryland, Virginia, North Carolina, &c.
146. Chart of New York Harbour
147. Nautical Directions to Sail in to the Harbour of New York, &c.
148. View of New York
149. Sketch of the Operations of His Majesty's Fleet & Army under the Command of Viscount Howe
150. The Phoenix and the Rose engaged by the enemy's fire and galleys.
151. References to Sketch of Operations of Earl Howe
152. The Phoenix and the Rose engaged by the enemy's fire and galleys on the 16 August 1776
153. Five Views on One Sheet
    i. View of the Highlands of Neversunk; ii. South Shore of Long Island; iii. New York with the Entrance of the North and East Rivers; iv. The Lighthouse on Sandy Hook; v. The Narrows
154. Chart of Western Passage into Long Island Sound
155. Chart of Hell Gate, Oyster Bay and Huntington Bay
156. Plan of Forts Montgomery and Clinton
157. Three Views Forts Montgomery and Clinton and St. Anthony's Nose
    i. Forts Montgomery and Clinton from the north; ii. Forts Clinton and Montgomery from the south; iii. Forts Clinton and Montgomery and St. Anthony's Nose from the south
158. Chart of Delawar Bay
159. Chart of Delawar River from Bombay Hook to Ridley Creek – Plan of Delawar River from Chester to Philadelphia
159*. Directions for Delawar Bay and River
160. Plan of the Environs of Philadelphia
161. Plan of the Posts of York and Gloucester
162. Chart of the Harbour of Charles Town in South-Carolina

163. Sketch of the Environs of Charlestown
164. Sketch of the Operations before Charlestown
165. Sketch of the Battle near Camden in South Carolina, 16 Augst. 1780
166. Chart of Port Royal in South Carolina
167. Chart of the Coast, Rivers and Inlets of Georgia
168. Chart of the Coast of Georgia
169. Chart of the Coast of West Florida from Bensecour River to the Bay of St. Joseph
170. Chart of the Gulf of Mexico from Pensacola to Apelousa River
171. Plan of the Harbour of St. Augustine
172. Chart of the Mississippi River from Iberville to Yazous
173. Chart of the Bay and Harbour of Pensacola
174. Chart of Port Royal and Kingston Harbours in Jamaica
175. Chart of Montego Bay – Port Antonio in the Island of Jamaica
176. Two Views of the Entrance of Havannah – The Harbour and Town of Havannah

Source: Henry Newton Stevens, *Catalogue of the Henry Newton Stevens Collection of the Atlantic Neptune* (London: Henry Stevens, Son & Stiles, 1937). Typescript at National Maritime Museum, Greenwich, London.

Quamphegon
Berwick
Canaan
Green Hill
Cocheco R
NORTHWOOD
DOVER
Building Yard
Sturgeon Cr.
Quakers M.
Frosts Hill
Scotland
Back R.
Two Miles
Streek
MADBURY
KITTERY
NOTTINGHAM
Saw M.
Mill
Sadleback Mt.
Oyster R
Upper Kittery
Spruce Cr. Mg
ERFIELD
LEE
DURHAM
Little B
NEWINGTON
Sulivans Mills
Nottm. Square
Gt. Bay
Greenland Bay
PORTSMOUTH
Little H.
PARISH
EPPING
NEW MARKET
Meet
Bridge
Frost Pt.
GREENLAND
RAYMOND
STRATHAM
3 Mills
RYE
EXETER
Rye Hr.
Joslings or Locks Pt.
BRENTWOOD
ESTER
POLIN
North Hill Mg
Philbrick or Rye Lodge
L. Boars Hd.
Exeter R
HAMPTON
Chester M.
Webster
L. River Rocks
Hamp.
Meet
Kingston M.
KINGSTON
KENSINGTON
HAMPTON FALLS
Gt. Boars Hd.
Derry Pond
Sunken Rocks
Hampton H.
Londonderry Mg
Southampton
Sea Brook
Round Rock
NEWTOWN
Powow R
Powow Hill
Iron Works
Almsbury M.
Salisbury M.
Ferry
River
PLAISTOW
Nicholls Cr.
Merrimack
NEWBURY PORT
NDHAM
Mt. Atkinson
HAMPSTEAD
Dalton
Pipe Stave Hill
NEWBURY PORT
HAVERHILL
West Parish
Pine I.
Mitchells Eddy
Griffins Fy.
Bradford
Old Town R.
Plum Island
NEW SALEM
Gages Fy.
Gages Neck
Clements
Mullikins Fy.
Gages Ferry
IPSWICH BAY
Rowley R.
Rowley M.
Spickett R
Saw M.
Peters Falls
Swans Fy.
Alewife Br.
Ipswich H.
Badwills Falls
Badwills Fy.
Lat. 42°41'2"
IPSWICH
Sand Downs
Ferry
Vemums Falls
Salt I.

NOTES

INTRODUCTION

1 Cited in William P. Cumming, *British Maps of Colonial America*, 56.
2 For an overview of Holland's surveys in Canada, see Don W. Thomson, *Men and Meridians: The History of Surveying and Mapping in Canada*, vol. I, *Prior to 1867*; for the New England survey, see Grace S. Machemer, "Headquartered at Piscataqua: Samuel Holland's Coastal and Inland Surveys, 1770–1774." Des Barres is covered in Thomson, *Men and Meridians*, and in a biography by G.N.D. Evans, *Uncommon Obdurate: The Several Public Careers of J.F.W. DesBarres*. The southern surveys have been better served by scholarly studies; see Louis De Vorsey Jr., ed., *De Brahm's Report of the General Survey in the Southern District of North America*, and "La Florida Revealed: The De Brahm Surveys of British East Florida, 1765–1771"; John D. Ware, *George Gauld: Surveyor and Cartographer of the Gulf Coast*.
3 P.J. Marshall, ed., *The Oxford History of the British Empire: The Eighteenth Century*. For the lack of attention to Canada in this volume, see Phillip A. Buckner, "Was There A 'British' Empire? *The Oxford History of the British Empire* from a Canadian Perspective," 117–19. For a review of the recent literature and debates, see Lisa Chilton, "Canada and the British Empire: A Review Essay." For the importance of Canada in the imperial structure, see Elizabeth Mancke, "Another British America: A Canadian Model for the Early British Empire."
4 The standard syntheses on the region for this period are J.M. Bumsted, "1763–1783: Resettlement and Rebellion"; and R. Cole Harris, *The Reluctant Land: Society, Space, and Environment in Early Canada*, 164–6.
5 Stephen J. Hornsby, *British Atlantic, American Frontier: Spaces of Power in Early Modern British America*.
6 Steven J. Harris, "Long-Distance Corporations, Big Sciences, and the Geography of Knowledge," 295.
7 John Gascoigne, *Science in the Service of Empire: Joseph Banks, the British State, and the Uses of Science in the Age of Revolution*. Surprisingly, the confluence of cartographic science and the navy is overlooked in N.A.M. Rodger, "Navies and the Enlightenment."
8 I am indebted to the phrase "shake-down" to one of the reviewers of the manuscript who wrote: "The Holland and Des Barres surveys might be considered a spatial 'shake-down,' leading to the eventual reorganization of imperial institutions."
9 J.B. Harley, *The New Nature of Maps: Essays in the History of Cartography*. For a sympathetic appreciation of Harley's work, see Matthew H. Edney, "The Origins and Development of J.B. Harley's Cartographic Theories." For a more critical view, see Barbara Belyea, "Images of Power: Derrida/Foucault/Harley." For another discussion of the

relationship between maps and power, see Denis Wood, *The Power of Maps*.

10 Harley, *New Nature of Maps*, 152. Such was Harley's emphasis on power, especially the power of the state, that cartographic historian J.H. Andrews, in a critical introduction to *New Nature of Maps*, summed up Harley's approach to the map with the somewhat facetious comment: "'Just as I thought: more glorification of state power.'" See Andrews, "Introduction: Meaning, Knowledge, and Power in the Map Philosophy of J.B. Harley," 32.

11 Jan Golinski, *Making Natural Knowledge: Constructivism and the History of Science*; David N. Livingstone, *Putting Science in its Place: Geographies of Scientific Knowledge*; Steven Shapin, "Placing the View from Nowhere: Historical and Sociological Problems in the Location of Science" and "Here and Everywhere: Sociology of Scientific Knowledge."

12 Charles W.J. Withers, *Placing the Enlightenment: Thinking Geographically about the Age of Reason*; David N. Livingstone and Charles W.J. Withers, eds, *Geography and Enlightenment*. For recent overviews of the Enlightenment, see Dorinda Outram, *Panorama of the Enlightenment*, and the various essays in Roy Porter, ed., *The Cambridge History of Science*, vol. 4: *Eighteenth-Century Science*.

13 An observation supported by Martin Brueckner and Michael Dettelbach in their comments on Charles Withers, *Placing the Enlightenment: Thinking Geographically about the Age of Reason* – Author Meets Critics Session at the 2008 Annual Meeting of the Association of American Geographers, Boston.

14 For a short but perceptive introduction to the burgeoning literature in Atlantic history, see Nancy L. Rhoden, "Introduction: Revisiting *The English Atlantic* in the Context of Atlantic History." For the relationship of Atlantic history to Canadian history, see the various essays in John G. Reid, et al., "Is There a 'Canadian' Atlantic World?"

15 For an attempt to understand the environmental history of the Atlantic, see W. Jeffery Bolster, "Putting the Ocean in Atlantic History: Maritime Communities and Marine Ecology in the Northwest Atlantic, 1500–1800."

16 The significance of *The Atlantic Neptune* has been completely missed by scholars; see for example, Charles W.J. Withers, "Where Was the Atlantic Enlightenment? – Questions of Geography."

17 Personal communication from Harold Borns, Jr., Research Professor and Professor Emeritus of Glacial and Quaternary Studies, University of Maine.

18 For an introduction to Latour's thinking, see Bruno Latour, *Reassembling the Social: An Introduction to Actor-Network Theory*.

19 David Lambert and Alan Lester, eds, *Colonial Lives Across the British Empire: Imperial Careering in the Long Nineteenth Century*, 1–31; Alan Lester, "Imperial Circuits and Networks: Geographies of the British Empire."

20 Tony Ballantyne, "Empire, Knowledge and Culture: From Proto-Globalization to Modern Globalization."

21 A.G. Doughty, "Notes and Documents: A New Account of the Death of Wolfe," 46.

CHAPTER ONE

1 The outstanding modern treatment of the war is Fred Anderson, *Crucible of War: The Seven Years' War and the Fate of Empire in British North America, 1754–1766*. The book provides essential context for this chapter.

2 R.C. Simmons and P.D.G. Thomas, eds, *Proceedings and Debates of the British Parliaments Respecting North America 1754–1783*, vol. I, *1754–1764*.

3 Douglas W. Marshall, "The British Engineers in America: 1755–1783."

4 Chris Tabraham and Doreen Grove, *Fortress Scotland and the Jacobites*; R.A. Skelton, "The Military Survey of Scotland 1747–1755"; Yolande O'Donoghue, *William Roy 1726–1790: Pioneer of the Ordnance Survey*.

5 Stanley McCrory Pargellis, *Lord Loudoun in North America*, 319; René Chartrand, *French Fortresses in North America 1535–1763*.

6 Pargellis, *Lord Loudoun*, 61–7; Alexander V. Campbell, "Atlantic Microcosm: The Royal American Regiment, 1755–1772."

7 Stanley McCrory Pargellis, ed., *Military Affairs in North America 1748–1765: Selected Documents from the Cumberland Papers in Windsor Castle*, 254.

8 F.J. Thorpe, "Holland, Samuel Johannes"; "Samuel Holland, Canada's First Surveyor-General, 1729–1801" at www.lowensteyn.com/Samuel_Holland/index.html; Holland to Pitt, 14 May 1789, quoted in Willis Chipman, "The Life and Times of Major Samuel Holland, Surveyor-General, 1764–1801," 77–9; "Memorial of Major Samuel Holland, Surveyor General of the Province of Quebec," 22 November 1783, NA AO 13/13, f.280–90.

9 William C. Lowe, "Lennox, Charles"; Stuart Reid, "Wolfe, James"; James Wolfe to his mother, 2 January 1753 in Beckles Willson, *The Life and Letters of James Wolfe*, 197–8.

10 Wolfe to 3rd Duke of Richmond, 28 July 1758, WSCRO Goodwood Ms. 223.

11 Holland to Sir Henry Moore, 9 October 1768, BL Papers relating to New York, Add.22679, f.50–1.

12 Holland to Hillsborough, 24 December 1769, CO 5/71, pt. 1; Holland to Moore, op. cit.

13 Pargellis, ed., *Military Affairs in North America*, 364.

14 "Memorial of Major Samuel Holland"; James Abercrombie to Earl of Loudoun, 29 November 1757, Loudoun Papers (American Series) LO 4915, Huntington Library.

15 J.S. McLennan, *Louisbourg from Its Foundation to Its Fall 1713–1758*, 286. For the best modern biography of Wolfe, see Stephen Brumwell, *Paths of Glory: The Life and Death of General James Wolfe*.

16 "Memorial of Major Samuel Holland."

17 Wolfe to Amherst, 25 July 1758, in Willson, *The Life and Letters of James Wolfe*, 381.

18 John Clarence Webster, ed., *Journal of William Amherst in America 1758–1760*, 31.

19 Wolfe to 3rd Duke of Richmond, 28 July 1758, WSCRO Goodwood Ms. 223.

20 René Chartrand, *Ticonderoga 1758*. None of Holland's plans and letters to the 3rd Duke of Richmond survive among the Goodwood Papers; they were most likely destroyed when Richmond House in London burnt in the late eighteenth century.

21 This famous encounter is described in Holland to Lieutenant governor Simcoe, 11 January 1792, reproduced in Chipman, "Life and Times of Major Samuel Holland," 18–19. The meeting has been discussed by several authors, notably R.A. Skelton, "Captain James Cook as a Hydrographer," 97–8; J.C. Beaglehole, *The Life of Captain James Cook*, 33–4; and Andrew David, ed., *The Charts & Coastal Views of Captain Cook's Voyages, Volume One: The Voyage of the Endeavour 1768–1771*, xix–xx.

22 John Knox, *An Historical Journal of the Campaigns in North America For the Years 1757, 1758, 1759, and 1760*, 1:262; Holland's maps of Fort Frederick are in the Abercromby Papers: Maps and Plans, Huntington Library. See also "A Sketch of St. John's Harbour and Part of the River," NA CO 700 New Brunswick 1 St. John River (? c.1760). The correct date should be 1758.

23 Holland to Simcoe in Chipman, "Life and Times of Major Samuel Holland," 19.

24 No manuscript or engraved copies of these maps survive. Andrew David suggests that only a few sheets were printed and were soon superseded by the updated map of 1760 printed by Jefferys; see David, ed., *Charts & Coastal Views*, xx.

25 W.A.B. Douglas, "Colvill (Colville), Alexander."

26 C.P. Stacey, *Quebec, 1759: The Siege and the Battle*.

27 Holland to Simcoe in Chipman, "Life and Times of Major Samuel Holland," 19.

28 Holland to Simcoe, 10 June 1792, in Doughty, "Notes and Documents," 47. Stacey does not consider this important letter in his account, but the 2002 edition does include a panoramic photograph of Anse au Foulon taken from the south shore; see Stacey, *Quebec, 1759*, 125.

29 Doughty, "Notes and Documents," 48.

30 Ibid., 51, 55; "Memorial of Major Samuel Holland."

31 Doughty, "Notes and Documents," 51.

32 "Memorial of Major Samuel Holland."

33 Ibid.
34 Evans, *Uncommon Obdurate*, 3–8; R.J. Morgan, "Des Barres, Joseph Frederick Wallet"; Jean-Marc Debard, "The Family Origins of Joseph Frederick Wallet DesBarres."
35 "Narrative of Colonel Joseph Frederick Wallet Des Barres," Public Archives of Nova Scotia MG1/1183/1. See also J.F.W. Des Barres, *A statement submitted by Lieutenant Colonel DesBarres, for consideration: respecting his services, from the year 1755, to the present time.*
36 William G. Godfrey, "Bastide, John Henry"; Evans, *Uncommon Obdurate*, 8.
37 John Clarence Webster, ed., *Re-Capture of St. John's Newfoundland, 1762: Journal of Lieut.-Col. William Amherst.*
38 "Narrative of Colonel Joseph Frederick Wallet Des Barres."
39 Ibid.
40 Spry to Honourable Commissioners of the Navy, 4 June 1763, NA ADM 106/1130/130.
41 F.M. Montrésor, "Notes and Documents: Captain John Montrésor in Canada"; R. Arthur Bowler, "Montresor (Montrésor), John."
42 G.D. Scull, "Lt. John Montresor's Journal of an Expedition in 1760 across Maine from Quebec."
43 Ibid., 32.
44 A copy of the map is in the Library of Congress. See also Stephen J. Hornsby and Micah A. Pawling, "British Survey the Interior."
45 Hornsby, *British Atlantic, American Frontier*, 214–23.
46 Adam Shortt and Arthur G. Doughty, eds, *Documents Relating to the Constitutional History of Canada 1759–1791*, Part I, 80.
47 Quoted in Gordon Shields, "General James Murray's Map of the St. Lawrence River Valley in 1761: A Cartographic Commentary," 30.
48 The best discussion of the Survey of Canada is James Gordon Shields, "The Murray Map Cartographically Considered: A Study of General James Murray's Survey of the St. Lawrence River Valley in 1761." See also Nathaniel N. Shipton, "General James Murray's Map of the St. Lawrence."
49 Douglas William Marshall, "The British Military Engineers 1741–1783: A Study of Organization, Social Origin, and Cartography," 209.
50 For early British mapping of North America, see Cumming, *British Maps of Colonial America*, and Jeffers Lennox, "An Empire on Paper: The Founding of Halifax and Conceptions of Imperial Space, 1744–55"; for French mapping, see Marcel Trudel, *Atlas de la Nouvelle-France: An Atlas of New France*, and Laurent Vidal and Émilie d'Orgeix, eds, *Les Villes Françaises du Nouveau Monde.*
51 J.B. Harley, "The Contemporary Mapping of the American Revolutionary War"; J.B. Harley, "Specifications for Military Surveys in British North America, 1750–75."
52 Shields, "Murray Map," 102–18.
53 Murray to Amherst, 15 November 1761, quoted in Shields, "Murray Map," 73.
54 Spry to Murray, 5 October 1762, quoted in Shields, "Murray Map," 90.
55 Murray to Amherst, 3 October 1762, quoted in Shields, "Murray Map," 91.
56 Murray to Amherst, 26 September 1762, quoted in Shields, "Murray Map," 89.
57 G.D. Scull, ed., *The Montresor Journals: Collections of the New York Historical Society for the Year 1881*, 116, 127.
58 Murray to Amherst, 3 October 1762, quoted in Shields, "Murray Map," 91.
59 Scull, ed., *Montresor Journals*, 135.
60 "General Murray's Report of the State of the Government of Quebec in Canada June 5th 1762," in Shortt and Doughty, eds, *Documents*, 50.
61 Murray to Amherst, 3 October 1762, quoted in Shields, "Murray Map," 91.
62 "Memorial of Major Samuel Holland."
63 P.D.G. Thomas, *British Politics and the Stamp Act Crisis: The First Phase of the American Revolution 1763–1767*, 21–8. See also Pargellis, *Lord Loudon*, 45–6, and Thomas Pownall, *The Administration of the Colonies.*
64 Peter Barber, "George III and His Geographical Collection"; Barber, "King George III's Topographical Collection: A Georgian View of

Britain and the World"; Jane Wess, "George III, Scientific Societies, and the Changing Nature of Scientific Collecting."

65 Earl of Ilchester, ed., *Henry Fox, First Lord Holland: His Family and Relations*, 2:268.

66 Bruno Latour, *Science in Action: How to Follow Scientists and Engineers Through Society*, 215–57.

67 Thomas C. Barrow, "Background to the Grenville Program, 1757–1763."

68 Jack M. Sosin, *Whitehall and the Wilderness: The Middle West in British Colonial Policy, 1760–1775*, 46.

69 Cited in ibid., 39.

70 Verner W. Crane, "Notes and Documents: Hints Relative to the Division and Government of the Conquered and Newly Acquired Countries in America," footnote 6 on 368; R.A. Humphreys, "Lord Shelburne and the Proclamation of 1763," 246; Franklin B. Wickwire, "John Pownall and British Colonial Policy"; and Edward J. Cashin, *Governor Henry Ellis and the Transformation of British North America*, 170–94.

71 Crane, "Hints." For the other papers submitted, see Shortt and Doughty, eds, *Documents*, 127–31.

72 Crane, "Hints," 371.

73 Ibid., 372.

74 Ibid.

75 Shortt and Doughty, eds, *Documents*, 142.

76 Ibid., note 1, 155.

77 Ibid., 163–8.

78 Duff Papers, National Maritime Museum, London, DUF/9. I would like to thank Gillian Hutchinson for this reference.

79 Ware, *George Gauld*, 14.

80 Derek Howse, *Greenwich Time and the Longitude*, 74; Jim Bennett, "The Travels and Trials of Mr Harrison's Timekeeper."

81 Hornsby, *British Atlantic, American Frontier*, 27.

82 John Mannion, Gordon Handcock, and Alan Macpherson, "The Newfoundland Fishery, 18th Century"; John J. Mannion, "Settlers and Traders in Western Newfoundland."

83 Mannion, et al., "Newfoundland Fishery."

84 Keith Matthews, *Lectures on the History of Newfoundland 1500–1830*, 103.

85 Jerry Bannister, *The Rule of the Admirals: Law, Custom, and Naval Government in Newfoundland, 1699–1832*; William H. Whiteley, "Governor Hugh Palliser and the Newfoundland and Labrador Fishery, 1764–1768."

86 Beaglehole, *Captain James Cook*, 60–98; William H. Whiteley, *James Cook in Newfoundland 1762–1767*; Whiteley, "James Cook and British Policy in the Newfoundland Fisheries, 1763–67"; James Cook, *James Cook Surveyor of Newfoundland Being a Collection of Charts of the Coasts of Newfoundland and Labradore, &c. Drawn from Original Surveys Taken by James Cook and Michael Lane*.

87 John Robson, *Captain Cook's World: Maps of the Life and Voyages of James Cook R.N.*, 0.18.

88 Neither Beaglehole nor Skelton recognized this vitally important qualification.

89 Spry to Honourable Commissioners of the Navy, 4 June 1763, NA ADM 106/1130/130.

90 Instructions given by Lord Colvill to the Captains under his Command, 25 October 1763, NA ADM 1/482/308.

91 Colvill to Stephens, 22 January 1764, NA ADM 1/482/388; Admiral's journal: Colvill, 22 January 1764, NA ADM 50/4.

92 Minutes of the Board of Admiralty, 8 December 1764, NA ADM 3/72.

93 Contemporary maps showing the distribution of British troops in the 1760s and early 1770s are in John W. Shy, *Toward Lexington: The Role of the British Army in the Coming of the American Revolution*.

94 Shields, "Murray Map," 84.

95 Hornsby and Pawling, "British Survey the Interior."

96 Shields, "Murray Map," 85.

97 Conrad E. Heidenreich and Françoise Noël, "French Interior Settlements, 1750s."

98 William Smith, *An Historical Account of the Expedition against the Ohio Indians, in the Year 1764*.

99 Lloyd Arnold Brown, *Early Maps of the Ohio Valley: A Selection of Maps, Plans, and Views made by Indians and Colonials from 1673 to 1783*, plates 38, 39, and 45; Edward G. Williams, ed., *The Orderly Book of Colonel Henry Bouquet's Expedition against the Ohio Indians 1764*, 8–10.

100 Brown, *Early Maps of the Ohio Valley*, 42, 43; Beverly W. Bond, Jr., ed., *The Courses of the Ohio River taken by Lt. T. Hutchins Anno 1776 and Two Accompanying Maps.*
101 Philip Pittman, *The Present State of the European Settlements on the Mississippi.*
102 Holland to Macleane, 31 July 1767, WCL Shelburne Papers 51/452.
103 "Memorial of Major Samuel Holland."
104 In Holland to Pownall, 6 April 1764, the address is given as "Richmond House," CO 323/17.
105 "Memorial of Major Samuel Holland."
106 Office of the Ordnance to Board of Trade, 17 February 1764, NA CO 323/17/79. For the history of the Drawing Room, see Douglas W. Marshall, "Military Maps of the Eighteenth Century and the Tower of London Drawing Room."
107 Samuel Holland, "Proposals for carrying on the General Survey of the American Colonies," received and read at the Board of Trade, 16 December 1763, NA CO 323/17/58–61.
108 *Journal of the Commissioners for Trade and Plantations from January 1759 to December 1763*, 423.
109 Board of Trade to the Privy Council, NA AO 3/140.
110 James Munro and Sir Almeric W. Fitzroy, eds, *Acts of the Privy Council of England: Colonial Series*, 4:620.
111 Munro and Fitzroy, eds, *Acts of the Privy Council*, 4:619; Privy Council Order, 10 February 1764, NA ADM 1/5166.
112 Simmons and Thomas, eds, *Proceedings and Debates*, 1:470–4.
113 "Memorial of Major Samuel Holland."
114 De Vorsey, ed., *De Brahm's Report*, and De Vorsey, "La Florida Revealed."
115 Cashin, *Governor Henry Ellis*, 86–8.
116 De Vorsey, ed., *De Brahm's Report*, 33.
117 Leonard Woods Labaree, ed., *Royal Instructions to British Colonial Governors 1670–1776*, 2:537–8.
118 Instructions given by Lord Colvill to the Captains under his Command, 25 October 1763, NA ADM 1/482/308.
119 Colin G. Calloway, *The Scratch of a Pen: 1763 and the Transformation of North America.*

CHAPTER TWO

1 Matthew H. Edney, "John Mitchell's Map of North America (1755): A Study of the Use and Publication of Official Maps in Eighteenth-Century Britain."
2 Admiralty to Navy Board, 24 February 1764, NMM ADM/A/2556; Minutes of the Board of Admiralty, 12 March 1764, NA ADM 3/71.
3 Admiralty to Navy Board, 12 March 1764, NMM ADM/A/2557.
4 Minutes of the Board of Admiralty, 14 March 1764, NA ADM 3/71; Andrew J. Wahll, ed., *Henry Mowat: Voyage of the Canceaux 1764–1776*, 1.
5 James P. Baxter, "A Lost Manuscript: Services of Henry Mowat, R.N.," 357.
6 Minutes of the Board of Admiralty, 18 April 1764, NA ADM 3/71.
7 Holland to Pownall, 6 April 1764, NA CO 323/17/81.
8 Stuart R.J. Sutherland, Pierre Tousignant, and Madeleine Dionne-Tousignant, "Haldimand, Sir Frederick," 888.
9 H.T. Holman, "Wright, Thomas."
10 "Memorial of Thomas Wright Deputy Surveyor of the Northern Dist. of America to the Right Honourable the Lords Commissioners for Trade and Plantations," NA CO 323/24, pt. 2/323–4; William P. Cumming, *The Southeast in Early Maps*, 290.
11 On Wright's map of Anticosti Island, which is reproduced in *The Atlantic Neptune*, are Cape Henry and Ellis Cove near West Point, which suggests Wright was making a point of commemorating Henry Ellis.
12 Shortt & Doughty, eds, *Documents*, 172–205.
13 Wahll, ed., *Voyage of the Canceaux*, 4–11.
14 Holland to Pownall, 20 August 1764, NA CO 323/18/20–1; Wahll, ed., *Voyage of the Canceaux*, 12–19.
15 Holland to Pownall, 20 August 1764, NA CO 323/18/25; Hilda Neatby, *Quebec: The Revolutionary Age 1760–1791*, 31.
16 Holland to Pownall, 20 August 1764, NA CO 323/18/25.

17 Des Barres to Gage, 26 June 1764, WCL Gage/America/20; Des Barres to Gage, 25 November 1764, WCL Gage/America/27; Holland to Pownall, 20 November 1764, NA CO 323/18/91–2.

18 Robert J. Hayward, "Collins, John."

19 Holland to Pownall, 10 November 1766, NA CO 323/24, pt. 1; Wahll, ed., *Voyage of the Canceaux*, 13 August 1764, 28.

20 Andrew Hill Clark, *Three Centuries and the Island: A Historical Geography of Settlement and Agriculture in Prince Edward Island, Canada*, 40.

21 Wahll, ed., *Voyage of the Canceaux*, 5 October 1764, 35; Holland to Pownall, 20 November 1764, NA CO 323/18/87–8.

22 Holland to Pownall, 20 November 1764, NA CO 323/18/90.

23 Ibid., 4 March 1765, NA CO 323/18/112; Des Barres to Gage, 25 November 1764, WCL Gage/America/27.

24 Holland to Pownall, 4 March 1765, NA CO 323/18/112; Wahll, ed., *Voyage of the Canceaux*, 27 October 1764, 37.

25 Holland to Pownall, 20 November 1764, NA CO 323/18/89–90; ibid., 4 March 1765, NA CO 323/18/116.

26 Ibid., 4 March 1765, NA CO 323/18/112–13; Holland, "Observations Made on the Islands of St. John, & Cape Britain, to Ascertain the Longitude & Latitude of Those Places; Agreeable to the Orders & Instructions of the Right Honorable The Lords Commissioners for Trade & Plantations," NA CO 323/24, pt. 1.

27 Holland to Pownall, 20 November 1764, NA CO 323/18/89.

28 "Meteorological Journal of the Weather on the Island of St. Johns, at Port Joy, from the 21st of December 1764 to April the 5th 1765," NA CO 323/18/118–29; Holland to Pownall, 4 March 1765, NA CO 323/18/112–17.

29 Wahll, ed., *Voyage of the Canceaux*, 9 April 1765, 40; ibid., 10 April 1765, 41.

30 Ibid., 14 April 1765, 41.

31 Ibid., 5 May 1765, 42; 24 May 1765, 43.

32 Ibid., 24 June 1765, 47.

33 "A PLAN of the ISLAND of ST. JOHN, in the PROVINCE of NOVA SCOTIA," NA CO 700/Prince Edward Is. 3.

34 Holland to Pownall, 6 October 1765, NA CO 323/18/194–5.

35 Wahll, ed., *Voyage of the Canceaux*, 1 September 1765, 54.

36 Ibid., 8–13 October 1765, 57–8; Holland to Pownall, 24 November 1765, NA CO 323/18/243.

37 "A Description of the Island of Cape Bretain," 10 November 1766, NA CO 5/70.

38 Holland to Pownall, 24 November 1765, NA CO 323/18/246.

39 Holland to Haldimand, 20 January 1768, BL Haldimand Papers Add. ms. 21728/f.211.

40 Holland to Pownall, 24 November 1765, NA CO 323/18/244. All documents written by Holland and Pownall concerning Cape Breton are reprinted in D.C. Harvey, ed., *Holland's Description of Cape Breton Island and Other Documents*.

41 Holland to Pownall, 6 October 1765, NA CO 323/18/194; Wahll, ed., *Voyage of the Canceaux*, 17 December 1765, 64.

42 Holland to Pownall, 24 November 1765, NA CO 323/18/245; Robert Fellows, "Sproule, George."

43 Holland to Pownall, 24 November 1765, NA CO 323/18/244; Wahll, ed., *Voyage of the Canceaux*, 26 February 1766, 66.

44 Holland to Pownall, 29 July 1766, NA CO 323/24, pt. 1/64.

45 Ibid., pt. 1/64–6.

46 "A Description of the Island of Cape Bretain," NA CO 5/70.

47 Holland to Pownall, 16 August 1766, NA CO 323/24, pt. 1.

48 Wahll, ed., *Voyage of the Canceaux*, 13 and 19 August 1766, 81–2; Holland to Pownall, 29 July 1766, NA CO 323/24, pt. 1/64–6; Holland to Gage, 10 September 1766, WCL Gage/America/57.

49 Holland to Pownall, 15 April 1767, NA CO 323/24, pt. 1.

50 Harvey, *Holland's Description*, 74–5.

51 Ibid., 80, 93.

52 Holland to Pownall, 29 July 1766, NA CO 323/24, pt. 1/65.

53 Holland to Hillsborough, 10 January 1767, NA CO 5/70.

54 Holland to Haldimand, 20 January 1768, BL Haldimand Papers Add. ms. 21728/f.213.

55 Glyn Williams, *Voyages of Delusion: The Quest for the Northwest Passage*.
56 Wahll, ed., *Voyage of the Canceaux*, 26 April–30 June 1767, 89–92.
57 Ibid., 3–31 July 1767, 93–5.
58 Holland to Pownall, 2 October 1767, NA CO 323/24, pt. 1.
59 Ibid.; Holland to Haldimand, 20 January 1768, BL Haldimand Papers Add. ms. 21728/f.212.
60 Holland to Pownall, 10 November 1766, NA CO 323/24, pt. 1.
61 Ibid., 2 October 1767, NA CO 323/24, pt. 1.
62 Holland to Haldimand, 20 January 1768, BL Haldimand Papers Add. ms. 21728/f.213.
63 Scull, ed., *Montresor Journals*, 342, 386–7; W.P. Cumming, "The Montresor-Ratzer-Sauthier Sequence of Maps of New York City, 1766–76."
64 Gage to Holland, 27 July 1767, WCL Gage/America/68.
65 Wahll, ed., *Voyage of the Canceaux*, 29 November–1 December 1767, 105.
66 Ibid., 4 December, 8 December 1767, 106–7.
67 Ibid., 14 January 1768, 108.
68 Holland to Hillsborough, 20 June 1768, NA CO 323/28.
69 Wahll, ed., *Voyage of the Canceaux*, 21 March–7 June 1768, 110–18.
70 Holland to Hillsborough, 20 June 1768, NA CO 323/28.
71 Ibid., 10 November 1768, NA CO 323/28.
72 Ibid.
73 Ibid.
74 Wahll, ed., *Voyage of the Canceaux*, 3–10 February 1769, 129.
75 Ibid., 3 November 1768, 128.
76 Holland to Hillsborough, 10 November 1768, NA CO 323/28.
77 "The Petition of Samuel Holland To The Right Honorable The Lords Commissioners for Trade and Plantations, &c, &c, &c," NA CO 323/29.
78 Wahll, ed., *Voyage of the Canceaux*, 5 June–29 October 1769, 137–51.
79 Harry Woolf, *The Transits of Venus: A Study of Eighteenth-Century Science*.
80 Fred Espenak, *Transits of Venus: Six Millennium Catalog: 2000 BCE to 4000 CE*.
81 Woolf, *Transits of Venus*, 188.
82 For the expedition by Wales and Dymond to Churchill on Hudson Bay, see Rita Griffin-Short, "The Transit of Venus."
83 Samuel Holland, "Astronomical Observations Made by Samuel Holland, Esquire, Surveyor-General of Lands for the Northern District of North-America; and Others of His Party. Communicated by the Astronomer Royal."
84 Holland to Hillsborough, 10 June 1769, NA CO 5/70/479.
85 Holland, "Astronomical Observations Made by Samuel Holland, Esquire, Surveyor-General …"; Thomas Wright, "An Account of an Observation of the Transit of Venus, Made at Isle Coudre Near Quebec."
86 Maskelyne to Board of Trade, 17 November 1769, NA CO 5/227.
87 Capt. Holland to His Excellency Sir Henry Moore, 9 October 1768, BL Papers Relating to New York 1764–1768 Add. 22679/50–1; Holland to Hillsborough, 10 November 1768, NA CO 323/28/122; Frederick W. Ricord and William Nelson, eds, *Documents Relating to the Colonial History of the State of New Jersey*, 9:581–2, 588–91, 622–5, 630–6.
88 Holland to Hillsborough, 15 October 1768, NA CO 5/70/11.
89 Ibid., 24 December 1769, NA CO 5/71, pt. 1.
90 Holland to Haldimand, 8 September 1769, BL Haldimand Papers Add. ms. 21679, f.68/69.
91 Brooke Hindle, *The Pursuit of Science in Revolutionary America 1735–1789*, 176.
92 Holland to Hillsborough, 24 December 1769, NA CO 5/71, pt. 1.
93 Ibid.
94 Ibid.
95 Holland to Hillsborough, 6 June 1770, NA CO 323/27.
96 Holland to Pownall, 2 October 1767, NA CO 323/24, pt. 1.
97 Holland to Hillsborough, 6 June 1770, NA CO 323/27.

98 "An Estimate of Population for the Year 1764" in *Censuses of Canada 1665 to 1871*, 4:62.
99 Graeme Wynn, "A Province Too Much Dependent on New England"; Graeme Wynn and Debra McNabb, "Pre-Loyalist Nova Scotia."
100 Admiral's journal: Colvill, 20 May 1764, NA ADM 50/4.
101 Ibid., 6 June 1764, NA ADM 50/4.
102 Des Barres to Gage, 26 June 1764, WCL Gage/America/20; ibid., 25 November 1764, WCL Gage/America/27.
103 Des Barres to Gage, 25 November 1764, WCL Gage/America/27.
104 Colvill to Stephens, 2 June 1765, NA ADM 1/482/451.
105 Des Barres to Colvill, 26 December 1764, NA ADM 1/482/435.
106 Admiral's journal: Colvill, 19 July 1766, NA ADM 50/4; Des Barres to Colvill, 23 July 1766, NA ADM 1/482/52.
107 Hood to Stephens, 5 September 1767, NA ADM 1/483/10–12.
108 Ibid.
109 Hood to Stephens, 5 August 1768, NA ADM 1/483/122.
110 Des Barres to Haldimand, 30 May 1768, BL Haldimand Papers Add 21728/284–7; Des Barres to Stephens, 13 November 1769, LAC Des Barres Papers Series 5, 52–5.
111 Admiralty Minutes, 24 March 1770, NA ADM 3/77.
112 Des Barres to Stephens, 13 November 1769, LAC Des Barres Papers Series 5, 52–5; Anon., "Biographical Memoir of John Knight Esq. Rear-Admiral of the White Squadron"; Susanna Fisher, "Knight, Sir John."
113 Log of HM Schooner *Diligent*, 1 June 1770–21 May 1771, NA ADM 52/1700.
114 Harold A. Innis, ed., *The Diary of Simeon Perkins 1766–1780*, 40–1.
115 Des Barres to Hood, 13 August 1770, LAC Des Barres Papers Series 5, 56–69.
116 Latour, *Science in Action*, 215–57.
117 Admiral's journal: Montagu, 22 November 1771, NA ADM 50/17; Log of HM Schooner *Diligent*, 1–5 October 1771, NA ADM 52/1700.
118 Montagu to Stephens, 9 April 1772, NA ADM 1/484/82; Admiral's journal: Montagu, 10 June 1772, NA ADM 50/17.
119 Admiral's journal: Montagu, 10 October 1772, NA ADM 50/17.
120 Ibid., 9 November 1772, NA ADM 50/17.
121 Holland to Haldimand, 16 March 1773, BL Haldimand Papers Add. ms. 21730/f.21; Wahll, ed., *Voyage of the Canceaux*, 4–6 November 1772, 209–10.
122 Admiral's journal: Montagu, 1 December 1772, NA ADM 50/17.
123 Admiralty Minutes, 3 March 1773, NA ADM 3/80; Philips to Des Barres, 15 February 1773, LAC Des Barres Papers Series 5, 91; Montagu to Stephens, 14 June 1773, NA ADM 1/484/228.
124 Admiral's journal: Montagu, 6 November 1773, NA ADM 50/17; Montagu to Stephens, 7 November 1773, NA ADM 1/484/268.
125 Log of HM Schooner *Diligent*, 20 May–12 June 1773, NA ADM 52/1700/3 and NA ADM 52/1700/4. The log ends on 12 June; the succeeding log, which covers the capture of the *Diligent* at Machias in 1775, is missing.
126 Admiralty Minutes, 14 April and 3 June 1774, NA ADM 3/80.
127 Graves to Stephens, 28 July 1775, in William Bell Clark, ed., *Naval Documents of the American Revolution*, 1:997.
128 Richard Candee, *"An Old Town By the Sea": Urban Landscapes and Vernacular Building in Portsmouth, NH 1660–1990*.
129 Holland to Hillsborough, 25 August 1770, CO 323/27. More generally, see Machemer, "Headquartered at Piscataqua."
130 Holland to Des Barres, 27 May 1771, LAC Des Barres Papers Series 5, 85; Holland to Pownall, 16 September 1771, NA CO 323/27; Holland to Pownall, 4 January 1772, NA CO 323/27; "A Plan of the River Connecticut from its source to the boundary line which divides the Provinces of New Hampshire and Massachusetts Bay," NA CO 700 Connecticut 3 Connecticut River; "A Plan of Winipissiokee or Richmond Lake including the New Settlements on Smith's Lake and the New Road from thence

130 to Rochester," NA CO 700 New Hampshire 22 Winnipesaukee Lake.
131 Nathaniel Bouton, ed., *Documents and Records relating to the Province of New-Hampshire, from 1764 to 1776*, 7:264, 268.
132 Holland to Hillsborough, 19 December 1770, NA CO 5/70.
133 "A Sketch of St. John's Harbour and Part of the River," NA CO 700 New Brunswick 1 St. John River.
134 Holland to Hillsborough, 15 June 1772, NA CO 5/70.
135 Ibid., 19 December 1770, NA CO 5/70.
136 Hillsborough to Holland, 1 April 1771, NA AO 3/140.
137 Joseph Williamson, "The Proposed Province of New Ireland," 1876, 7:201–6; Williamson, "The Proposed Province of New Ireland," 1904, 1:147–57. Williamson was apparently unaware of Holland's original proposal for the new province.
138 Holland to Pownall, 16 September 1771, NA CO 323/27.
139 Labaree, ed., *Royal Instructions*, 440.
140 Paul W. Wilderson, *Governor John Wentworth & the American Revolution: The English Connection*, 181; Bouton, ed., *Documents and Records*, 293–4.
141 Holland to Pownall, 4 January 1772, NA CO 323/27.
142 Holland to Dartmouth, 10 November 1772, NA CO 323/27.
143 Holland to Pownall, 15 June 1772, NA CO 323/27.
144 Holland to Dartmouth, 18 April 1774, NA CO 5/75.
145 Julian W. Green, "A Map of New Hampshire – Spanning the Revolution."
146 Holland to Pownall, 15 June 1772, NA CO 323/27.
147 Holland to Dartmouth, 10 November 1772, NA CO 323/27.
148 Holland to Haldimand, 15 May 1773, BL Haldimand Papers Add. ms. 21730/f.47; Wright to Dartmouth, 10 October 1773, NA CO 5/228.
149 Holland to Haldimand, 2 July 1773, BL Haldimand Papers Add. ms. 21730/f.123.
150 Ibid., 15 May 1773 BL Haldimand Papers Add. ms. 21730/f.47.
151 Wahll, ed., *Voyage of the Canceaux*, 3–24 June 1773, 217–20.
152 Wentworth to the Right Honorable the Lord Commissioner of His Majesty's Treasury, 26 November 1773, Transcript of Wentworth Letter Book M#204.00 NHARM.
153 Ibid.
154 Holland to Dartmouth, 28 October 1773, NA CO 5/75.
155 Ibid.
156 Holland to Pownall, 14 April 1774, NA CO 5/75.
157 Ibid.; J.F.W. Des Barres, "A Chart of the Harbour of Boston," *Atlantic Neptune*.
158 Holland to Haldimand, 13 June 1774, BL Haldimand Papers Add. ms. 21731/f.176.
159 Admiral's journal: Montagu, 13 May 1774, NA ADM 50/17.
160 Ibid., 15 and 28 May 1774, NA ADM 50/17.
161 Bernard Bailyn, *The Ordeal of Thomas Hutchinson*, 271; Wahll, ed., *Voyage of the Canceaux*, 11 June 1774, 238.
162 Wahll, ed., *Voyage of the Canceaux*, 7 June, 19 July, and 19 August 1774, 238, 240, 242.
163 Holland to Haldimand, 13 June 1774, BL Haldimand Papers Add. ms. 21731/f.176.
164 Holland to Pownall, 20 December 1774, NA CO 323/29; Wahll, ed., *Voyage of the Canceaux*, 25 August 1774, 242.
165 Holland to Pownall, 22 August 1774, NA CO 323/29; Admiral's journal: Graves, 3 September 1774, NA ADM 50/9.
166 Graves to Stephens, 8 January 1775, in Clark, ed., *Naval Documents*, 1:60.
167 "Proceedings & Observations of Capt. Holland for Settling the 42nd. degree of North Latitude on Delaware River" enclosed in Holland to Pownall, 20 December 1774, NA CO 323/29; NA CO 323/29 Delaware River 1774.
168 Holland to Dartmouth, 27 May 1775, NA CO 5/75.
169 Ibid., 20 September 1775, NA CO 5/75.
170 Ibid.
171 Graves to Holland, 12 September 1775, in Clark, ed., *Naval Documents*, 2:81–2; Graves to Stephens, 26 September 1775, in ibid., 2:210–11.
172 Vandeput to Graves, 9 October 1775, in Clark, ed., *Naval Documents*, 2:381.

173 Graves to Vandeput, 11 November 1775, in Clark, ed., *Naval Documents*, 2:982–3.

174 Memorial of Major Samuel Holland, Surveyor General of the Province of Quebec, 22 November 1783 NA AO 13/13, f.287; Memorial of Major Samuel Holland, 29 April 1789, NA AO 13/13.

175 NA AO 3/140; "The Petition of Samuel Holland To The Right Honorable The Lords Commissioners for Trade and Plantations, &c, &c, &c," NA CO 323/29.

176 Logbook of HMS *Asia*, 15 November 1775, NMM ADM/L/A/195.

177 Holland's accounts for the survey, prepared in London, were dated 24 December 1775 (NA AO 3/140).

178 "The Petition of Samuel Holland To The Right Honorable The Lords Commissioners for Trade and Plantations, &c, &c, &c," NA CO 323/29.

179 Holland to Simcoe, 11 January 1792, quoted in Chipman, "Life and Times of Major Samuel Holland," 18.

180 For further discussion of the making of space, see John Law, "Materialities, Spatialities, Globalities."

CHAPTER THREE

1 Arthur Herbert Basye, *The Lords Commissioners of Trade and Plantations Commonly known as the Board of Trade 1748–1782*.

2 Oliver Morton Dickerson, *American Colonial Government 1696–1765*, 39–47.

3 Steven G. Greiert, "The Earl of Halifax and the Settlement of Nova Scotia, 1749–1753."

4 Bernard Bailyn, *Voyagers to the West: A Passage in the Peopling of America on the Eve of the Revolution*, 29–36.

5 *Council of Trade and Plantations 1696–1782: British History Online*. www.british-history.ac.uk.

6 Wickwire, "John Pownall and British Colonial Policy," 544; Franklin B. Wickwire, *British Subministers and Colonial America 1763–1783*, 69–72.

7 Wickwire, "John Pownall and British Colonial Policy," 547.

8 N.A.M. Rodger, *The Admiralty*.

9 David Syrett, *Admiral Lord Howe: A Biography*, 33–4.

10 Mary Blewitt, *Surveys of the Seas: A Brief History of British Hydrography*, 29.

11 Rodger, *Admiralty*, 69; see also Franklin B. Wickwire, "Admiralty Secretaries and the British Civil Service."

12 Mowat to Colvill, 14 October 1764, NA ADM 1/482/407.

13 Special Minutes of the Admiralty Board, 9 March 1765, NA ADM 3/353.

14 Holland to Pownall, November 1764, NA CO 323/18/106–11; Special Minutes of the Admiralty Board, 9 March 1765, NA ADM 3/353; Stephens to Pownall, 11 April 1765, NA CO 323/18/269; Colvill to Stephens, 30 June 1765, NA ADM 1/482/461; Admiral's journal: Colvill, 13 June 1765, NA ADM 50/4.

15 Colvill to Mowat, 9 November 1764, NA ADM 1/482/407.

16 Holland to Hillsborough, 10 January 1767, NA CO 5/70.

17 Montagu to Stephens, 22 September 1771, NA ADM 1/484/53.

18 Montagu to Sandwich, 12 June 1772, NA ADM 1/484/96.

19 Neil R. Stout, *The Royal Navy in America, 1760–1775: A Study of the Enforcement of British Colonial Policy in the Era of the American Revolution*.

20 Barrow, "Background to the Grenville Program."

21 Hood to Stephens, 28 March 1768, NA ADM 1/483; Wahll, ed., *Voyage of the Canceaux*, 22–26 October, 7 November 1767, 101–3.

22 Hood to Stephens, 30 May 1768, 15 June 1768, NA ADM 1/483.

23 Holland to Hillsborough, 20 June 1768, NA CO323/28.

24 Admiral's journal: Montagu, 2 October 1771, NA ADM 50/17.

25 Hood to Des Barres, 8 December 1773, LAC Des Barres Papers Series 5, 100.

26 Log of HM Schooner *Diligent*, 1 June 1770–21 May 1771, NA ADM 52/1700/2; Log of HM Schooner *Diligent*, 1 June 1771–31 May 1772, NA ADM

26. ...52/1700/1; Log of HM Schooner *Diligent*, 1 June 1772–31 May 1773, NA ADM 52/1700/3.
27. For discussion of the speed of communication and the integration of the Atlantic world, see Ian K. Steele, *The English Atlantic 1675–1740: An Exploration of Communication and Community*.
28. Holland to Pownall, 10 November 1766, NA CO 323/24, pt. 1.
29. Ibid.
30. Ibid.
31. *Journal of the Commissioners for Trade and Plantations from January 1764 to December 1767*, 374.
32. Holland to Pownall, 2 October 1767, NA CO 323/24, pt. 1.
33. Holland to Gage, 15 November 1767, WCL Gage/America/72.
34. Holland to Pownall, 20 December 1774, NA CO 323/29.
35. Beaglehole did not consider this context; see Beaglehole, *Life of Captain James Cook*, 126–7.
36. Holland to Pownall, 20 November 1764, NA CO 323/18/91.
37. Ibid., 10 November 1766, NA CO 323/24, pt. 1.
38. Ibid.
39. "A List of Men Employed upon the Survey with Capt. Holland in No. America" enclosed in Mowat to Montagu, 3 May 1772, NA ADM 1/484/99A.
40. David Dundas: "each surveyor was attended by a Non-Commissioned Officer and 6 Soldiers as Assistants – One carried the Instrument – Two measured with the Chain. Two for the fore, and back stations – *One* as Batman." Two soldiers with flags marked the fore and back stations; in O'Donoghue, *William Roy*, 10.
41. Holland to Pownall, 15 June 1772, NA CO 323/27; Holland to Haldimand, 2 July 1773, BL Haldimand Papers Add. ms. 21730/f.127–8; Holland to Gage, 20 May 1773, WCL Gage Papers America/118; Gage to Barrington, 4 March 1772, 2 June 1773, and Dartmouth to Gage, 3 February 1773, in Clarence Edwin Carter, ed., *The Correspondence of General Thomas Gage with the Secretaries of State, and with the War Office and Treasury 1763–1775*, 600, 644, 155.
42. "Journal, Portsmouth New Hampshire, 1773" and "Journal, Portsmouth New Hampshire, 1774," NA AO 3/140.
43. "Journal, Quebec, November 1769" and "Journal, Portsmouth, New Hampshire, 1772," NA AO 3/140.
44. Minutes of the Board of Admiralty, 11 April 1764, NA ADM 3/71.
45. Colvill to Mowat, 9 November 1764, NA ADM 1/482/407.
46. Mowat to Montagu, 3 May 1772, NA ADM 1/484/101.
47. Admiral's journal: Colvill, 12 May, 2 June 1764, NA ADM 50/4.
48. Admiral's journal: Colvill, 6 December 1764, 29 May 1766, NA ADM 50/4.
49. Des Barres to Stephens, 10 November 1769, LAC Des Barres Papers Series 5, 52.
50. "State and Condition of His Majesty's Ships and Vessels employed under the comand [sic] of Comodore [sic] Gambier," NA ADM 1/483.
51. Stout, *Royal Navy in America*, 138–9; Colvill to Stephens, 30 June 1765, NA ADM 1/482/461.
52. Colvill to Stephens, 9 September 1764, NA ADM 1/482/386.
53. Des Barres to Gambier, 9 September 1770, LAC Des Barres Papers Series 5; see also Hood to Stephens, 5 September 1767, NA ADM 1/483/10.
54. Des Barres to Hood, 13 August 1770, LAC Des Barres Papers Series 5, 56; Gambier to Stephens, 13 June 1771, NA ADM 1/484/407.
55. Montagu to Stephens, 28 June 1773, NA ADM 1/148/229.
56. For comparison, James Cook, who surveyed the west coast of Newfoundland in the mid-1760s, never had to contend with such conditions. Like other members of the Newfoundland squadron, Cook sailed for England in the fall and did not return until the following spring.
57. Hood to Stephens, 10 September 1770, NA ADM 1/483/302.
58. Hillsborough to Holland, 1 March 1769, NA CO 5/70.
59. Holland to Hillsborough, 10 June 1769, NA CO 5/70.

60 "State of Expences for the Survey of the Northern District of North America, incurred from 24th Decr. 1764 to 24th Decr. 1765," NA AO 3/140/122.
61 "State of Expences for the Survey of the Northern District of North America, incurred from the 25th December, 1774, to the 24th December, 1775, Inclusive," NA AO 3/140/19.
62 "State of Expences, for the Survey of the Northern District, incurred from the 25th Decr 1765 to 24th Decr 1766, Inclusive," NA AO 3/140/52; "State of Expences for the Survey of the Northern District of North America, incurred from 25th December, 1767 to 24th 1768, Inclusive," NA AO 3/140/111.
63 Holland to Pownall, 10 November 1766, NA CO 323/24, pt. 1.
64 N.A.M. Rodger, *Articles of War: The Statutes which Governed Our Fighting Navies 1661, 1749 and 1886*.
65 Wahll, ed., *Voyage of the Canceaux*.
66 DesBarres to Hood, 13 August 1770, LAC DesBarres Series 5, 65–6.
67 Log of HM Schooner *Diligent*, 1 June 1770–21 May 1771, NA ADM 52/1700.
68 Wahll, ed., *Voyage of the Canceaux*, passim.
69 Wahll, ed., *Voyage of the Canceaux*, 10 October 1767, 100.
70 Admiral's journal: Colvill, 16, 18, 21, and 27 May 1766, NA ADM 50/4.
71 Colvill to Stephens, 9 September 1764, NA ADM 1/482/386.
72 "Court Martial onboard the *Romney* in Halifax Harbour the 20 of November 1764," NA ADM 1/5302 pt. 2, 497.
73 "Minutes at a Court martial on the *Romney* in Halifax Harbour the 20 Nov. 1764," NA ADM 1/5302 pt. 2, 505.
74 Colvill to Stephens, 30 November 1764, NA ADM 1/482/418.
75 Admiral's journal: Colvill, 28 April 1765, NA ADM 50/4.
76 "Minutes at a Court martial on the *Romney* in Halifax Harbour the 20 Nov. 1764," NA ADM 1/5302 pt. 2, 500.
77 Ibid., 507–8.
78 Richard Sorrenson, "The Ship as a Scientific Instrument in the Eighteenth Century."
79 Withers, *Placing the Enlightenment*, 88–9.
80 Memorial of Samuel Holland to the Board of Trade, 12 March 1764, NA AO 3/140.
81 Minutes of the Board of Admiralty, 24 February 1764, NA ADM 3/71; Admiralty to Navy Board, 24 February 1764, NMM ADM/A/2556.
82 Navy Board to Stephens, 2 March 1764, NMM ADM/B/174.
83 Admiralty to Navy Board, 3 March 1764, NMM ADM/A/2557.
84 A.P. McGowan, "Captain Cook's Ships."
85 Navy Board to Stephens, 9 March 1784, NMM ADM/B/174; Minutes of the Board of Admiralty, 12 March 1764, NA ADM 3/71; Colvill to Mowat, 9 November 1764, NA ADM 1/482/407. See also David Lyon, *The Sailing Navy List: All the Ships of the Royal Navy – Built, Purchased and Captured – 1688–1860*, 212.
86 Mowat to Colvill, 14 October 1764, NA ADM 1/482/406.
87 Holland to Pownall, 24 November 1765, NA CO 323/118/244.
88 Ibid., 2 October 1767, NA CO 323/24, pt. 1.
89 Colvill to Stephens, 30 June 1765, NA ADM 1/482/461.
90 Wahll, ed., *Voyage of the Canceaux*, 26 August 1765, 53. The *Jupiter* was also referred to as a sloop.
91 Holland to Pownall, 29 July, 16 August 1766, NA CO 323/24, pt. 1; Wahll, ed., *Voyage of the Canceaux*, 13 August 1766, 81.
92 Holland to Pownall, 2 October 1767, NA CO 323/24, pt. 1.
93 Colvill to Stephens, 18 July 1764, NA ADM 1/482/356.
94 Admiral's journal: Colvill, 20, 23, and 26 June 1766, NA ADM 50/4.
95 Hood to Stephens, 5 September 1767, NA ADM 1/483/12.
96 Ibid., 5 May 1768, NA ADM 1/483/79; ibid., 5 August 1768, NA ADM 1/483/122.
97 Des Barres to Stephens, 10 November 1769, LAC Des Barres Papers Series 5; Admiralty Minutes, 24

98 Des Barres to Hood, 13 August 1770, LAC Des Barres Papers Series 5, 63–5.

99 Des Barres, "Minute for a Letter to Com[modo]re Gambier," 9 September 1770, LAC Des Barres Papers Series 5, 76–82.

100 Admiral's journal: Montagu, 14 June 1773, NA ADM 50/17; Montagu to Stephens, 14 June 1773, NA ADM 1/484/228.

101 Robson, *Captain Cook's World*, 1.20.

102 Association Salomon, *Le mystère Lapérouse ou le rêve inachevé d'un roi*.

103 Holland to Pownall, 24 November 1765, NA CO 323/18/243.

104 Wahll, ed., *Voyage of the Canceaux*, 21 June 1766, 78.

105 Holland to Pownall, 4 March 1765, NA CO 323/18/116–17.

106 For another account of coastal surveying, see Andrew David, "Vancouver's Survey Methods and Surveys."

107 Des Barres to Colvill, 27 May 1765, NA ADM 1/482/452–3.

108 Ibid.; for Holland, see "A Description of the Island of Cape Bretain," NA CO 5/70.

109 Des Barres to Colvill, 27 May 1765, NA ADM 1/482/452–3.

110 Bruno Latour, *Pandora's Hope: Essays on the Reality of Science Studies*, 24–79.

111 Des Barres to Colvill, 27 May 1765, NA ADM 1/482/452–3.

112 Ibid.

113 Des Barres, "Minute for a Letter to Com[modo]re Gambier," 9 September 1770, LAC Des Barres Papers Series 5, 76.

114 Log of HM Schooner *Diligent*, 23 August 1772, NA ADM 52/1700.

115 Holland, Captain, R. Brocklesby, and George Derbage, "Observations Made on the Islands of Saint John and Cape Briton, to Ascertain the Longitude and Latitude of Those Places, Agreeable to the Orders and Instructions of the Right Honourable the Lords Commissioners for Trade and Plantations."

116 E.G.R. Taylor, *Navigation in the Days of Captain Cook*, 2; Gloria Clifton, *Directory of British Scientific Instrument Makers 1550–1851*, 131, 253.

117 Thomas Wright and Astronomer Royal, "Immersions and Emersions of Jupiter's First Satellite, Observed at Jupiter's Inlet, on the Island of Anticosti, North America, by Mr. Thomas Wright, Deputy Surveyor General of Lands for the Northern District of America; And the Longitude of the Place, Deduced from Comparison with Observations Made at the Royal Observatory at Greenwich, by the Astronomer Royal," 190.

118 Holland, "Astronomical Observations Made by Samuel Holland …," 247. See also Samuel Holland, "A Letter to the Astronomer Royal, from Samuel Holland, Esq., Surveyor General of the Lands for the Northern District of America, Containing Some Eclipses of Jupiter's Satellites, Observed Near Quebec," 173.

119 Holland, "A Letter to the Astronomer Royal …," 173.

120 Ibid., 174.

121 Albert Van Helden, "Longitude and the Satellites of Jupiter."

122 Randall C. Brooks, "Time, Longitude Determination and Public Reliance upon Early Observatories," 165–70. Chabert's observations on latitude are listed in "Latitudes" in LAC Des Barres Papers Series 5, 30.

123 Eric G. Forbes, "The Foundation and Early Development of the Nautical Almanac"; D.H. Sadler, *Man Is Not Lost: A Record of Two Hundred Years of Astronomical Navigation with The Nautical Almanac 1767–1967*; Mary Croarken, "Tabulating the Heavens: Computing the *Nautical Almanac* in 18th-Century England."

124 "Journal, Perth Amboy, New Jersey, 1775," NA AO 3/140.

125 "General Surveys in North America, for the Northern District, thereof, in Account Current with Samuel Holland. Esq., Surveyor General of Said District," 24th December 1763 to 24th December 1764, NA AO 3/140; Holland, Brocklesby, and Derbage, "Observations Made …" Also in

125 Holland, "Observations Made on the Islands …," NA CO 323/24, pt. 1.
126 Holland, Brocklesby, and Derbage, "Observations Made …"
127 Holland, "Observations Made on the Islands …," NA CO 323/24, pt. 1.
128 David Cordingly, *Capt. James Cook Navigator*, 74.
129 David Penney, "Thomas Mudge and the Longitude: A Reason to Excel."
130 Holland to Pownall, 2 October 1767, NA CO 323/24, pt. 1.
131 Samuel Holland, "A Letter to the Astronomer Royal …," 174.
132 Brooks, "Time, Longitude Determination and Public Reliance," 170; Wright and Astronomer Royal, "Immersions and Emersions of Jupiter's First Satellite …"
133 Holland, "Astronomical Observations, made by Samuel Holland … for ascertaining the Longitude of Sundry Places in said District," 2 May 1773, NA CO 5/70.
134 Samuel Holland and Astronomer Royal, "Observations of Eclipses of Jupiter's First Satellite Made at the Royal Observatory at Greenwich, Compared with Observations of the Same, Made by Samuel Holland Esquire, Surveyor General of Lands for the Northern District of America, and Others of His Party, in Several Parts of North America, and the Longitudes of the Places Thence Deduced, by the Astronomer Royal."
135 Des Barres to Hood, 13 August 1770, LAC Des Barres Papers Series 5, 59.
136 "Goods ship'd for Mr Desbarres of Hallifax," LAC Des Barres Papers Series 5, 125; Roy L. Bishop, "J.F.W. DesBarres: An Eighteenth-Century Nova Scotia Observatory."
137 "Astronom: Observat: By Major Holland and his Assistants" LAC Des Barres Papers Series 5, 30–3.
138 Ibid.
139 Anthony Giddens, *The Consequences of Modernity*, 18–19.
140 Latour, *Science in Action*, 215–57. See also David Turnbull, "Travelling Knowledge: Narratives, Assemblage and Encounters."
141 Minutes of the Board of Admiralty, 20 April 1764, NA ADM 3/71.
142 "State of Expences, for the Survey of the Northern District of North America, incurred from the 25th December 1764 to 24th December 1765, Inclusive," NA AO 3/140/63.
143 "State of Expences, for the Survey of the Northern District, incurred from the 25th Decr. 1765 to 24th Decr. 1766, Inclusive," NA AO 3/140/52.
144 "State of Expences for the Survey of the Northern District of North America, incurred from 24th Decr. 1766 to 24th Decr. 1767 inclusive," NA AO 3/140/53.
145 "State of Expences, for the Survey of the Northern District of North America, incurred from 25th Decr. 1767 to 24th Decr. 1768 inclusive," NA AO 3/140/4.
146 "State of Expences, for the Survey of the Northern District of North America, incurred from 25th Decr. 1768 to 24th Decr. 1769 inclusive," NA AO 3/140/5.
147 "State of Expences, for the Survey of the Northern District of North America, incurred from 25th Decr. 1769 to 24th Decr. 1770 inclusive," NA AO 3/140/101, and succeeding years to 1773.
148 Gambier to Stephens, 13 June 1771, ADM 1/484/407.
149 Holland to Pownall, 24 November 1765, NA CO 323/118/244.
150 Biographical Sketch, MG100/134/22 Public Archives of Nova Scotia.
151 Latour, "Centres of Calculation," in *Science in Action*, 215–57.
152 John Montrésor, "Montresor's Journal," 345.
153 Fannie Hardy Eckstorm, *History of the Chadwick Survey*, 13.
154 "A Description of the Island of Cape Bretain," NA CO 5/70.
155 Holland, "A Description of the Island of Cape Britain Relative to the Plan surveyed Agreeable to the Orders and I[n]structions of the Right Honorable the Lords Commissioners for Trade and Plantation," NA CO 323/24, pt. 1.
156 Holland to Hillsborough, 10 November 1768, NA CO 323/28.
157 Latour, *Reassembling the Social*.

CHAPTER FOUR

1. Edney, "John Mitchell's Map."
2. Pascal Geneste, "The Archivist and the Hydrographer: The Origins of the French Hydrographic Service in the Dépôt des Cartes et Plans de la Marine, 1720–1850."
3. For all his interest in maps as text, Brian Harley failed to consider maps as part of a larger corpus of texts; see Harley, *New Nature of Maps*.
4. Samuel Holland, "Proposals for carrying on the General Survey of the American Colonies," received and read at the Board of Trade, 16 December 1763, NA CO 323/17/58–61.
5. Holland to Pownall, 20 December 1774, NA CO 323/29.
6. For example, "A Plan of the Island of Cape Bretain [Breton] Reduced from the Large Survey," Gage Papers, WCL; another copy survives in the Hydrographic Office, Taunton.
7. Des Barres to Stephens, 10 November 1769, LAC Des Barres Papers Series 5, 52.
8. Clinton Papers. Clinton Map 32, WCL.
9. Clinton Papers, Clinton Map 29, WCL.
10. Maps K. Top. 120.18 BL.
11. Hydrographic Office, z52/1.
12. Des Barres to Colvill, 27 May 1765, ADM 1/482/452–4.
13. Harley, et al., eds., *Mapping the American Revolutionary War*, 26.
14. Des Barres to Colvill, 26 December 1764, NA ADM 1/482/435.
15. Des Barres to Hood, 13 August 1770, LAC Des Barres Papers Series 5, 56–69.
16. Holland to Pownall, 15 June 1772, NA CO 323/27.
17. Inscription on manuscript map of Narragansett Bay to Hampton, NH, Number 9, Press 73, Hydrographic Office, Taunton.
18. For example, see "A Plan of the Island of Cape Britain [Breton] Reduced from the Large survey," Gage Papers, WCL.
19. Holland to Pownall, 7 October 1765, NA CO 323/18/238–40.
20. Holland to Hillsborough, 24 December 1769, NA CO 5/7, pt. 1.
21. Holland to Pownall, 16 September 1771, NA CO 323/27.
22. Dartmouth to Holland, 4 November 1772, NA AO 3/140.
23. Holland to Dartmouth, 2 May 1773, NA CO 5/228; Holland to Pownall, 14 April 1774, NA CO 5/75.
24. Holland to Pownall, 20 December 1774, NA CO 323/29.
25. Holland to Haldimand, 4 July 1775, BL Haldimand Papers Add. ms. 21731/f.220–1.
26. "Instructions to General Murray" in Shortt and Doughty, eds, *Documents*, 194.
27. Report of the Board of Trade, 23 March 1764, BL Hardwicke Papers Add. ms. 35914/f.75.
28. Clark, *Three Centuries and the Island*, 45; "Explanation Refering to the Townships" enclosed in Holland to Pownall, 6 October 1765, NA CO 323/18/226.
29. Holland to Pownall, 15 October 1768, NA CO 5/70.
30. Cited in David Nasaw, "AHR Roundtable: Historians and Biography."
31. Paul Carter, *The Road to Botany Bay: An Exploration of Landscape and History*, 1–33, quote from 9.
32. The following discussion of place names is based on Samuel Holland, "A Plan of the Island of St. John, in the Province of Nova Scotia … 19th September, 1765," NA CO 700/PEI/3, Alan Rayburn, *Geographical Names of Prince Edward Island*, and extensive research on the Internet into aristocratic family trees.
33. Holland's Report on St. John's Island, 5 October 1765, NA CO 323/18.
34. Ibid.
35. Bunbury was usually called Charles; see William James Smith, ed., *The Grenville Papers: Being the Correspondence of Richard Grenville Earl Temple, K.G., and the Right Hon: George Grenville, Their Friends and Contemporaries*, 2:186, note 1.
36. Smith, ed., *Grenville Papers*, 2:208.
37. J.M. Bumsted, *Land, Settlement, and Politics on Eighteenth-Century Prince Edward Island*, 18.
38. Smith, ed., *Grenville Papers*, 2:179.
39. Ibid., 2:222.

40 Ian R. Christie, "A Vision of Empire: Thomas Whately and the Regulations Lately Made Concerning the Colonies."
41 George Paston, *Little Memoirs of the Eighteenth Century*, 57–116.
42 Paul David Nelson, *William Tryon and the Course of Empire: A Life in British Imperial Service*, 11.
43 "A Plan of the Island of Cape Britain [Breton] Reduced from the Large Survey," Gage Papers, WCL.
44 Hydrographic Office z52/1.
45 Hydrographic Office z52/4 and "King's Bay [and] Lunenburg" in Des Barres, *The Atlantic Neptune*.
46 Harley, "Power and Legitimation in the English Geographical Atlases of the Eighteenth Century," 112.
47 Harley, "Silences and Secrecy: The Hidden Agenda of Cartography in Early Modern Europe," 99.
48 Harley, "New England Cartography and the Native Americans," 180–1.
49 "A Plan of the Island of St. John in the Province of Nova Scotia … 19th September, 1765," NA CO 700/PEI/3.
50 Rayburn, *Geographical Names*, 2. No systematic study has been made of the survival of Holland's place names on Cape Breton Island but my familiarity with the island suggests few have been preserved.
51 "Capt. Holland's Account of the Island of St. John's" and "Explanation Refering to the Townships" enclosed in Holland to Pownall, 6 October 1765, NA CO323/18/211–17, 218–27, and 228–37; Harvey, *Holland's Description*; Holland to Pownall, 20 August 1764, NA CO 323/18/20–7; Peter Frederick Haldimand, "A Description of the Magdalen Island &c" enclosed in Holland to Pownall, 6 October 1765, NA CO 323/18/196–210; Des Barres to Colvill, 23 July 1766, NA ADM 1/482/522.
52 "A Plan of the Island of St. John in the Province of Nova Scotia … 19th September, 1765," NA CO 700/PEI/3.
53 "A Plan of the Island of Cape Britain [Breton] Reduced from the Large Survey," Gage Papers, WCL.
54 Harvey, *Holland's Description*, 92.
55 "Meteorological Journal of the Weather on the Island of St. Johns; at Port Joy, from 21st of December 1764 to April the 5th 1765," NA CO 323/18/118–29; "Journal of the Weather from the 8th October 1764 to the 20th Decemr foll[owin]g before the Barometer was set up," NA CO 323/18/130–1.
56 William Wales, "Journal of a Voyage, Made by Order of the Royal Society, to Churchill River, on the North-West Coast of Hudson's Bay; Of Thirteen Months Residence in That Country; and of the Voyage Back to England; In the Years 1768 and 1769."
57 Harley, "Maps, Knowledge, and Power," 67–9.
58 Harley, "Text and Contexts in the Interpretation of Early Maps," 46.
59 Harley, "Silences and Secrecy," 104–5.
60 Holland to Pownall, 24 November 1765, NA CO 323/18.
61 "A Description of the Island of Cape Britain," NA CO 5/70.
62 For discussion of the agricultural potential of Cape Breton and the subsequent history of settlement, see Stephen J. Hornsby, *Nineteenth-Century Cape Breton: A Historical Geography*.
63 Felix Driver and Luciana Martins, "John Septimus Roe and the Art of Navigation, c.1815–1830," 158.
64 Law, "Materialities, Spatialities, Globalities"; Latour, *Science in Action*.
65 Harley, "Text and Contexts."

CHAPTER FIVE

1 For a geographic overview, see Hornsby, *British Atlantic, American Frontier*, 126–79.
2 For a general discussion of this expansion, see Bailyn, *Voyagers to the West*, 7–28; for the migration to Nova Scotia, see Wynn and McNabb, "Pre-Loyalist Nova Scotia."
3 Amherst to Egremont, 27 January 1763, quoted in Marshall, *The Making and Unmaking of Empires*, 117.
4 Shortt and Doughty, eds, *Documents*, 166.
5 *Journal of the Commissioners for Trade and Plantations from January 1764 to December 1767*, 49, 53, 69.
6 *Journal of the Commissioners for Trade and Plantations from January 1764 to December 1767*, 6.

7 *Journal of the Commissioners for Trade and Plantations from January 1764 to December 1767*, 64.
8 Des Barres to Hood, 13 August 1770, LAC Des Barres Papers Series 5, 56–9.
9 The best discussion of the early British period is in Bumsted, *Land, Settlement, and Politics*.
10 Hornsby, *British Atlantic, American Frontier*, 73–88.
11 Duncan Campbell, *History of Prince Edward Island*, 10–12. See also Bailyn, *Voyagers to the West*, 364–71.
12 Hardwicke Papers, BL 35.914.
13 Cited in Campbell, *History of Prince Edward Island*, 15–16.
14 *Journal of the Commissioners for Trade and Plantations from January 1764 to December 1767*, 277–8.
15 Holland to Pownall, 16 August 1766, NA CO 323/24 pt. 1.
16 Ibid., 10 November 1766 NA CO 323/24, pt. 1.
17 *Journal of the Commissioners for Trade and Plantations from January 1764 to December 1767*, 374.
18 Ibid., 400–6.
19 Ibid., 413.
20 Holland to Haldimand, 20 January 1768, BL Haldimand Papers Add. ms. 21728/f.210.
21 "Explanation Referring to the Townships," NA CO 323/18/213.
22 Holland to Haldimand, 20 January 1768, BL Haldimand Papers Add. ms. 21728/f.210.
23 Ibid., 16 March 1773, BL Haldimand Papers Add. ms. 21730/f.20/21.
24 Ibid., 8 September 1769, BL Haldimand Papers Add. ms. 21679/f.68/69.
25 Holland to Peter Hunter, 12 March 1786, NA AO 13/13/f.293.
26 Holland to Haldimand, 16 March 1773, BL Haldimand Papers Add. ms. 21730/f.20/21.
27 Estimate of Lands & Leases contained in the Claim filed by S. Holland, NA AO 13/13/f.306.
28 Abstract from the Acct. of Major Samuel Holland's Losses, NA AO 13/13/f.308.
29 Schedule or State of Major Samuel Holland's Losses, 22 April 1776, NA AO 13/13/f.289–90.
30 Wentworth to Holland, 20 March 1775, NHARM, Gov. John Wentworth Letter Book M 204.00, f.70.
31 Schedule or State of Major Samuel Holland's Losses, 22 April 1776, NA AO 13/13/f.290.
32 A right to land was a share in a township.
33 Schedule or State of Major Samuel Holland's Losses, 22 April 1776, NA AO 13/13/f.290; Wentworth refers to the "new townships" in his letter to Holland, 20 March 1775.
34 "Memorial of Des Barres, 1784, stating his services," LAC Des Barres Papers Series 5, 678–84; J.F.W. DesBarres, "'Mr Desbarres – Description of Nova Scotia,' ca. 1763."
35 Wynn, "A Province Too Much Dependent on New England."
36 "Memorial of Des Barres, 1784, stating his services," LAC Des Barres Papers Series 5, 678–84; John V. Duncanson, *Falmouth – A New England Township in Nova Scotia*, 24, 144; "Estimate of Estates belonging to Lieutenant Governor Desbarres Situated in the Provinces of Nova Scotia & New Brunswick," LAC Des Barres Papers Series 5.
37 "Estates and Property seized by Mr. Sparrow, in Nova Scotia for a presumed Debt of £6041," LAC Des Barres Papers Series 5. The extent of the Falmouth property is here listed at 4,195 acres, and the date of Des Barres's first acquisition of land as 1761.
38 "Estimate of Estates belonging to Lieutenant Governor Desbarres …," LAC Des Barres Papers Series 5.
39 "Memorial of Des Barres, 1784," LAC Des Barres Papers Series 5, 678–84. "Estimate of Estates belonging to Lieutenant Governor Desbarres …," LAC Des Barres Papers Series 5.
40 "Estates and Property, seized by Mr. Sparrow …," LAC Des Barres Papers Series 5.
41 Ibid.
42 Frank H. Patterson, *History of Tatamagouche, Nova Scotia*, appendix A.
43 Evans, *Uncommon Obdurate*, 27.
44 "Estimate of Estates belonging to Lieutenant Governor Desbarres …," LAC Des Barres Papers Series 5; "Information respecting the estates of Memramcook, Peticodiack, and some rights of parcels of land near Cumberland in Nova Scotia obtained on the spot by Capt. John Macdonald and

transmitted to his friend Lt. Gov. Desbarres the proprietor thereof," LAC Des Barres Papers Series 5, 60; list of estates in "The Plaintiffs Case," LAC Des Barres Papers Series 5.

45 "Estates and Property seized by Mr. Sparrow …," LAC Des Barres Papers Series 5.

46 "Memorial of Des Barres, 1784," LAC Des Barres Papers Series 5, 678–84.

47 "Return of State of Falmouth 1770," reprinted in Duncanson, *Falmouth*, 68.

48 "Information respecting the estates of Memramcook …," LAC Des Barres Papers Series 5, 37.

49 "Memorial of Des Barres, 1784," LAC Des Barres Papers Series 5, 680.

50 Ibid., 680–1.

51 For Macdonald, see Bumsted, *Land, Settlement and Politics*, 55–60, and F.L. Pigot, "MacDonald of Glenaladale, John."

52 "Information respecting the estates at Minudie or the Elysian Fields–and Macan & Napan obtained on the spot by Captain John Macdonald and transmitted to his friend Lt. Gov. Desbarres the proprietor thereof," LAC Des Barres Papers Series 5, 17.

53 "Estates and Property seized by Mr. Sparrow …," LAC Des Barres Papers Series 5; "Estimate of Estates belonging to Lieutenant Governor Desbarres …," LAC Des Barres Papers Series 5.

54 "Information respecting the estates at Minudie or the Elysian Fields–and Macan & Napan …," LAC Des Barres Papers Series 5, 54.

55 Captain John Macdonald, "Conditions of Settlement at Tatamagouche, Nova Scotia, 1795," xxvii.

56 "Estimate of Estates belonging to Lieutenant Governor Desbarres …," LAC Des Barres Papers Series 5; Des Barres – Memramcook: Stephen Millidge, 27 June 1792, Public Archives of Nova Scotia MG100/134/24.

57 "Information respecting the estates of Memramcook …," LAC Des Barres Papers Series 5, 10.

58 "Information respecting the estates at Minudie or the Elysian Fields–and Macan & Napan …," LAC Des Barres Papers Series 5, 33.

59 R. Cole Harris, "The Simplification of Europe Overseas."

60 "Information respecting the estates at Minudie or the Elysian Fields–and Macan & Napan …," LAC Des Barres Papers Series 5, 69.

61 Graeme Wynn, "Late Eighteenth Century Agriculture on the Bay of Fundy Marshlands"; Wynn, "A Province Too Much Dependent on New England."

62 Holland to Haldimand, 16 March 1773, BL Haldimand Papers Add. ms. 21730 f.21.

63 Ibid., 15 May 1773, BL Haldimand Papers Add. ms. 21730 f.47.

CHAPTER SIX

1 G.R. Crone, *Maps and Their Makers*, 147.

2 J.B. Harley, "The Bankruptcy of Thomas Jefferys: An Episode in the Economic History of Eighteenth Century Map-making."

3 Laurence Worms, "The Maturing of British Commercial Cartography: William Faden (1749–1836) and the Map Trade"; Susanna Fisher, *The Makers of the Blueback Charts: A History of Imray Laurie Norie & Wilson Ltd*, 4–5.

4 Paper used for some *Atlantic Neptune* charts carries the watermark "Smith 1802." See Edwin B. Newman and Augustus P. Loring, "Some Notes on the Paper of the *Atlantic Neptune*."

5 For discussion of the Luttrell family in parliament, see Clive Wilkinson, *The British Navy and the State in the Eighteenth Century*, 181–99.

6 Luttrell to Des Barres, 1 February 1773, LAC Des Barres Papers Series 5, 87–90.

7 N.A.M. Rodger, *The Insatiable Earl: A Life of John Montagu, 4th Earl of Sandwich*.

8 Thomas Spry to Des Barres, 7 January 1774, LAC Des Barres Papers Series 5, 606–9.

9 Hood to Des Barres, 4 September 1770, LAC Des Barres Papers Series 5, 72.

10 Julian Gwyn, *Ashore and Afloat: The British Navy and the Halifax Naval Yard before 1820*.

11 Hood to Des Barres, 8 December 1773, LAC Des Barres Papers Series 5, 100.

12 Cook, *James Cook Surveyor of Newfoundland*.

13 Admiralty Minutes, 25 May 1774, NA ADM 3/80.
14 Simmons and Thomas, *Proceedings and Debates*, 5:508–9, 520–1.
15 Stephens to Des Barres, 18 March 1775, LAC Des Barres Papers Series 5, 131.
16 Des Barres, *The Atlantic Neptune*.
17 "A Map of That Part of the Sea Coast of Nova Scotia comprehended between the Entrance to St. Margaret's Bay and Canso," circa 1770, UK Hydrographic Office z52/1.
18 Walter K. Morrison, "Cape Breton Maps in *The Atlantic Neptune*: A Holland-DesBarres Connection," 119–20. The third variant of the chart of "Port Hood" in the Henry Stevens Newton Collection at the National Maritime Museum bears an imprint date of 4 March 1776.
19 Luttrell to Des Barres, 1 February 1773, LAC Des Barres Papers Series 5, 87–90. For Luttrell's later opposition to Sandwich, see Wilkinson, *British Navy and the State*, 181–99.
20 Howe to Des Barres, 3 March 1775, LAC Des Barres Papers Series 5, 118.
21 "White Haven" in Des Barres, *Atlantic Neptune*.
22 Syrett, *Admiral Lord Howe*.
23 Blewitt, *Surveys of the Seas*, 29.
24 "Estimate of the Expense attending General Surveys of His Majesty's Dominions in North America for the Year 1777," NA CO5/78.
25 "Estimate of the Expense attending General Surveys of His Majesty's Dominions in North America for the Year 1778," NA CO5/79; "Estimate of the Expense attending General Surveys of His Majesty's Dominions in North America for the Year 1779," NA CO5/80; "Estimate of the Expense attending General Surveys of His Majesty's Dominions in North America for the Year 1780," NA CO5/81.
26 Title page of volume I of *The Atlantic Neptune*.
27 Des Barres to Sandwich, 1 May 1778, Sandwich Papers NMM SAN/F/14/3.
28 Title page to the volume of North America charts in *The Atlantic Neptune*.
29 Latour, *Science in Action*, 227.
30 "Estimate of the Expense attending General Surveys of His Majesty's Dominions in North America for the Year 1777," NA CO5/78; "Estimate of the Expense attending General Surveys of His Majesty's Dominions in North America for the Year 1778," NA CO5/79; "Estimate of the Expense attending General Surveys of His Majesty's Dominions in North America for the Year 1779," NA CO5/80; "Estimate of the Expense attending General Surveys of His Majesty's Dominions in North America for the Year 1780," NA CO5/81.
31 Mary Sponberg Pedley, *The Commerce of Cartography: Making and Marketing Maps in Eighteenth-Century France and England*, 99. Of particular importance is Ian Maxted, "The London Book Trades, 1775–1800: A Topographical Guide." Also relevant is Laurence Worms, "The Book Trade at the Royal Exchange."
32 Addresses on Spry to Des Barres, 7 January 1774, LAC Des Barres Papers Series 5, 609; and Des Barres to Howe, 17 February 1777, LAC Des Barres Papers Series 5, 156.
33 Address on Des Barres to Charles Wellbore, 3 July 1782, LAC Des Barres Papers Series 5, 333.
34 Tessa Murdoch, *The Quiet Conquest: The Huguenots 1685 to 1985*, III, 161–75.
35 Ian Maxted, "The London Book Trades 1775–1800: A Preliminary Checklist of Members. Names N–O," *Exeter Working Papers in Book History*. One of the several views of Boston in the *Neptune* was engraved by James Newton; see Donald H. Cresswell, ed., *The American Revolution in Drawings and Prints*, 179–80.
36 Ian Maxted, "The London Book Trades 1775–1800: A Preliminary Checklist of Members. Names A," *Exeter Working Papers in Book History*.
37 Ian Maxted, "The London Book Trades 1775–1800: A Preliminary Checklist of Members. Names D," *Exeter Working Papers in Book History*.
38 Ian Maxted, "The London Book Trades 1775–1800: A Preliminary Checklist of Members. Names B," *Exeter Working Papers in Book History*. Bensley lived at Bolt Court from 1785 to 1789, which suggests that this illustrated text was published between those years.
39 Tally based on imprint dates of charts in the Henry Newton Stevens Collection of *The Atlantic*

Neptune held in the National Maritime Museum, London. I would like to thank Gillian Hutchinson, Curator of the History of Cartography at the National Maritime Museum, for drawing my attention to the imprint dates.

40 R.S.J. Clarke, "The Irish Charts in 'Le Neptune Francois': Their Sources and Influence."

41 For the most detailed discussion of the complexities of the publishing history of *The Atlantic Neptune*, see Henry Newton Stevens, *Catalogue of the Henry Newton Stevens Collection of the Atlantic Neptune*.

42 *A Map of the Island of St. John in the Gulf of St. Laurence Divided into Counties and Parishes and the Lots, as granted by Government* (London: R. Sayer and J. Bennett, 6 April 1775).

43 Walter K. Morrison, "The Cartographical Revolution of 1775," 84.

44 "Nautical Chart of the South Coast of Nova Scotia from Sambro Light House to Rugged Islands," Hydrographic Office, Taunton z52/4.

45 Walter K. Morrison, "The 'Modern' Mapping of Nova Scotia," 33.

46 Hank and Jan Taft, and Curtis Rindlaub, *A Cruising Guide to the Maine Coast*, 311. I would like to thank Max Edelson for this reference.

47 Samuel Eliot Morison, *The Story of Mount Desert Island*, 24.

48 Wahll, ed., *Voyage of the Canceaux*, 219.

49 Taft, Taft, and Rindlaub, *Cruising Guide*, 311.

50 "A Plan of the Sea Coast from Cape Elizabeth … to St. John's River," BL K. Top 120.18.

51 Augustus P. Loring, "The Atlantic Neptune."

52 For examples of their work, see Conrad Graham, *Mont Royal – Ville Marie: Vues et plans anciens de Montréal – Early Plans and Views of Montreal*, 42–4.

53 British Library, *The American War of Independence 1775–1783*, 62; Cresswell, ed., *The American Revolution in Drawings and Prints*, 178–9.

54 For military topographic drawing, see Bruce Robertson, "Venit, Vidit, Depinxit: The Military Artist in America."

55 For a general discussion of naval drawing, see John O. Sands, "The Sailor's Perspective: British Naval Topographic Artists."

56 Richard L. Raymond, *J.F.W. DesBarres: Views and Profiles*

57 Hornsby, *British Atlantic, American Frontier*, 216–17; John E. Crowley, "A Visual Empire: Seeing the British Atlantic World from a Global British Perspective"; Crowley, "'Taken on the Spot': The Visual Appropriation of New France for the Global British Landscape."

58 Holland's role in providing the surveys that Des Barres published is frequently overlooked; see, for example, Joseph G. Garver, *Surveying the Shore: Historic Maps of Coastal Massachusetts 1600–1930*, 26–7.

59 Derek Howse and Michael Sanderson, *The Sea Chart*, 104–5.

60 Cook, *James Cook Surveyor of Newfoundland*.

61 David, ed., *Voyage of the Endeavour*; David, ed., *The Charts and Coastal Views of Captain Cook's Voyages, Volume Two: The Voyage of the Resolution and Adventure 1772–1775*.

62 Graves to Stephens, 4 December 1775, in Clark, ed., *Naval Documents*, 2:1266.

63 Stephens to Graves, 7 September 1775, in Clark, ed., *Naval Documents*, 2:708.

64 Stephens to Des Barres, 25 October 1775, LAC Des Barres Papers Series 5, 135.

65 Des Barres to Howe, 14 September 1776, LAC Des Barres Papers Series 5, 145.

66 Ibid., 17 February 1777 LAC Des Barres Papers Series 5, 155–6.

67 Des Barres to Gambier, 4 November 1777, LAC Des Barres Papers Series 5, 174.

68 William Knox to Des Barres, 19 December 1776, LAC Des Barres Papers Series 5, 149. For Knox, see Jack P. Greene, "The Deeper Roots of Colonial Discontent: William Knox's Structural Explanation for the American Revolution."

69 Gage to Holland, 27 July 1767, WCL Gage/America/68.

70 Holland to Haldimand, 31 January 1778, BL Haldimand Papers Add. ms. 21732/f.10-11.

71 David Syrett, "Home Waters or America? The Dilemma of British Naval Strategy in 1778."

72 DesBarres to Sandwich, 1 May 1778, Sandwich Papers NMM SAN/F/14/3.

73 Des Barres to Stephens, 8 May 1779, LAC Des Barres Papers Series 5, 182.
74 Ibid., 20 July 1779, LAC Des Barres Papers Series 5, 183–5.
75 Ibid., 24 July 1779 LAC Des Barres Papers Series 5, 187–9.
76 Stephens to Des Barres, 3 July 1781, LAC Des Barres Papers Series 5, 300–1.
77 Des Barres to Stephens, 9 October 1784, LAC Des Barres Papers Series 5, 362–4.
78 W.H. Lyttleton to Des Barres, 20 December 1775, LAC Des Barres Papers Series 5, 137–8.
79 Lady William Campbell to Des Barres, 7 February 1777, LAC Des Barres Papers Series 5, 151–2.
80 W. Eden to Des Barres, 30 April 1780, LAC Des Barres Papers Series 5, 201.
81 Correspondence to Des Barres, LAC Des Barres Papers Series 5.
82 Pedley, *Commerce of Cartography*, 119–55, particularly 131.
83 Pedley, *Commerce of Cartography*, 144.
84 Reproduced in Barbara Backus McCorkle, *New England in Early Printed Maps 1513 to 1800: An Illustrated Carto-Bibliography*, 289. For Laurie and Whittle, see Fisher, *Makers of the Blueback Charts*, 59.
85 Two manuscript charts with soundings by Lieutenant Knight are in the Library of Congress, "Part of Long Island; and the coast eastward to the shoals of Nantucket," circa 1778, and "Buzzards Bay and shoals of Nantucket," 1778.
86 For example, "Steel's New and Correct Chart of North America," 3rd edn. (London: Steel, 1813).
87 For the need to study circulation of knowledge, see James A. Secord, "Knowledge in Transit."
88 A.R.T. Jonkers, "Parallel Meridians: Diffusion and Change in Early-Modern Oceanic Reckoning," note 66.
89 Morrison, "Cartographical Revolution," and "'Modern' Mapping of Nova Scotia."
90 David Philip Miller, "Joseph Banks, Empire, and 'Centers of Calculation' in Late Hanoverian London," 33.
91 Robert Darnton, "What is the History of Books?" 67.

CHAPTER SEVEN

1 Imperial circuits and networks are explored in Lambert and Lester, *Colonial Lives across the British Empire: Imperial Careering in the Long Nineteenth Century*, 1–31.
2 James S. Leamon, *Revolution Downeast: The War for American Independence in Maine*, 67–69; John Howard Ahlin, *Maine Rubicon: Downeast Settlers during the American Revolution*, 11–26.
3 Knight to Graves, 10 August 1775, in Clark, ed., *Naval Documents*, 1:1108; Ahlin, *Maine Rubicon*, 23–4; Stephen Jones, "Historical Account of Machias, Maine," Maine Historical Society. Unfortunately, the log of the *Diligent* is missing from 1 June 1773, no doubt because it was taken by the rebels in 1775, and so the exact series of events preceding the capture can not be reconstructed.
4 James Warren to Samuel Adams, 4 August 1775, in Clark, ed., *Naval Documents*, 1:1059.
5 Graves to Stephens, 17 August 1775, in Clark, ed., *Naval Documents*, 1:1166.
6 James Lyon to James Otis, 19 January 1776, in Clark, ed., *Naval Documents*, 3:854.
7 Fisher, "Knight, Sir John," 910. This useful biography is marred by confusion over the dates of the Machias affair.
8 See John Knight, "Buzzards Bay and shoals of Nantucket" (1778) and "Part of Long Island; and the coast eastward to the shoals of Nantucket" (1778?), Library of Congress. These maps were originally part of Admiral Howe's map collection. See also "A Chart of New York Harbour with the soundings views of land marks and nautical directions for the use of pilotage composed from surveys and observations of Lieutenants John Knight[,] John Hunter of the Navy & others" published by Des Barres, 19 May 1779. For the French threat, see Piers Mackesy, *The War for America 1775–1783*, 216–19.

9 Fisher, "Knight, Sir John"; *Naval Chronicle* 11 (1804): 425–31, Robert Gardiner, ed., *Navies and the American Revolution 1775–1783*, 77–8, 84–7.

10 N.A.M. Rodger, *The Wooden World: An Anatomy of the Georgian Navy*, 275–94.

11 St. Vincent cited in M.K. Barritt, *Eyes of the Admiralty: J.T. Serres An Artist in the Channel Fleet*, 45.

12 Graves to Mowat, 6 October 1775, in Clark, ed., *Naval Documents*, 2:324–6.

13 Wahll, ed., *Voyage of the Canceaux*, 8 April–16 May 1775, 260–7.

14 Mowat to the people of Falmouth, 16 October 1775, in Clark, ed., *Naval Documents*, 2:471.

15 Wahll, ed., *Voyage of the Canceaux*, 21 October 1775, 320.

16 Baxter, "A Lost Manuscript," 367.

17 Mowat to Sandwich, 20 January 1778, NMM Sandwich Papers F/12/46.

18 For an eyewitness account, see John Calef, *The Siege of Penobscot*.

19 Baxter, "A Lost Manuscript," 368.

20 Charles E. Banks, "The Burial Place of Captain Henry Mowat."

21 *Canceaux* Ship's Muster, 30 August 1771, NA ADM 36/8516.

22 John Bassett Moore, ed., *International Adjudications Ancient and Modern: History and Documents*, 452.

23 NA CO 700 New Brunswick 11 Passamaquoddy Bay 1772; T. Hurd, "Bay of Fundy, Campobello Island to Beaver Harbour," HO 284/1; "Bay of Fundy, Mace[s] Bay to St. John," HO 284/2; "Coast westward of Penobscot Bay," HO 289; and "Coast of North America from Penobscot to St. Johns," HO E468.

24 *Canceaux* Ship's Muster, 6 July 1774, NA ADM 36/8518.

25 Andrew David, "Hurd, Thomas Hannaford."

26 Chipman, "Life and Times of Major Samuel Holland," 78–9.

27 J.B. Harley, "The Spread of Cartographical Ideas between the Revolutionary Armies," 71–3.

28 Cumming, "The Montresor-Ratzer-Sauthier Sequence," 59. Holland's map of New York City is reproduced in Paul E. Cohen and Robert T. Augustyn, *Manhattan in Maps 1527–1995*, 82–3. The British Headquarters Map, depicting Manhattan Island at a large-scale, and reproduced in Cohen and Augustyn, may also have been by Holland. See also "A plan of Fort Montgomery & Fort Clinton" published by Des Barres 1 January 1779.

29 Holland to Haldimand, 31 January 1778, BL Haldimand Papers Add. ms. 21732/f10–11.

30 Ibid., 19 October 1778 BL Haldimand Papers Add. ms. 21732/f137.

31 The official correspondence between Haldimand and Holland concerning this trip is reproduced in Chipman, "Life and Times of Major Samuel Holland," 45–52; see also E.A. Cruikshank, ed., *The Settlement of the United Empire Loyalists on the Upper St. Lawrence and Bay of Quinte in 1784: A Documentary Record*.

32 Louis Gentilcore and Kate Donkin, *Land Surveys of Southern Ontario: An Introduction and Index to the Field Notebooks of the Ontario Land Surveyors 1784–1859*. The townships are shown on R. Louis Gentilcore, Don Measner, and David Doherty, "The Coming of the Loyalists." For dates of the original township surveys, see W.G. Dean, ed., *Economic Atlas of Ontario*, plate 99.

33 E.A. Cruikshank, ed., *The Correspondence of Lieut. Governor John Graves Simcoe*, 3:191.

34 E.A. Cruikshank, ed., *The Correspondence of Lieut. Governor John Graves Simcoe*, 1:347, 397.

35 Reproduced in Derek Hayes, *Historical Atlas of Canada*, 131.

36 Reproduced in Raymond Litalien, Jean-François Palomino, and Denis Vaugeois, *Mapping a Continent: Historical Atlas of North America, 1492–1814*, 242.

37 Clark, *Three Centuries and the Island*, 42–8.

38 Harvey, *Holland's Description*.

39 "A Plan of the Island of Cape Breton …," Gage Papers, WCL.

40 D.N. Jeans, "Territorial Divisions and the Locations of Towns in New South Wales, 1826–1842," 244; Jeans, "The Impress of Central Authority upon the Landscape: Southeastern Australia 1788–1850," 5–7.

41 Cruikshank, *Correspondence of Lieut. Governor John Graves Simcoe*, 1:146–7.

42 Holland to John Graves Simcoe, 10 June 1792, reproduced in Doughty, "Notes and Documents," 46.
43 *Canceaux* Ship's Muster, NA ADM 36/8518.
44 Fellows, "Sproule, George."
45 Robert Fellows, "The Loyalists and Land Settlement in New Brunswick, 1783–1790"; Ronald Rees, *Land of the Loyalists: Their Struggle to Shape the Maritimes*.
46 Fellows, "Sproule," 774.
47 Holman, "Wright, Thomas."
48 David Demeritt, "Representing the 'True' St. Croix: Knowledge and Power in the Partition of the Northeast."
49 Morgan, "Des Barres, Joseph Frederick Wallet"; Evans, *Uncommon Obdurate*; Debard, "The Family Origins of Joseph Frederick Wallet DesBarres."
50 William Dyott in Brian Tennyson, ed., *Impressions of Cape Breton*, 58.
51 Christopher Terrell, "A Sequel to *The Atlantic Neptune* of J.F.W. DesBarres: The Story of the Copperplates."
52 Sir Archibald Day, *The Admiralty Hydrographic Service 1795–1919*.
53 For an attractive account of the Vancouver expedition, see Robin Fisher, *Vancouver's Voyage: Charting the Northwest Coast, 1791–1795*; for a more critical account, see Daniel W. Clayton, *Islands of Truth: The Imperial Fashioning of Vancouver Island*, and "On the Colonial Genealogy of George Vancouver's Chart of the North-West Coast of North America."

BIBLIOGRAPHY

GUIDES TO SOURCES

Brun, Christian, compiler. *Guide to the Manuscript Maps in the William L. Clements Library.* Ann Arbor: University of Michigan, 1959.

Cock, Randolph, and N.A.M. Rodger, eds. *A Guide to the Naval Records in the National Archives of the UK.* London: Institute of Historical Research, 2006.

David, A.C.F. *Supplement to Geographical Index: North & South America, West Indies & Polar Regions for Items relating to Canadian and adjacent waters recorded up to 1850 and held in the Hydrographic Office, Taunton.* Taunton: UK Hydrographic Office, 1993.

Harley, John Brian, and Minda C. Phillips. *Manuscript Maps Relating to North America and the West Indies*, Part I: *The Revolutionary Era in the Public Record Office London.* East Ardsley: EP Microform, 1974.

Peckham, Howard H., compiler. *Guide to the Manuscript Collections in the William L. Clements Library.* Ann Arbor: University of Michigan Press, 1942.

Stevens, Henry Newton. *Catalogue of the Henry Newton Stevens Collection of the Atlantic Neptune.* London: Henry Stevens, Son & Stiles, 1937. Typescript at National Maritime Museum, Greenwich, London.

MANUSCRIPT SOURCES

British Library (BL)
    Haldimand Papers
    Hardwicke Papers

National Archives UK (NA)
    Admiralty Papers
    Audit Office Papers
    Colonial Office Papers

National Maritime Museum, London (NMM)
    Duff Papers
    Navy Board Papers
    Sandwich Papers
    Henry Newton Stevens Collection

UK Hydrographic Office (HO)
    Des Barres charts
    Holland charts
    Hurd charts

West Sussex County Record Office (WSCRO)
    Goodwood Papers

William L. Clements Library, University of Michigan (WCL)
    Gage Papers
    Shelburne Papers

Huntington Library, San Marino, California (HL)
  Abercromby Papers
  Loudoun Papers

Library of Congress, Geography and Map Division (LC)
  Knight charts

Maine Historical Society (MHS)
  Stephen Jones, "Historical Account of Machias, Maine."

New Hampshire Archives and Records Management (NHARM)
  Governor John Wentworth Letter Book (transcript)

Library and Archives Canada (LAC)
  Des Barres Papers

Public Archives and Records Office of Prince Edward Island (PAROPEI)
  Thomas Wright Field Book

PRINTED PRIMARY SOURCES

Anon. "Biographical Memoir of John Knight Esq. Rear-Admiral of the White Squadron." *Naval Chronicle* 11 (1804): 425–31.

Baxter, James P. "A Lost Manuscript: Services of Henry Mowat, R.N." *Collections and Proceedings of the Maine Historical Society*, Second Series, vol. II (1891): 345–75.

Bond Jr., Beverly W., ed. *The Courses of the Ohio River taken by Lt. T. Hutchins Anno 1776 and Two Accompanying Maps.* Cincinnati, OH: Historical and Philosophical Society of Ohio, 1942.

Bouton, Nathaniel, ed. *Documents and Records Relating to the Province of New-Hampshire, from 1764 to 1776*, vol. VII. Nashua, NH: Orren C. Moore, 1873.

Brown, Lloyd Arnold. *Early Maps of the Ohio Valley: A Selection of Maps, Plans, and Views made by Indians and Colonials from 1673 to 1783.* Pittsburgh: University of Pittsburgh Press, 1959.

Calef, John. *The Siege of Penobscot.* New York: New York Times, 1971, originally published in London, 1781.

Carter, Clarence Edwin, ed. *The Correspondence of General Thomas Gage with the Secretaries of State, and with the War Office and Treasury 1763–1775*, vol. II. New Haven: Yale University Press, 1933.

Censuses of Canada 1665 to 1871, vol. IV. Ottawa: Printed by I.B. Taylor, 1876.

Clark, William Bell, ed. *Naval Documents of the American Revolution*, vols. I–III. Washington, DC: Naval History Division, 1964–1968.

Cook, James. *James Cook Surveyor of Newfoundland Being a Collection of Charts of the Coasts of Newfoundland and Labradore, &c. Drawn from Original Surveys Taken by James Cook and Michael Lane* (London: Thomas Jefferys, 1769–1770). Reproduced in Facsimile from the Copy in the Library of the University of California at Los Angeles with an Introductory Essay by R.A. Skelton. San Francisco: David Magee, 1965.

Council of Trade and Plantations 1696–1782: British History Online. www.british-history.ac.uk.

Crane, Verner W. "Notes and Documents: Hints Relative to the Division and Government of the Conquered and Newly Acquired Countries in America." *Mississippi Valley Historical Review* 8, 4 (1922): 367–73.

Cresswell, Donald H., compiler. *The American Revolution in Drawings and Prints.* Washington, DC: Library of Congress, 1975.

Cruikshank, E.A., ed. *The Correspondence of Lieut. Governor John Graves Simcoe*, vol. I, *1789–1793.* Toronto: Ontario Historical Society, 1923.

– *The Correspondence of Lieut. Governor John Graves Simcoe*, vol. III, *1794–1795.* Toronto: Ontario Historical Society, 1925.

– *The Settlement of the United Empire Loyalists on the Upper St. Lawrence and Bay of Quinte in 1784: A Documentary Record.* Toronto: Ontario Historical Society, 1934.

David, Andrew, ed. *The Charts and Coastal Views of Captain Cook's Voyages, Volume One: The Voyage of the Endeavour 1768–1771.* London: Hakluyt Society, 1988.

- *The Charts and Coastal Views of Captain Cook's Voyages, Volume Two: The Voyage of the Resolution and Adventure 1772–1775.* London: Hakluyt Society, 1992.
- *The Charts and Coastal Views of Captain Cook's Voyages, Volume Three: The Voyage of the Resolution and Discovery 1776–1780.* London: Hakluyt Society, 1997.

Des Barres, Joseph F.W. *A statement submitted by Lieutenant Colonel DesBarres, for consideration: respecting his services, from the year 1755, to the present time: in the capacity of an officer and engineer during the war of 1756: the utility of his surveys and publications of the coasts and harbours of North America, intituled, The Atlantic neptune: and his proceedings and conduct as lieutenant governor and commander in chief of His Majesty's colony of Cape Breton.* London: 1796?

- *Charts of the Coast and Harbours in the Gulph & River of St. Lawrence.* 1778.
- *Charts of the Coast and Harbours of New England.*
- *Charts of several Harbours, and divers parts of the Coast of North America, from New York south Westwards to the Gulph of Mexico.* 1778.
- *The Atlantic Neptune, Published For the Use of the Royal Navy of Great Britain*, vol. I, 1777.
- "'Mr Desbarres – Description of Nova Scotia,' ca. 1763." *Nova Scotia Historical Review* 6, 2 (1986): 105–19.

De Vorsey Jr., Louis, ed. *De Brahm's Report of the General Survey in the Southern District of North America.* Columbia, SC: University of South Carolina Press, 1971.

Doughty, A.G. "Notes and Documents: A New Account of the Death of Wolfe." *Canadian Historical Review* 4, 1 (1923): 45–55.

Eckstorm, Fannie Hardy. *History of the Chadwick Survey.* 1926.

Harvey, D.C., compiler. *Holland's Description of Cape Breton Island and Other Documents.* Halifax, NS: Public Archives of Nova Scotia, 1935.

Holland, Captain, R. Brocklesby, and George Derbage. "Observations Made on the Islands of Saint John and Cape Briton, to Ascertain the Longitude and Latitude of Those Places, Agreeable to the Orders and Instructions of the Right Honourable the Lords Commissioners for Trade and Plantations." *Philosophical Transactions* 58 (1768): 46–53.

Holland, Samuel. "A Letter to the Astronomer Royal, from Samuel Holland, Esq. Surveyor General of Lands for the Northern District of America, Containing Some Eclipses of Jupiter's Satellites, Observed Near Quebec." *Philosophical Transactions* 64 (1774): 171–6.

- "Astronomical Observations Made by Samuel Holland Esquire, His Majesty's Surveyor General of Lands for the Northern District of North America, for Ascertaining the Longitude of Several Places in the Said District." *Philosophical Transactions* 64 (1774): 182–3.
- "Astronomical Observations Made by Samuel Holland, Esquire, Surveyor-General of Lands for the Northern District of North-America; and Others of His Party. Communicated by the Astronomer Royal." *Philosophical Transactions* 59 (1769): 247–52.
- and Astronomer Royal. "Observations of Eclipses of Jupiter's First Satellite Made at the Royal Observatory at Greenwich, Compared with Observations of the Same, Made by Samuel Holland Esquire, Surveyor General of Lands for the Northern District of America, and Others of His Party, in Several Parts of North America, and the Longitudes of the Places Thence Deduced, by the Astronomer Royal." *Philosophical Transactions* 64 (1774): 184–9.

Ilchester, Earl of, ed. *Henry Fox, First Lord Holland His Family and Relations*, vol. II. New York: Charles Scribner's Sons, 1920.

Innis, Harold A., ed. *The Diary of Simeon Perkins 1766–1780.* Toronto: Champlain Society, 1948.

Journal of the Commissioners for Trade and Plantations from January 1759 to December 1763. London: H.M.S.O., 1935.

Journal of the Commissioners for Trade and Plantations from January 1764 to December 1767. London: H.M.S.O., 1936.

Journal of the Commissioners for Trade and Plantations from January 1767 to December 1775. London: H.M.S.O., 1937.

Knox, John. *An Historical Journal of the Campaigns in North America for the Years 1757, 1758, 1759, and 1760*, vol. I. Toronto: Champlain Society, 1914.

Labaree, Leonard Woods, ed. *Royal Instructions to British Colonial Governors 1670–1776*, vol. II. New York: D. Appleton-Century, 1935.

Macdonald, Captain John. "Conditions of Settlement at Tatamagouche, Nova Scotia, 1795." *Report of the Department of Public Archives for the Year 1945*. Ottawa: 1946, xxvii–xliii.

McCorkle, Barbara Backus. *New England in Early Printed Maps 1513 to 1800: An Illustrated Carto-Bibliography*. Providence, RI: John Carter Brown Library, 2001.

Montrésor, John. "Montresor's Journal." *Collections of the Maine Historical Society* 1 (1831): 341–57.

Moore, John Bassett, ed. *International Adjudications Ancient and Modern: History and Documents*. Modern Series, vol. I. New York: Oxford University Press, 1929.

Munro, James, and Sir Almeric W. Fitzroy, eds. *Acts of the Privy Council of England: Colonial Series*, vol. IV, *A.D. 1745–1766*. Buffalo, NY: William S. Hein, 2004; originally published in London: H.M.S.O., 1911.

Pargellis, Stanley McCrory, ed. *Military Affairs in North America 1748–1765: Selected Documents from the Cumberland Papers in Windsor Castle*. Archon Books, 1969.

Pittman, Philip. *The Present State of the European Settlements on the Mississippi*. Gainesville: University of Florida Press, 1973; originally published 1770.

Pownall, Thomas. *The Administration of the Colonies*. London: Printed for J. Wilkie, 1764.

Ricord, Frederick W., and William Nelson, eds. *Documents Relating to the Colonial History of the State of New Jersey*, Vol. IX, *1757–1767*. Newark, NJ: Daily Advertiser Printing House, 1885.

Scull, G.D., ed. *The Montresor Journals: Collections of the New York Historical Society for the Year 1881*. New York: New York Historical Society, 1882.

– "Lt. John Montresor's Journal of an Expedition in 1760 across Maine from Quebec." *New England Historical and Genealogical Register* 36 (January 1882): 29–36.

Shortt, Adam, and Arthur G. Doughty, eds. *Documents Relating to the Constitutional History of Canada, 1759–1791*. Ottawa: J. De L. Taché, 1918.

Simmons, R.C., and P.D.G. Thomas, eds. *Proceedings and Debates of the British Parliaments Respecting North America 1754–1783*. Millwood, NY: Kraus International Publications, 1982–86.

Smith, William. *An Historical Account of the Expedition against the Ohio Indians, in the Year 1764*. Philadelphia: W. Bradford, 1765.

Smith, William James, ed. *The Grenville Papers: Being the Correspondence of Richard Grenville Earl Temple, K.G., and the Right Hon. George Grenville, Their Friends and Contemporaries*, vol. II. London: John Murray, 1852.

Tennyson, Brian, ed. *Impressions of Cape Breton*. Sydney, NS: University College of Cape Breton Press, 1986.

Wahll, Andrew J., transcriber and annotator. *Henry Mowat: Voyage of the Canceaux 1764–1776*. Bowie, MD: Heritage Books, 2003.

Wales, William. "Journal of a Voyage, Made by Order of the Royal Society, to Churchill River, on the North-West Coast of Hudson's Bay; Of Thirteen Months Residence in That Country; and of the Voyage Back to England; In the Years 1768 and 1769." *Philosophical Transactions* 60 (1770): 100–6.

Webster, John Clarence, ed. *Re-Capture of St. John's Newfoundland, 1762: Journal of Lieut.-Col. William Amherst*. Self-published, 1928.

– *Journal of William Amherst in America 1758–1760*. London: Butler & Tanner, 1927.

Williams, Edward G., ed. *The Orderly Book of Colonel Henry Bouquet's Expedition against the Ohio Indians 1764*. Privately printed, 1960.

Willson, Beckles. *The Life and Letters of James Wolfe*. London: William Heinemann, 1909.

Wright, Thomas. "An Account of an Observation of the Transit of Venus, Made at Isle Coudre Near Quebec." *Philosophical Transactions* 59 (1769): 273–80.

– and Astronomer Royal. "Immersions and Emersions of Jupiter's First Satellite, Observed at Jupiter's Inlet, on the Island of Anticosti, North America, by Mr. Thomas Wright, Deputy Surveyor General of Lands for the Northern District of America; And the Longitude of the Place, Deduced from Comparison with Observations Made at the Royal Observatory at Greenwich, by the Astronomer Royal." *Philosophical Transactions* 64 (1774): 190–3.

SECONDARY SOURCES

Ahlin, John Howard. *Maine Rubicon: Downeast Settlers during the American Revolution.* Calais, ME: Calais Advertiser Press, 1966.

Anderson, Fred. *Crucible of War: The Seven Years' War and the Fate of Empire in British North America, 1754–1766.* New York: Alfred A. Knopf, 2000.

Andrews, J.H. "Introduction: Meaning, Knowledge, and Power in the Map Philosophy of J.B. Harley." In *The New Nature of Maps: Essays in the History of Cartography*, by J.B. Harley, 1–32. Baltimore: Johns Hopkins University Press, 2001.

Association Salomon. *Le mystère Lapérouse ou le rêve inachevé d'un roi.* Édition de Conti, 2008.

Bailyn, Bernard. *Voyagers to the West: A Passage in the Peopling of America on the Eve of the Revolution.* New York: Alfred A. Knopf, 1986.

– *The Ordeal of Thomas Hutchinson.* Cambridge, MA: Harvard University Press, 1974.

Baker, William Avery. "Vessel Types of Colonial Massachusetts." *Publications of the Colonial Society of Massachusetts* 52 (1980): 3–29.

Ballantyne, Tony. "Empire, Knowledge and Culture: From Proto-Globalization to Modern Globalization." In *Globalization in World History*, edited by A.G. Hopkins, 115–40. London: Pimlico, 2002.

Banks, Charles E. "The Burial Place of Captain Henry Mowat." *Collections and Proceedings of the Maine Historical Society*, Second Series, IX (1898): 308–11.

Bannister, Jerry. *The Rule of the Admirals: Law, Custom, and Naval Government in Newfoundland, 1699–1832.* Toronto: University of Toronto Press, 2003.

Barber, Peter. "George III and His Geographical Collection." In *The Wisdom of George the Third*, edited by Jonathan Marsden, 263–89. London: Royal Collection Enterprises, 2005.

– "King George III's Topographical Collection: A Georgian View of Britain and the World." In *Enlightenment: Discovering the World in the Eighteenth Century*, edited by Kim Sloan, 158–65. London: British Museum Press, 2003.

Barritt, M.K. *Eyes of the Admiralty: J.T. Serres An Artist in the Channel Fleet.* London: National Maritime Museum, 2008.

Barrow, Thomas C. "Background to the Grenville Program, 1757–1763." *William and Mary Quarterly* 22, 1 (1965): 93–104.

Basye, Arthur Herbert. *The Lords Commissioners of Trade and Plantations Commonly Known as the Board of Trade 1748–1782.* New Haven: Yale University Press, 1925.

Beaglehole, J.C. *The Life of Captain James Cook.* Stanford: Stanford University Press, 1974.

Belyea, Barbara. "Images of Power: Derrida/Foucault/Harley." *Cartographica* 29, 2 (1992): 1–9.

Bennet, Jim. "The Travels and Trials of Mr Harrison's Timekeeper." In *Instruments, Travel and Science: Itineraries of Precision from the Seventeenth to the Twentieth century*, edited by Marie-Noëlle Bourguet, Christian Licoppe, and H. Otto Sibum, 75–95. London: Routledge, 2002.

Bishop, Roy L. "J.F.W. DesBarres: An Eighteenth-Century Nova Scotia Observatory." In *Profiles of Science and Society in the Maritimes Prior to 1914*, edited by Paul A. Bogaard, 65–81. Fredericton, NB: Acadiensis Press, 1990.

Blewitt, Mary. *Surveys of the Seas: A Brief History of British Hydrography.* London: MacGibbon & Kee, 1957.

Bolster, W. Jeffrey. "Putting the Ocean in Atlantic History: Maritime Communities and Marine

Ecology in the Northwest Atlantic, 1500–1800." *American Historical Review* 113, 1 (2008): 19–47.

Bowler, R. Arthur. "Montresor (Montrésor), John." In *Dictionary of Canadian Biography*, vol. IV, *1771 to 1800*, edited by Francess G. Halpenny, 552–3. Toronto: University of Toronto Press, 1979.

British Library. *The American War of Independence 1775–1783*. London: British Library, 1975.

Brooks, Randall C. "Time, Longitude Determination and Public Reliance upon Early Observatories." In *Profiles of Science and Society in the Maritimes Prior to 1914*, edited by Paul A. Bogaard, 163–92. Fredericton, NB: Acadiensis Press, 1990.

Brown, Lloyd Arnold. *Early Maps of the Ohio Valley: A Selection of Maps, Plans, and Views made by Indians and Colonials from 1673 to 1783*. Pittsburgh: University of Pittsburgh Press, 1959.

Brumwell, Stephen. *Paths of Glory: The Life and Death of General James Wolfe*. Montreal and Kingston: McGill-Queen's University Press, 2006.

Buckner, Phillip A. "Was There A 'British' Empire? *The Oxford History of the British Empire* from a Canadian Perspective." *Acadiensis* 32, 1 (2002): 110–28.

Bumsted, J.M. "1763–1783: Resettlement and Rebellion." In *The Atlantic Region to Confederation: A History*, edited by Phillip A. Buckner and John G. Reid, 156–83. Toronto: University of Toronto Press, 1994.

– *Land, Settlement, and Politics on Eighteenth-Century Prince Edward Island*. Kingston and Montreal: McGill-Queen's University Press, 1987.

Calloway, Colin G. *The Scratch of a Pen: 1763 and the Transformation of North America*. Oxford: Oxford University Press, 2006.

Campbell, Alexander V. "Atlantic Microcosm: The Royal American Regiment, 1755–1772." In *English Atlantics Revisited: Essays Honouring Professor Ian K. Steele*, edited by Nancy L. Rhoden, 284–309. Montreal: McGill-Queen's University Press, 2007.

Campbell, Duncan. *History of Prince Edward Island*. Charlottetown: Bremner Brothers, 1875.

Candee, Richard. *"An Old Town By the Sea": Urban Landscapes and Vernacular Building in Portsmouth, NH 1660–1990*. Portsmouth, NH: Vernacular Architecture Field Guide, 1992.

Carter, Paul. *The Road to Botany Bay: An Exploration of Landscape and History*. New York: Knopf, 1988.

Cashin, Edward J. *Governor Henry Ellis and the Transformation of British North America*. Athens: University of Georgia Press, 1994.

Chartrand, René. *French Fortresses in North America 1535–1763*. Oxford: Osprey Publishing, 2005.

– *Ticonderoga 1758*. Oxford: Osprey Publishing, 2000.

Chilton, Lisa. "Canada and the British Empire: A Review Essay." *Canadian Historical Review* 89, 1 (2008): 89–95.

Chipman, Willis. "The Life and Times of Major Samuel Holland, Surveyor-General, 1764–1801." *Papers and Records – Ontario Historical Society* 21 (1924): 11–90.

Christie, Ian R. "A Vision of Empire: Thomas Whately and the Regulations Lately Made Concerning the Colonies." *English Historical Review* 113, 451 (1998): 300–20.

Clark, Andrew Hill. *Three Centuries and the Island: A Historical Geography of Settlement and Agriculture in Prince Edward Island, Canada*. Toronto: University of Toronto Press, 1959.

Clarke, R.S.J. "The Irish Charts in 'Le Neptune Francois': Their Sources and Influence." *Map Collector* 30 (March 1985): 10–14.

Clayton, Daniel. *Islands of Truth: The Imperial Fashioning of Vancouver Island*. Vancouver, BC: UBC Press, 2000.

– "On the Colonial Genealogy of George Vancouver's Chart of the North-West Coast of North America." *Ecumene* 7, 4 (2000): 371–401.

Clifton, Gloria. *Directory of British Scientific Instrument Makers 1550–1851*. London: Zwemmer, 1995.

Cohen, Paul E. and Robert T. Augustyn, *Manhattan in Maps 1527–1995*. New York: Rizzoli, 1997.

Cordingly, David. *Capt. James Cook Navigator.* London: National Maritime Museum, 1988.

Croarken, Mary. "Tabulating the Heavens: Computing the *Nautical Almanac* in 18th-Century England." *IEEE Annals of the History of Computing* 25, 3 (2003): 48–61.

Crone, G.R. *Maps and Their Makers.* London: Hutchinson, 1953.

Crowley, John E. "A Visual Empire: Seeing the British Atlantic World from a Global British Perspective." In *The Creation of the British Atlantic World*, edited by Elizabeth Mancke and Carole Shammas, 283–303. Baltimore: Johns Hopkins University Press, 2005.

– "'Taken on the Spot': The Visual Appropriation of New France for the Global British Landscape." *Canadian Historical Review* 86, 1 (2005): 1–28.

Cumming, William P. *The Southeast in Early Maps.* Third edition, revised and enlarged by Louis De Vorsey, Jr. Chapel Hill: University of North Carolina Press, 1998.

– "The Colonial Charting of the Massachusetts Coast." *Publications of the Colonial Society of Massachusetts* 52 (1980): 67–118.

– "The Montresor-Ratzer-Sauthier Sequence of Maps of New York City, 1766–76." *Imago Mundi* 31 (1979): 55–65.

– *British Maps of Colonial America.* Chicago: University of Chicago Press, 1974.

Darnton, Robert. "What Is the History of Books?" *Daedalus* 111, 3 (1982): 65–83.

David, Andrew. "Hurd, Thomas Hannaford." In *Oxford Dictionary of National Biography.* Oxford: Oxford University Press, 2004.

– "Vancouver's Survey Methods and Surveys." In *From Maps to Metaphors: The Pacific World of George Vancouver*, edited by Robin Fisher and Hugh Johnson, 51–69. Vancouver, BC: UBC Press, 1993.

Day, Sir Archibald. *The Admiralty Hydrographic Service 1795–1919.* London: H.M.S.O., 1967.

Dean, W.G., ed., and G.J. Matthews, cartographer, *Economic Atlas of Ontario.* Toronto: University of Toronto Press for the Government of Ontario, 1969.

Debard, Jean-Marc. "The Family Origins of Joseph Frederick Wallet DesBarres: A Riddle Finally Solved." *Nova Scotia Historical Review* 14, 2 (1994): 108–22.

Demeritt, David. "Representing the 'True' St. Croix: Knowledge and Power in the Partition of the Northeast." *William and Mary Quarterly* 54, 3 (1997): 515–48.

De Vorsey Jr., Louis. "La Florida Revealed: The De Brahm Surveys of British East Florida, 1765–1771." In *Pattern and Process: Research in Historical Geography*, edited by Ralph E. Ehrenberg, 87–102. Washington, DC: Howard University Press, 1975.

Dickerson, Oliver Morton. *American Colonial Government 1696–1765.* New York: Russell & Russell, 1962.

Douglas, W.A.B. "Colvill (Colville), Alexander." In *Dictionary of Canadian Biography*, vol. III, *1741–1770*, edited by Francess G. Halpenny, 131–3. Toronto: University of Toronto Press, 1974.

Driver, Felix, and Luciana Martins. "John Septimus Roe and the Art of Navigation, c. 1815–1830." *History Workshop Journal* 54 (2002): 144–61.

Duncanson, John V. *Falmouth – A New England Township in Nova Scotia 1760–1965.* Windsor, ON: 1965.

Edney, Matthew H. "John Mitchell's Map of North America (1755): A Study of the Use and Publication of Official Maps in Eighteenth Century Britain." *Imago Mundi* 60, 1 (2008): 63–85.

– "The Origins and Development of J.B. Harley's Cartographic Theories." *Cartographica* 40, 1–2 (2005): 1–143.

Espenak, Fred. *Transits of Venus: Six Millennium Catalog: 2000 BCE to 4000 CE.* Greenbelt, MD: NASA/Goddard Space Flight Center, ND, at http://sunearth.gsfc.nasa.gov/eclipse/transit/catalog/VenusCatalog.html

Evans, G.N.D. *Uncommon Obdurate: The Several Public Careers of J.F.W. DesBarres.* Toronto: University of Toronto Press, 1969.

Fellows, Robert. "Sproule, George." In *Dictionary of Canadian Biography*, vol. V, *1801–20*, edited by Francess G. Halpenny, 773–4. Toronto: University of Toronto Press, 1983.

– "The Loyalists and Land Settlement in New Brunswick, 1783–1790." *Canadian archivist/Archiviste canadien* 2, 2 (1971): 5–15.

Fisher, Robin. *Vancouver's Voyage: Charting the Northwest Coast, 1791–1795.* Vancouver, BC: Douglas & McIntyre, 1992.

Fisher, Susanna. "Knight, Sir John." In *Oxford Dictionary of National Biography.* Oxford: Oxford University Press, 2004.

– *The Makers of the Blueback Charts: A History of Imray Laurie Norie & Wilson Ltd.* St. Ives: Regatta Press, 2001.

Forbes, Eric G. "The Foundation and Early Development of the Nautical Almanac." *Journal of the Institute of Navigation* 18, 4 (1965): 391–401.

Gardiner, Robert, ed. *Navies and the American Revolution 1775–1783.* Annapolis, MD: Naval Institute Press, 1996.

Garver, Joseph G. *Surveying the Shore: Historic Maps of Coastal Massachusetts 1600–1930.* Beverly, MA: Commonwealth Editions, 2006.

Gascoigne, John. *Science in the Service of Empire: Joseph Banks, the British State, and the Uses of Science in the Age of Revolution.* Cambridge: Cambridge University Press, 1998.

Geneste, Pascal. "The Archivist and the Hydrographer: The Origins of the French Hydrographic Service in the Dépôt des Cartes et Plans de la Marine, 1720–1850." In *Science and the French and British Navies, 1700–1850*, edited by Pieter van de Merwe, 34–47. Greenwich: National Maritime Museum, 2003.

Gentilcore, Louis, and Kate Donkin. *Land Surveys of Southern Ontario: An Introduction and Index to the Field Notebooks of the Ontario Land Surveys 1784–1859.* Supplement to *Canadian Cartographer* 10 (1973).

Gentilcore, R. Louis, Don Measner, and David Doherty. "The Coming of the Loyalists." In *Historical Atlas of Canada II: The Land Transformed, 1800–1891*, edited by R. Louis Gentilcore, plate 7. Toronto: University of Toronto Press, 1993.

Giddens, Anthony. *The Consequences of Modernity.* Stanford: Stanford University Press, 1990.

Godfrey, William G. "Bastide, John Henry." In *Dictionary of Canadian Biography*, vol. III, *1741 to 1770*, edited by Francess G. Halpenny, 32–4. Toronto: University of Toronto Press, 1974.

Golinski, Jan. *Making Natural Knowledge: Constructivism and the History of Science.* Cambridge: Cambridge University Press, 1998.

Graham, Conrad. *Mont Royal – Ville Marie: Vues et plans anciens de Montréal – Early Plans and Views of Montreal.* Montreal: McCord Museum of Canadian History, 1992.

Green, Julian W. "A Map of New Hampshire – Spanning the Revolution." *Dartmouth College Library Bulletin* 16, 2 (1976): 71–8.

Greene, Jack P. "The Deeper Roots of Colonial Discontent: William Knox's Structural Explanation for the American Revolution." In *Understanding the American Revolution: Issues and Actors.* Charlottesville, VA: University Press of Virginia, 1995.

Greiert, Steven G. "The Earl of Halifax and the Settlement of Nova Scotia, 1749–1753." *Nova Scotia Historical Review* 1 (1981): 4–23.

Griffin-Short, Rita. "The Transit of Venus." *The Beaver* 83, 2 (April/May 2003): 8–12.

Gwyn, Julian. *Ashore and Afloat: The British Navy and the Halifax Naval Yard before 1820.* Ottawa: University of Ottawa Press, 2004.

Harley, J.B. *The New Nature of Maps: Essays in the History of Cartography.* Baltimore: Johns Hopkins University Press, 2001.

– "Text and Contexts in the Interpretation of Early Maps." In *The New Nature of Maps*, 34–49.

– "Maps, Knowledge, and Power." In *The New Nature of Maps*, 52–81.

– "Silences and Secrecy: The Hidden Agenda of Cartography in Early Modern Europe." In *The New Nature of Maps*, 84–107.

– "Power and Legitimation in the English Geographical Atlases of the Eighteenth Century." In *The New Nature of Maps*, 110–47.

- "New England Cartography and the Native Americans." In *The New Nature of Maps*, 170–95.
- "The Contemporary Mapping of the American Revolutionary War." In *Mapping the American Revolutionary War*, edited by J.B. Harley, Barbara Bartz Petchenik, and Lawrence W. Towner, 1–44.
- "The Spread of Cartographical Ideas between the Revolutionary Armies." In *Mapping the American Revolutionary War*, edited by Harley, Petchenik, and Towner 45–78.
- "Specifications for Military Surveys in British North America, 1750–75." In *International Geography, 1972*, 22nd International Geographical Congress, 424–5. Toronto: University of Toronto Press, 1972.
- "The Bankruptcy of Thomas Jefferys: An Episode in the Economic History of Eighteenth Century Map-making." *Imago Mundi* 20 (1966): 27–48.

Harley, J.B., Barbara Bartz Petchenik, and Lawrence W. Towner, eds. *Mapping the American Revolutionary War.* Chicago: University of Chicago Press, 1978.

Harris, R. Cole. *The Reluctant Land: Society, Space, and Environment in Canada before Confederation.* Vancouver, BC: UBC Press, 2008.
- "The Simplification of Europe Overseas." *Annals of the Association of American Geographers* 67, 4 (1977): 469–83.

Harris, Steven J. "Long-Distance Corporations, Big Sciences, and the Geography of Knowledge." *Configurations* 6, 2 (1998): 269–304.

Hayes, Derek. *Historical Atlas of Canada.* Vancouver, BC; Douglas & McIntyre, 2002.

Hayward, Robert J. "Collins, John." In *Dictionary of Canadian Biography*, volume IV, *1771 to 1800*, edited by Francess G. Halpenny, 161–2. Toronto: University of Toronto Press, 1979.

Heidenreich, Conrad E., and Françoise Noël. "French Interior Settlements, 1750s." In *Historical Atlas of Canada*, vol. I: *From the Beginning to 1800*, edited by R. Cole Harris, plate 41. Toronto: University of Toronto Press, 1987.

Helden, Albert Van. "Longitude and the Satellites of Jupiter." In *The Quest for Longitude*, edited by William J.H. Andrewes, 86–100. Cambridge, MA: Collection of Historical Scientific Instruments, Harvard University, 1996.

Hindle, Brooke. *The Pursuit of Science in Revolutionary America 1735–1789.* Chapel Hill: University of North Carolina Press, 1956.

Holman, H.T. "Wright, Thomas." In *Dictionary of Canadian Biography*, vol. V, *1801 to 1820*, edited by Francess G. Halpenny, 873–4. Toronto: University of Toronto Press, 1983.

Hornsby, Stephen J. *British Atlantic, American Frontier: Spaces of Power in Early Modern British America.* Hanover, NH: University Press of New England, 2005.
- *Nineteenth-Century Cape Breton: A Historical Geography.* Montreal and Kingston: McGill-Queen's University Press, 1992.
- and Micah A. Pawling, "British Survey the Interior." In *Historical Atlas of Maine*, edited by Stephen J. Hornsby and Richard W. Judd. Orono, ME: University of Maine Press, 2015.

Howse, Derek. *Greenwich Time and the Longitude.* Greenwich: National Maritime Museum, 1997.
- and Michael Sanderson, *The Sea Chart.* New York: McGraw-Hill, 1973.

Humphreys, R.A. "Lord Shelburne and the Proclamation of 1763." *English Historical Review* 49, 194 (1934): 241–64.

Jeans, D.N. "The Impress of Central Authority upon the Landscape: Southeastern Australia 1788–1850." In *Australian Space, Australian Time: Geographical Perspectives*, edited by J.M. Powell and M. Williams, 1–17. Melbourne: Oxford University Press, 1975.
- "Territorial Divisions and the Locations of Towns in New South Wales, 1826–1842." *Australian Geographer* 10 (1967): 243–55.

Jonkers, A.R.T. "Parallel Meridians: Diffusion and Change in Early-Modern Oceanic Reckoning." In *Noord-Zuid in Oostindisch Perspectif*, edited by Jan Parmentier, 17–42. The Hague: Walburg, 2005.

Kernaghan, Lois. "A Man and His Mistress: J.F.W. DesBarres and Mary Cannon." *Acadiensis* 1, 1 (1981): 23–42.

Lambert, David, and Alan Lester, eds. *Colonial Lives across the British Empire: Imperial Careering in the Long Nineteenth Century.* Cambridge: Cambridge University Press, 2006.

Lande, Lawrence Montague. *The 3rd Duke of Richmond: A Study in Early Canadian History.* Montreal: 1956.

Latour, Bruno. *Reassembling the Social: An Introduction to Actor-Network Theory.* Oxford: Oxford University Press, 2005.

– *Pandora's Hope: Essays on the Reality of Science Studies.* Cambridge, MA: Harvard University Press, 1999.

– *Science in Action: How to Follow Scientists and Engineers Through Society.* Cambridge, MA: Harvard University Press, 1987.

Law, John. "Materialities, Spatialities, Globalities." Lancaster: Centre for Science Studies, Lancaster University, 1999.

Leamon, James S. *Revolution Downeast: The War for American Independence in Maine.* Amherst: University of Massachusetts Press, 1993.

Lennox, Jeffers. "An Empire on Paper: The Founding of Halifax and Conceptions of Imperial Space, 1744–55." *Canadian Historical Review* 88, 3 (2007): 373–412.

Lester, Alan. "Imperial Circuits and Networks: Geographies of the British Empire." *History Compass* 3 (2005): 1–18.

Litalien, Raymond, Jean-François Palomino, and Denis Vaugeois. *Mapping a Continent: Historical Atlas of North America, 1492–1814.* Sillery, QC: Septentrion, 2007.

Livingstone, David N. *Putting Science in its Place: Geographies of Scientific Knowledge.* Chicago: University of Chicago Press, 2003.

– and Charles W.J. Withers, eds. *Geography and Enlightenment.* Chicago: University of Chicago Press, 1999.

Loring, Augustus P. "*The Atlantic Neptune*." In *Seafaring in Colonial Massachusetts*, edited by Frederick S. Allis, Jr., 119–30. Boston: Colonial Society of Massachusetts, 1980.

Lowe, William C. "Lennox, Charles." In *Oxford Dictionary of National Biography.* Oxford: Oxford University Press, 2004.

Lyon, David. *The Sailing Navy List: All the Ships of the Royal Navy – Built, Purchased and Captured – 1688–1860.* London: Conway Maritime Press, 1993.

McGowan, A.P. "Captain Cook's Ships." *Mariner's Mirror* 65, 2 (1979): 109–18.

McLennan, J.S. *Louisbourg from Its Foundation to Its Fall 1713–1758.* London: Macmillan, 1918.

Machemer, Grace S. "Headquartered at Piscataqua: Samuel Holland's Coastal and Inland Surveys, 1770–1774." *Historical New Hampshire* 57, 1 & 2 (2002): 4–25.

Mackesy, Piers. *The War for America 1775–1783.* Lincoln: University of Nebraska Press, 1993.

Mancke, Elizabeth. "Another British America: A Canadian Model for the Early British Empire." *Journal of Imperial and Commonwealth History* 25, 1 (1997): 1–36.

Mannion, John J. "Settlers and Traders in Western Newfoundland." In *The Peopling of Newfoundland: Essays in Historical Geography*, edited by John J. Mannion, 234–78. St. John's: Memorial University of Newfoundland, 1977.

– Gordon Handcock, and Alan Macpherson. "The Newfoundland Fishery, 18th Century." In *Historical Atlas of Canada*, vol. I: *From the Beginning to 1800*, edited by R. Cole Harris, plate 25. Toronto: University of Toronto Press, 1987.

Marshall, Douglas William. "Military Maps of the Eighteenth Century and the Tower of London Drawing Room." *Imago Mundi* 32 (1980): 22–44.

– "The British Military Engineers 1741–1783: A Study of Organization, Social Origin, and Cartography." PhD dissertation, University of Michigan, 1976.

– "The British Engineers in America: 1755–1783." *Journal of the Society for Army Historical Research* 51 (1973): 155–63.

Marshall, P.J. *The Making and Unmaking of Empires: Britain, India, and America c.1750–1783.* Oxford: Oxford University Press, 2005.

– ed. *The Oxford History of the British Empire: The Eighteenth Century.* Oxford: Oxford University Press, 1998.

Matthews, Keith. *Lectures on the History of Newfoundland 1500–1830.* St. John's: Breakwater Books, 1988.

Maxted, Ian. "The London Book Trades, 1775–1800: A Topographical Guide." *Exeter Working Papers in Book History,* at www.bookhistory.blogspot.com/2007/01/streets-introduction.html, accessed on 10 February 2009.

Miller, David Philip. "Joseph Banks, Empire, and 'Centers of Calculation' in Late Hanoverian London." In *Visions of Empire: Voyages, Botany, and Representation of Nature,* edited by David Philip Miller and Peter Hanns Reill, 21–37. Cambridge: Cambridge University Press, 1996.

Montrésor, F.M. "Notes and Documents: Captain John Montrésor in Canada." *Canadian Historical Review* 5 (1924): 336–40.

Morgan, R.J. "Des Barres, Joseph Frederick Wallet." In *Dictionary of Canadian Biography,* vol. VI, *1821 to 1835,* edited by Francess G. Halpenny, 192–7. Toronto: University of Toronto Press, 1987.

Morison, Samuel Eliot. *The Story of Mount Desert Island.* Boston: Little, Brown, 1960.

Morrison, Walter K. "Cape Breton Maps in *The Atlantic Neptune*: A Holland-DesBarres Connection." *Nova Scotia Historical Review* 10, 2 (1990): 111–23.

– "The 'Modern' Mapping of Nova Scotia." *Map Collector* 18 (1982): 28–34.

– "The Other Revolution in 1775." In *Proceedings of the Annual Conference of the Association of Canadian Map Libraries* 9 (1975): 59–79. Reprinted, with modifications as "The Cartographical Revolution in 1775" in *Explorations in the History of Canadian Mapping: A Collection of Essays,* edited by Barbara Farrell and Aileen Desbarats, 75–88. Ottawa: Association of Canadian Map Libraries and Archives, 1988.

Murdoch, Tessa. *The Quiet Conquest: The Huguenots 1685 to 1985.* London: Museum of London, 1985.

Nasaw, David. "AHR Roundtable: Historians and Biography." *American Historical Review* 114, 3 (2009): 577–8.

Neatby, Hilda. *Quebec: The Revolutionary Age 1760–1791.* Toronto: McClelland & Stewart, 1966.

Nelson, Paul David. *William Tryon and the Course of Empire: A Life in British Imperial Service.* Chapel Hill: University of North Carolina Press, 1990.

Newman, Edwin B. and Augustus P. Loring, "Some Notes on the Paper of the *Atlantic Neptune.*" *American Neptune* 46, 3 (1986): 173–8.

O'Donoghue, Yolande. *William Roy 1726–1790: Pioneer of the Ordnance Survey.* London: British Museum Publications, 1977.

Outram, Dorinda. *Panorama of the Enlightenment.* Los Angeles: Getty Publications, 2006.

Pargellis, Stanley McCrory. *Lord Loudon in North America.* New Haven: Yale University Press, 1933.

Paston, George. *Little Memoirs of the Eighteenth Century.* New York: E.P. Dutton, 1901.

Patterson, Frank H. *History of Tatamagouche, Nova Scotia.* Belleville, ON: Mika Publishing, 1973; originally published 1917.

Pedley, Mary Sponberg. *The Commerce of Cartography: Making and Marketing Maps in Eighteenth-Century France and England.* Chicago: University of Chicago Press, 2005.

Penney, David. "Thomas Mudge and the Longitude: A Reason to Excel." In *The Quest for Longitude,* edited by William J.H. Andrewes, 294–310. Cambridge, MA: Collection of Historical Scientific Instruments, Harvard University, 1996.

Pigot, F.L. "MacDonald of Glenaladale, John." In *Dictionary of Canadian Biography,* vol. V, *1801 to 1820,* edited by Francess G. Halpenny, 514–17. Toronto: University of Toronto Press, 1983.

Porter, Roy, ed. *The Cambridge History of Science,* vol. 4, *Eighteenth-Century Science.* Cambridge: Cambridge University Press, 2003.

Rayburn, Alan. *Geographical Names of Prince Edward Island*. Ottawa: Energy, Mines and Resources Canada, 1973.

Raymond, Richard L. *J.F.W. DesBarres: Views and Profiles*. Halifax, NS: Dalhousie Art Gallery, 1982.

Rees, Ronald. *Land of the Loyalists: Their Struggle to Shape the Maritimes*. Halifax, NS: Nimbus Publishing, 2000.

Reid, John G., with H.V. Bowen and Elizabeth Mancke. "Is There a 'Canadian' Atlantic World?" *International Journal of Maritime History* 21, 1 (2009): 1–33.

Reid, Stuart. "Wolfe, James." In *Oxford Dictionary of National Biography*. Oxford: Oxford University Press, 2004.

Rhoden, Nancy L. "Introduction: Revisiting *The English Atlantic* in the Context of Atlantic History." In *English Atlantics Revisited: Essays Honouring Professor Ian K. Steele*, edited by Nancy L. Rhoden, xiii–xxxviii. Montreal and Kingston: McGill-Queen's University Press, 2007.

Robertson, Bruce. "Venit, Vidit, Depinxit: The Military Artist in America" in *Views and Visions: American Landscape before 1830*, edited by Edward Nygren with Bruce Robertson, 83–103. Washington, DC: Corcoran Gallery of Art, 1986.

Robson, John. *Captain Cook's World: Maps of the Life and Voyages of James Cook R.N.* Auckland: Random House New Zealand, 2000.

Rodger, N.A.M. "Navies and the Enlightenment." In *Science and the French and British Navies, 1700–1850*, edited by Pieter van der Merwe, 5–23. Greenwich: National Maritime Museum, 2003.

– *The Insatiable Earl: A Life of John Montague, 4th Earl of Sandwich*. New York: Norton, 1994.

– *The Wooden World: An Anatomy of the Georgian Navy*. London: Collins, 1986.

– *Articles of War: The Statutes which Governed Our Fighting Navies 1661, 1749 and 1886*. Havant: Kenneth Mason, 1982.

– *The Admiralty*. Lavenham: Terence Dalton, 1979.

Sadler, D.H. *Man Is Not Lost: A Record of Two Hundred Years of Astronomical Navigation with The Nautical Almanac 1767–1967*. London: H.M.S.O., 1968.

Sands, John O. "The Sailor's Perspective: British Naval Topographic Artists." In *Background to Discovery: Pacific Exploration from Dampier to Cook*, edited by Derek Howse, 185–200. Berkeley, CA: University of California Press, 1990.

Secord, James A. "Knowledge in Transit." *Isis* 95 (2004): 654–72.

Shapin, Steven. "Placing the View from Nowhere: Historical and Sociological Problems in the Location of Science." *Transactions of the Institute of British Geographers* NS 23 (1998): 5–12.

– "Here and Everywhere: Sociology of Scientific Knowledge." *Annual Review of Sociology* 21 (1995): 289–321.

Shields, Gordon. "General James Murray's Map of the St. Lawrence River Valley in 1761: A Cartographic Commentary." *Association of Canadian Map Libraries Bulletin* 48 (1983): 30–5.

Shields, James Gordon. "The Murray Map Cartographically Considered: A Study of General James Murray's Survey of the St. Lawrence River Valley in 1761." MA thesis, Queen's University, 1980.

Shipton, Nathaniel N. "Samuel Holland's Plan of Cape Breton." *Canadian Cartographer* 5, 2 (1968): 81–9.

– "General James Murray's Map of the St Lawrence." *Cartographer* 4, 2 (1967): 93–101.

Shy, John W. *Toward Lexington: The Role of the British Army in the Coming of the American Revolution*. Princeton, NJ: Princeton University Press, 1965.

Skelton, R.A. "The Military Survey of Scotland 1747–1755." *Scottish Geographical Magazine* 83, 1 (1967): 1–12.

– "Captain James Cook as a Hydrographer." *Mariner's Mirror* 40, 2 (1954): 92–119.

Sorrenson, Richard. "The Ship as a Scientific Instrument in the Eighteenth Century." *Osiris* 2nd series, 11 (1996): 221–36.

Sosin, Jack M. *Whitehall and the Wilderness: The Middle West in British Colonial Policy, 1760–1775*. Lincoln: University of Nebraska Press, 1961.

Stacey, C.P. Edited and with new material by Donald E. Graves. *Quebec, 1759: The Siege and the Battle.* Toronto: Robin Brass Studio, 2002, originally published 1959.

Steele, Ian K. *The English Atlantic 1675–1740: An Exploration of Communication and Community.* New York: Oxford University Press, 1986.

Stout, Neil R. *The Royal Navy in America, 1760–1775: A Study of Enforcement of British Colonial Policy in the Era of the American Revolution.* Annapolis, MD: Naval Institute Press, 1973.

Sutherland, Stuart R.J., Pierre Tousignant, and Madeleine Dionne-Tousignant. "Haldimand, Sir Frederick." In *Dictionary of Canadian Biography*, vol. V, *1801 to 1820*, edited by Francess G. Halpenny, 887–904. Toronto: University of Toronto Press, 1983.

Syrett, David. *Admiral Lord Howe: A Biography.* Annapolis, MD: Naval Institute Press, 2006.

– "Home Waters or America? The Dilemma of British Naval Strategy in 1778." *Mariner's Mirror* 77, 4 (1991): 365–77

– and R.L. DiNardo. *The Commissioned Sea Officers of the Royal Navy 1660–1815.* Aldershot: Scolar Press for the Navy Records Society, 1994.

Tabraham, Chris, and Doreen Grove. *Fortress Scotland and the Jacobites.* London: B.T. Batsford, 1995.

Taft, Hank and Jan, and Curtis Rindlaub. *A Cruising Guide to the Maine Coast.* Peaks Island, ME: Diamond Pass Publishing, 2002.

Taylor, E.G.R. *Navigation in the Days of Captain Cook.* London: National Maritime Museum, 1975.

Terrell, Christopher. "A Sequel to *The Atlantic Neptune* of J.F.W. DesBarres: The Story of the Copperplates." *The Map Collector* 72 (1995): 2–9.

Thomas, P.D.G. *British Politics and the Stamp Act Crisis: The First Phase of the American Revolution, 1763–1767.* Oxford: Clarendon Press, 1975.

Thomson, Don W. *Men and Meridians: The History of Surveying and Mapping in Canada*, vol. I, *Prior to 1867.* Ottawa: Queen's Printer, 1966.

Thorpe, F.J. "Holland, Samuel Johannes." In *Dictionary of Canadian Biography*, vol. V, *1801 to 1820*, edited by Francess G. Halpenny, 425–9. Toronto: University of Toronto Press, 1983.

Trudel, Marcel. *Atlas de la Nouvelle-France: An Atlas of New France.* Québec: Les Presses de l'Université Laval, 1968.

Turnbull, David. "Travelling Knowledge: Narratives, Assemblage and Encounters" in *Instruments, Travel and Science*, edited by Marie-Noëlle Bourguet, Christian Licoppe, and H. Otto Sibum, 273–94. London: Routledge, 2002.

Vidal, Laurent, and Émilie d'Orgeix, eds. *Les Villes Françaises du Nouveau Monde.* Paris: Somogy, 1999.

Ware, John D. Revised and completed by Robert R. Rea. *George Gauld: Surveyor and Cartographer of the Gulf Coast.* Gainesville/Tampa, FL: University Presses of Florida, 1982.

Wess, Jane. "George III, Scientific Societies, and the Changing Nature of Scientific Collecting." In *The Wisdom of George the Third*, edited by Jonathan Marsden, 313–30. London: Royal Collection Enterprises, 2005.

Whiteley, William H. *James Cook in Newfoundland 1762–1767.* St. John's: Newfoundland Historical Society, 1975.

– "James Cook and British Policy in the Newfoundland Fisheries, 1763–67." *Canadian Historical Review* 54, 3 (1973): 245–72.

– "Governor Hugh Palliser and the Newfoundland and Labrador Fishery, 1764–1768." *Canadian Historical Review* 50, 2 (1969): 141–63.

Wickwire, Franklin B. *British Subministers and Colonial America 1763–1783.* Princeton, NJ: Princeton University Press, 1966.

– "Admiralty Secretaries and the British Civil Service." *Huntington Library Quarterly* 28, 3 (1965): 235–54.

– "John Pownall and British Colonial Policy." *William and Mary Quarterly* 20, 4 (1963): 543–54.

Wilderson, Paul W. *Governor John Wentworth and the American Revolution: The English Connection.*

Hanover, NH: University Press of New England, 1994.
Wilkinson, Clive. *The British Navy and the State in the Eighteenth Century.* Woodbridge: Boydell Press, 2004.
Williams, Glyn. *Voyages of Delusion: The Quest for the Northwest Passage.* New Haven: Yale University Press, 2003.
Williamson, Joseph. "The Proposed Province of New Ireland." *Collections of the Maine Historical Society* Third Series, vol. I, 147–57. Portland: Maine Historical Society, 1904.
– "The Proposed Province of New Ireland." *Collections of the Maine Historical Society*, vol. VII, 201–6. Bath: Maine Historical Society, 1876.
Withers, Charles W.J. "Where Was the Atlantic Enlightenment? – Questions of Geography." In *The Atlantic Enlightenment*, edited by Susan Manning and Francis D. Cogliano, 37–60. Aldershot: Ashgate, 2008.
– *Placing the Enlightenment: Thinking Geographically about the Age of Reason.* Chicago: University of Chicago Press, 2007.

Wood, Denis. *The Power of Maps.* New York: Guilford Press, 1992.
Woolf, Harry. *The Transits of Venus: A Study of Eighteenth-Century Science.* Princeton, NJ: Princeton University Press, 1959.
Worms, Laurence. "The Maturing of British Commercial Cartography: William Faden (1749–1836) and the Map Trade." *Cartographic Journal* 41, 1 (2004): 5–11.
– "The Book Trade at the Royal Exchange." In *The Royal Exchange*, edited by Ann Saunders, 209–26. London: London Topographical Society, 1997.
Wynn, Graeme. "A Province Too Much Dependent on New England." *Canadian Geographer* 31, 2 (1987): 98–113.
– "Late Eighteenth Century Agriculture on the Bay of Fundy Marshlands." *Acadiensis* 8 (1979): 80–9.
– and Debra McNabb. "Pre-Loyalist Nova Scotia" in *Historical Atlas of Canada*, vol. I: *From the Beginning to 1800*, edited by R. Cole Harris, plate 31. Toronto: University of Toronto Press, 1987.

INDEX

Italics indicate figures

Abercromby, Maj. Gen. James, 16, 19
Admiralty, British 9, 31, 33–8, 89–90, *90*; and *The Atlantic Neptune*, 164–72, 175, 190–3, 196; chain of command, 87–95, 117–18; collection of geographic and hydrographic information by, 19, 33–4, 37, 87–8; colonial defence, 31; Hydrographic Office, 4, 19, 171, 205; surveys of, 4, 33–8, *35*, 43, 45–7. *See also* General Survey of North America; Survey of Nova Scotia
American Atlas, 177
American Revolution. *See* War of American Independence
Amherst, Lord Jeffrey, 16, 25, 26, 32, 134; promoted to commander-in-chief, 19
army, British, surveys of, 4, 5, 8, 11, 12–23, 31, *35*, 38–9, 43; cooperation with navy, 19; Royal American Regiment (60th Regiment of Foot), 13–14. *See also* Survey of Canada
astronomical observation, 3, 6, 18–19, 52, 68, 62–4, 80, 86, 111–12. *See also* survey instruments, methods and equipment; Transit of Venus
Atlantic Neptune, The, 4–6, 8, 9, 86, 122, 161, *162*, 163–97; accuracy of maps, 181–4; Admiralty financial support for, 164–72; in Canadian historiography, 4; coastal views, 185–8; commemoration of patrons, 165–72; copper plates from, 210–11; copying of by other publishers, 194–5; credit to other surveyors, 172, 188; distribution of, 190–4; engravers and printers, 171, 173–5; and meeting of Samuel Holland and Joseph Des Barres, 69–71, 124–5; organization into volumes, 176; patronage for, 164–72, 194; publication of, 172–90; revision of place names, 167–72; scale of maps, 177–81; seabed description, 184–5, *186–7*; use in War of American Independence, 190–3

Banks, Joseph, 5
Blaskowitz, Charles, 62, 95, 99; survey of New England, 72, *74*, 80, 81
Board of Ordnance 12–13, 30; cartographic resources of, 40; Corp of Engineers, 12–13
Board of Trade and Plantations, 4, 5, 8–9, 31, 32–3, 37, *89*; chain of command, 87, 88–9, 93–5, 117–18; colonial administration, 88–9; and settlement on St. John's Island, 125–7, 147, 150–2; surveys of, 4, 5, *35*, 39–43, 45, 57, 59, 65, 142. *See also* General Survey of North America
Borthwick, Robert, 96–7
Bouquet, Henry, 13, 38–9, 134
British Atlantic World, 6–8, 117–18, 195–7
British Empire, 3–8, 85–6, 117–18, 198; imperial government chain of command, 87–95; and scientific leadership, 3, 5–6, 11, 63–4, 85–6. *See also* British North America; British North America, surveying and mapping of; Seven Years War; Treaty of Paris (1763)
British North America, 30; colonial governance and administration, 4–5, 30–3; Indian land, 32–3, 145, 147; need for survey, 30–4; political unrest, 7, 72, 73, 81–5, *82*, 89, 93; reorganization of, 30–3;

settlement policy 4–6, 30–3, 43, 48, 144–6, 147–8. *See also* British North America, survey and mapping of

British North America, surveying and mapping of, 3–9, 25–30, 33–43, *35*, 67–8, 85–6, 87–8, 117–18; and colonial settlement, 4, 5, 6, 9, 43, 144–6, 147–8; and political power, 6–7, 9, 145–6; and role of military, 4, 5–6; scientific equipment and techniques, 5–6, 34–6, 41–2, 43, 85–6. *See also* General Survey of North America, Survey of Nova Scotia

Cabinet, 30–1, 90
Canada, defined, 9, 125
Canada, Survey of. *See* Survey of Canada
Canceaux, Armed Vessel, 46–7, *82*, 101–3. *See also* General Survey of the Northern District; Holland, Samuel; Mowat, Henry
Cape Breton Island, 32–3, 54–7, *56*, 117; coal, 54, 57; coastal soundings, 56; cod fishery, 54, 55, 57; division into counties, parishes and townships, 135–7, *136*; naming of civil division and geographic features, 135–7, 140–2. *See also* Cape Breton Island, survey of
Cape Breton Island, survey of, 3–9, 54–7, *56*, 93–4, 95, 208; report on, 57, 142–4
Carleton, Gen. Guy, 15, 134; as Governor of Quebec, 62
Carter, Paul, 128
cartography, 3, 6–9, 119–20, 145–6, 163, 195–7. *See also The Atlantic Neptune*; maps and survey plans; survey instruments, methods and equipment
chronometer, 14, 35–6, 111. *See also* Harrison, John
Collins, John, 49
Colvill, Rear Adm. Alexander, 19, 47, 91, 133–4, 167; and survey of Nova Scotia, 19, 37–8, 66–7, 68
Connecticut, survey of, 81
Connecticut River, 73, 74, 81
Cook, James, *35*, 58, 85–6; mapping of St. Lawrence River, 20; Newfoundland marine survey, 36–7, 188–90; and Samuel Holland, 17–19, 85–6; voyage to the Pacific, 3, 4, 5, 6, 94–5
Corp of Engineers, 12–13. *See also* Board of Ordnance

Des Barres, Joseph Frederick Wallet, 3–9, 13, 19, *24*, *35*; Acadian tenants, 158–60; as assistant engineer in the Royal American Regiment, 23–5; background of, 23–5; death, 210; expedition to Quebec, 23–4; as lieutenant governor of Nova Scotia and Prince Edward Island, 210; mapping of the St. Lawrence, 23–4; meeting with Samuel Holland, 69–71, 124–5; in Newfoundland, 24–5; as proprietor in Nova Scotia, 156–62; *156*, *157*, *158*; in Seven Years War, 23–5, *24*; Survey of Nova Scotia, 9, 37–8, 65–72. *See also The Atlantic Neptune*; Survey of Nova Scotia
De Brahm, William Gerard, *35*, 42–3
Derbage, George, 95
Diligent, schooner, 68, 103. *See also* Survey of Nova Scotia
Dispatch, 66–7, 68, 103. *See also* Survey of Nova Scotia
Durrell, Vice Adm. Phillip, 19, 20, 123, 133–4

Egmont, John Perceval, earl of, 46, 90, 132; and proposal for settlement of St. John's Island, 149–51
Ellis, Henry, 32, 42, 47

fisheries, 31, 36–7, 54, 55, 57, 143, 149; and French fishing rights, 36–7
Florida and Gulf Coast, 30, 31, 32–3, 43; survey of Gulf Coast, 34–6, *35*, 38, 43
forts: Amherst, 50, *51*; Carillon (Ticonderoga), 16; Frederick, 17
Fox, Henry, 31, 132
France, 3, 11, 30. *See also* Seven Years War; Treaty of Paris (1763)
Fusier, Lt. Lewis, 26, *27*

Gage, Gen. Thomas, 5, 29, 37–8, 59–61, 82, 134
Gauld, George, 34–6, *35*
General Survey of North America (1764–1770), 3–9, *35*, 39–43, 45–86, 117–18, 147; budget, 41–2, 59; division into Northern and Southern districts, 42; impact on later surveys, 6, 118, 145–6; proposal for, 41–2; scale and scope of survey, 41, 43, 45–6. *See also The Atlantic Neptune*; General Survey of

the Northern District; General Survey of the Southern District

General Survey of the Northern District (1764–1770), 3–9, *35*, 39–43, 45–86; appointment of surveyor general, 42; in Canadian historiography, 4; chain of command, 87–95, 117–18; communication, 93–4; impact on later surveys, 6, 118, 145–6; proposal for, 41–2; reports on, 142–4; role of military, 5; survey teams, 95–6, 97–8. *See also* Admiralty, British; *The Atlantic Neptune*; Board of Trade and Plantations; General Survey of North America; survey instruments, methods and equipment

General Survey of the Southern District, *35*, 42–3; and appointment of surveyor general, 42–3. *See also The Atlantic Neptune;* De Brahm, William Gerard; Florida and Gulf Coast; General Survey of North America

George III, 8, 28, 29, 30, 39

Goldfrap, George, 95, 99

Gordon, Capt. Henry, 39

Grant, James, *74*, 78, 80, 81, 95, 99

Graves, Capt. Thomas, 36, 83–6

Gulf Coast. *See* Florida and the Gulf Coast

Haldimand, Lt. Frederic, Jr., 13, 26, *27*, 47–8, 50, 52, 54, 55, 82, 134

Halifax, George Montagu Dunk, earl of, 33, 88, *89*, 132

Halifax, Nova Scotia, 16–19, 24, *66*, 67–8, 88

Harley, J.B., 7, 140–2, 144

Harrison, John, 3, 35

Hillsborough, Willis Hill, earl of, 33, 57, 64, 65, 78, 88–9, *129*, 130

Holland, Samuel, 3–9, 13–25, *14*, *15*, *35*, *209*; appointment as surveyor general of the Northern District, 42, 46; appointment as surveyor general of Quebec, 42, 205; and assault on Louisbourg, 16–17; and attack on Quebec, 20–3; background of, 13; death, 208; at Fort Frederick 17; and General Survey of North America, 39–42; and James Cook, 17, 18, 19; mapping of province of New York, 14–16, *15*, 64; meeting with Des Barres, 69–71, 124–5; in Perth Amboy, 83–5; promotion to captain and appointment as acting chief engineer, 22–3; property in Province of New Hampshire, 153–6, *153*, *155*; property in Province of New York, 153, 154–6, *155*; property on St. John's Island, 152–3, *152*, *155*; as proprietor, 147–56, *155*, 161–2; in Seven Years War, 13–25; survey of boundary between New York and New Jersey, 64; survey of boundary between New York and Pennsylvania, 83; and Survey of Canada, 26–30; survey of Cape Breton Island, 54–7; survey of New England, 65–85; survey of New Hampshire, 72–80; survey of St. John's Island, 49–54, 177; surveys of Gulf of St. Lawrence, 18–20, 46–65, *60*, 70–1; surveys of Loyalist townships in Upper Canada, 206–8, *206*, *207*; in War of American Independence, 205

Hood, Samuel, Commodore, 68, 93, 165, 170, 201

Howe, Lord George, 16, 17, 170–2

Hurd, Thomas, 5–6, 204–5, 211

Hutchins, Ens. Thomas, *35*, 39

Jefferys, Thomas, 42, 163

Knight, Lt. John, 5, 68–9, 72, 97, 110, 199–201, *201*; capture at Machias, 200

Labrador. *See* Newfoundland and Labrador

Lane, Michael, *35*, 37

Latour, Bruno, 8, 31, 71, 116, 172

Lloyd, William, 96

Louisbourg, 54–5, *54*; assault on and siege of, 16–17, *18*; French surrender, 17; plan of, *18*

Loudoun, Lord, Comm. in Chief, 13–16

Luttrell, James, 97, 164, 170

Magdalen Islands, 54, 142, *182*

Maine, District of, 75, 80–1, 142; separation from Massachusetts, 75–8

maps and survey plans, 119–42, 143–6; manuscript plans, 120–5; scale and accuracy of maps and plans 119, 120–1, 123–4; and soundings, 121–3; and survey descriptions and reports, 119–20. *See also* survey descriptions and reports

marine survey. *See* survey, hydrographic and soundings
Maskelyne, Nevil, Astronomer Royal at Greenwich, 63, 64
Massachusetts, 73, 75–8; political unrest in, 72, 73, 81–4; survey of, 81–3
Military Survey of Scotland, 12–13
Mississippi River, survey of, *35*, 38–9, 43
Monckton, Brig. Gen. Robert, 19, 134
Montagu, Rear Adm. John, 72, 82, 91–2
Montreal, attack on, 25–6
Montrésor, Lt. John, *26, 27*, 38, 60; expedition from Quebec to Maine, 25–6, 116; and survey of Canada, *26, 27*, 29, 30
Mowat, Lt. Henry, 46–7, 61, 65, 72, 80–1, 91–3, 96, 110, 201–4; seizure of vessels, 92–3; in War of American Independence, 201–4
Murray, James, 25, 26, 29–30, 38, 47, 48–9, 134; attack on Quebec, 21–3; promoted to Brig. Gen., 19

Native Americans, 115–17, *116;* Mi'kmaq, 115–16, 117; and place names, 140, 142, 144–5; as source of local information, 115–17. *See also* Proclamation of 1763
navigation, 31, 35–6, 92–3
navy, British, 5, 6, 19, 31 43, 92–3; and illegal trade, 92–3. *See also* Admiralty, British
New England, survey of, 65–85, *74, 75*, 95; and timber resources, 75
New Hampshire, Province of, 72–80, *73, 77, 79*
New Ireland, 75–8, *77*
New York, Province of, 14–16, *15*, 60–1, 64, 81, 201; boundary with New Jersey, 62, 64; boundary with Pennsylvania, 83
Newfoundland and Labrador, 36–7, 43, 188–90; marine survey of 36–7, 43
Northwest Passage, 57–8, *59*
Nova Scotia, 32–3, 36–7, 43, 75, 88; naming of civil divisions and geographic features, 137, *138–9*, 140–2. *See also* Survey of Nova Scotia
Nova Scotia, Survey of. *See* Survey of Nova Scotia

Observation Cove, 50, *51*, 135
Ohio River, survey of, *35*, 38–9, 43

Peach, Lt. Joseph, 26, *27*, *35*, 38
Perkins, Simeon, 69–70, *69*
Piscataqua River, 73
Pitt, William, 8, 11, 28, 30, 39
Pittman, Ens. Philip, 26, *27, 35*, 39
place names, 6, 9, 127–37, 145–6, 167–70, 172; of Cape Breton, 135–7, *136;* French and Acadian 142–3, 144–5; Native American, 140, 142, 144–5; of Nova Scotia, 135, 137, *138–9;* persistence on the landscape, 140–2; of St. John's Island, 127–35, *128, 130, 133*
Pontiac's Rebellion, 32, 38–9
Portsmouth, New Hampshire, 72–3, *73*
Pownall, John, 32, 89
Prince Edward Island. *See* St. John's Island
Pringle, John, Lt. Col., 52, 54, 55, 56, 57, 58–9, 95
Privy Council, 30, 46; role in colonial administration, 30
Proclamation of 1763, 48, 145, 147–8

Quebec, province of, 9; British attack on, 19–23, *22*, 25; French settlement in *20*, 25–6, *28*, 32–3, 47, 48–9, *59*, 61–2; map of seigneuries, 61–2. *See also* Survey of Canada

Rhode Island, Province of, survey of, 81–3
Richmond, Charles Lennox, 3rd duke of, 13–14, 17, 39, 129, 130–1
Rittenhouse, David, 64
Royal Observatory, 112
Royal Proclamation, 33, 88
Royal Society, 3, 63–4, 118

Sable Island, 37, 66, 68, *106*
Saint John River, survey of, *35*, 38, 62, 75–8, *76–7*, 80
St. John's Island, 17, 32–3, 49–54; British settlement of, 149–53; division into counties, parishes and townships, 125–7, *127, 128;* naming of civil divisions and geographic features, 127–35, *127, 128, 130, 133*, 140–2. *See also* St. John's Island, survey of
St. John's Island, survey of (1764–65), 3, 4, 49–54, *53*, 91, *106, 122, 127*, 208; map in *American Atlas*, 177, report on survey, 142–4

St. Lawrence, Gulf of, and St. Lawrence River, 18–19, 20, 23–4, 45–65, *60*

St. Lawrence, Gulf of, surveys, 17, 20, 46–65, *60, 63, 64*, 80; soundings, 62, 65

Sandwich, John Montagu, 4th earl of, 164–5, *165*, 167–70, *168–9*, 172

Seven Years War (1756–1763), 3, 8, 11–25; and development of military engineering, 11, 12. *See also* Treaty of Paris (1763)

Shelburne, William Petty, earl of, 32–3, 39–40, 45

Simcoe, Capt. John, 17–19

Spain, 3, 11, 30. *See also* Seven Years War; Treaty of Paris (1763)

Sproule, George, 55, 56, 57, 58, 59, 61, 62, 95, 208–9, 210; survey of New England, 72, *74*, 78, 80, 81

Spry, Capt. Lt. William, 26, *27*, 29–30, 165

Spry, Cdre. Richard, 37, 133–4

Survey of Canada (1760–1763), 8, 25–30, *27, 28*; area surveyed, 26; census of parishes, 26–7, *28*; scale of, *27*; techniques used, 26, 27–8

survey descriptions and reports, 119–20, 142–6; and natural resources, 143

survey, hydrographic and soundings, 3, 5–6, 8, 56, 67–9, 70–1, 86, 87–8, 96–8, 121–3, *122*; charts in *The Atlantic Neptune*, 184–5, *186–7, 189*; methods, 109–10; need for survey, 33–8; vessels used for, 103–5. *See also* Mowat, Henry; Survey of Nova Scotia

survey instruments, methods and equipment, 17–19, 105–14, *110*, 117–18; astronomical clock, *111*, 111–12; astronomical observation, 6, 18–19, 62–4, 68, 86, 110–12; establishment of latitude and longitude, 3, 6, 18–19, 35–6, 110–14, *113*; plane table, 17–18, 105–7; quadrants, 110–12, 113–14; telescopes, 111–12, 113, 114; triangulation, 105–9, *106*, 114; use of local informants, 114–17, *114*; use of pilots and guides, 114–17. *See also* survey vessels

Survey of Nova Scotia (1764–1775), 3–11, 25, 37–8, 43, 65–72, *67, 104*, 106–10; in Canadian historiography, 4; chain of command, 93, 117–8; survey teams, 96–8. *See also The Atlantic Neptune*; Des Barres, Joseph Frederick Wallet; survey, hydrographic and soundings

survey personnel, 87–101, chain of command, 87–95; compensation of, 98–9; desertion by, 100–1; discipline and punishment of, 98–101; naval detachment, 96; survey teams, *12*, 95–8

survey vessels, 101–5, *102, 104*. *See also Canceaux; Diligent; Dispatch*

timber resources, 75, 78, 80–1, 143

transit of Venus, 3, 62–4, *63*, 112

Treaty of Paris (1763), 3, 11, 30, 36

War of American Independence, 4, 5, 72, 79–80, 83–4, 124–5, 199–205; and need for maps, 4–5, 79, 175–6, 190–3

War Office, 31, 33. *See also* army, British

Wentworth, Gov. John, 72–3, *74*–5, 78, 80–1

Wheeler, Thomas, *74*, 81, 95

Wolfe, James, Lt. Col., 13, 16–17, 134; attack on Quebec, 20–2; promoted to Maj. Gen., 19

Wright, Thomas, 95, 208–10; survey in the Gulf of St Lawrence, 47–8, 50, 52, 54, 55, 57, 58–9, 63–4; survey of New England, 73, *74*, 80

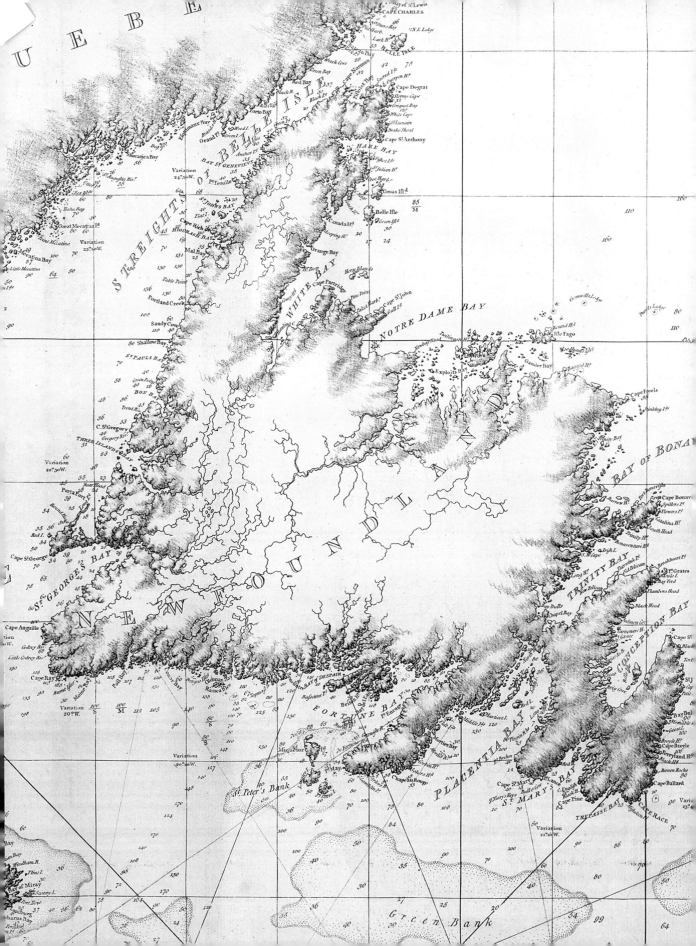